Engineering Thermodynamics

D. B. Spalding
Professor of Heat Transfer

E. H. Cole
Senior Lecturer

Imperial College of Science and Technology
University of London

J. A. Dallender, *Lecturer, Imperial College of Science and Technology,*
assisted with the preparation of the third edition

EDWARD ARNOLD

First Published 1958
by Edward Arnold (Publishers) Ltd
25 Hill Street W1X 8LL

Reprinted 1959, 1961, 1963, 1964
Second edition 1966
Reprinted 1967
Third edition 1973

Boards edition ISBN 0 7131 3298 1
Paper edition   ISBN 0 7131 3299 X

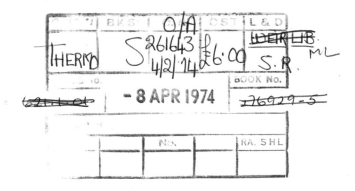
Text set in 10/12pt. Monotype Times New Roman, printed by photolithography,
and bound in Great Britain at The Pitman Press, Bath

Engineering Thermodynamics

Preface to the Third Edition

The preparation of this new edition results primarily from the introduction of the Système International d'Unités.

The order of presentation and development of the subject matter remain as in the second edition; the text, figures, tables, worked examples and end-of-chapter problems have been revised to accommodate the use of SI units.

In Chapter 2 we have retained a general treatment of mechanical units, but with the emphasis now switched from British to SI units. We have done this not only to facilitate understanding but also because students and practising engineers will still need to interpret and use material presented in diverse unit systems; in particular our treatment allows the constant g_c in Newton's Second Law of Motion to be correctly assigned to suit the various unit systems encountered in practice. The tables of conversion factors have been retained, but with priority now given to SI units.

Similarly, in Chapters 3, 4, and 5, on work, temperature, and heat, priority has been given to the use of SI units; but where appropriate, the corresponding British units are also discussed.

In addition to the incorporation of the revisions necessitated by the use of SI units, the opportunity has been taken to re-write some of the other parts of the text; also a new page layout has been adopted.

The steam tables included in Appendix B are extracts based on the U.K. Steam Tables in SI units (1970) prepared by the United Kingdom Committee on the Properties of Steam. We acknowledge with thanks permission to use these data.

South Kensington, 1973

D.B.S.
E.H.C.

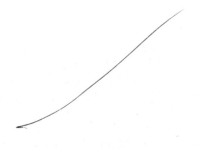

Preface to the Second Edition

The purpose of the present book is the same as that of the first edition: to present the fundamentals of classical thermodynamics, to students of all branches of engineering, in a manner that combines rigour with practical utility. The treatment is also unchanged: it rests on the laws governing the heat and work interactions between a system and its surroundings. The system is always large enough, i.e. contains a sufficiently large number of individual molecules, for heat and work to be easily distinguishable; no attempt is made to present the laws as consequences of the microscopic structure of matter. On this foundation, we erect the conventional framework: the properties energy, enthalpy and entropy are defined; we describe the relations between them for various substances of importance in power-plant engineering; and we present the analytical techniques by which the engineer employs the laws and the properties in the prediction of plant performance.

Some new material has been added, particularly in Chapter 2 (Mechanical Quantities and Their Units), which has been completely rewritten, and in Chapter 13 (Entropy), which now includes a section entitled "Understanding Entropy". Other material has been redistributed: the section on systems in motion has been transferred from Chapter 8 to Chapter 7, and the treatment of reversible and irreversible engines has been rearranged in Chapters 11 and 12. Further, the opportunity has been taken to bring our unit abbreviations and symbols more closely into line with the recommendations of the British Standards Institution. The remaining changes, though of detail rather than essence, are quite extensive. The whole book has been reset, with larger type and more generous margins; many of the figures have been enlarged, while others have been revised and redrawn; the English has been polished; and we have eliminated all the printer's and other errors to which readers of the 1958 edition have helpfully drawn our attention.

We have been assisted by many of our colleagues, who have participated in the use of the book as an undergraduate text; the comments of Dr. J. R. Singham have been particularly valuable to us. Miss E. M. Archer, Librarian, Mechanical Engineering Department, City and Guilds College, has given much assistance in the revision and extension of the list of references, and Miss M. P. Steele has checked some of the proofs; we wish

Preface

to thank them for their help. In the preface to the first edition we acknowledged out debt to Professor Keenan's *Thermodynamics* of 1941; this indebtedness has not diminished with the years. We also wish to record our gratitude to the following persons and organisations, who have assisted us with the provision of diagrams and other material: The Science Museum (Fig. 1.1 and Fig. 1.2); The General Electric Company Ltd. (Fig. 1.3); D. Napier & Son Ltd. (Fig. 1.4); Ruston & Hornsby Ltd. (Fig. 1.5); Hawker-Siddeley Group Ltd. (Fig. 1.6); Budenburg Gauge Co. Ltd. and the Society of Instrument Technology (Fig. 2.4); Dobbie McInnes Ltd. (Fig. 3.9); Griffin & George Ltd. (Fig. 16.6 and 16.7); Professor J. H. Keenan, Professor F. G. Keyes and John Wiley & Sons, Inc. (Extract from *Thermodynamic Properties of Steam*); British Standards Institution (Tables A.1 to A.14 inclusive).

D.B.S.

E.H.C.

South Kensington, 1966.

Contents

1 Introductory Survey

Historical Introduction

Mechanical power as the basis of civilized life

Man is physically a weak animal; yet he dominates the globe. How has this come about? Man's dominion rests on his ability to control forces far greater than those which his own muscles can exert. Without this control, neither agriculture nor urban life could have developed; for water had to be pumped to irrigate the land and supply the towns; food and building materials required transportation; corn must be ground and wood sawn. Later, when valuable ores were discovered beneath the surface of the earth, these too had to be raised and processed; the deeper the mines became, the more difficult grew the task of keeping them free of water. Man's ability to provide sufficient power for these and similar purposes set limits to the rate of growth of civilization, as indeed it still does.

Provision of this power is nowadays one of the main tasks of the mechanical engineer.

Sources of power

Animals, wind and water. When civilization began, during the sixth millennium BC, the only available sources of power were animal: heavy tasks were performed by gangs of slaves, or by domestic animals. This division of labour was a prerequisite for the establishment of collective life.

The wind was the next source of power, at first used only for driving sailing vessels. Not until the tenth century AD, however, was the wind harnessed, by means of windmills, to mechanical tasks such as milling and sawing.

Water-wheels were invented earlier, probably in primitive form before the beginning of the Christian era. For centuries they remained man's main source of mechanical power. But animals were slow and winds uncertain; water power was not always found in the right places. Particularly for mining, more concentrated, reliable and disposable sources of power were needed and urgently looked for throughout the Middle Ages.

The beginnings of steam power. The seventeenth century saw the rise of modern science in Europe. Observations were recorded and communicated; novel experiments were planned and executed; above all, quantitative measurements were systematically carried out.

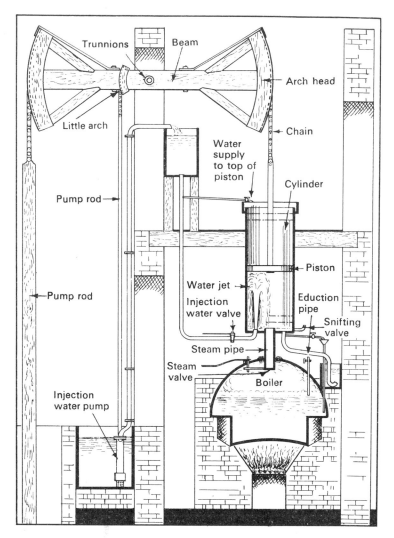

Fig. 1.1 Newcomen's pumping engine, 1712.

This diagram shows a section through a typical "atmospheric" pumping engine, as built by Newcomen, with the piston in the middle of the downward or working stroke. Steam, generated at atmospheric pressure in the boiler, fills the cylinder during the upward stroke of the piston. The steam valve is closed at the end of the stroke, and the steam is condensed by a jet of cold water; this reduces the pressure under the piston. The atmospheric pressure acting on the top of the piston forces it down (hence the name "atmospheric" engine); this constitutes the working stroke. The piston is raised again by the overbalancing weight of the pump-rods, when steam is again admitted to the cylinder.

Much study was devoted to the properties of air and water. With the invention of the piston-and-cylinder air pump by von Guericke in Germany in 1654, the properties of gases under pressure, and of vacua, could be investigated. Notable discoveries were made by Boyle and Hooke in England. Even earlier, Porta in Italy and de Caus in France had shown how steam could be made to raise water by pressure or by condensation. By the end of the century, Savery had managed to combine these processes into a practical means of pumping water.

A most important advance was made by Newcomen in 1712: he combined the piston and cylinder of the air pump with the use of steam (Fig. 1.1). A vertical cylinder was filled with steam from a boiler; cold water was then injected into the cylinder, causing the steam to condense and its pressure to fall; the atmospheric pressure pushing on top of the piston forced it down into the cylinder, thereby moving a beam which drove a water pump. This was the first practical steam engine. Its immediate effect was to solve the problem of draining the mines; of more far-reaching significance however was the fact that man now had a source of power of unprecedented magnitude, dependent neither on flesh and blood nor on the vagaries of weather and geography.

Power developments in the eighteenth century. The importance of the Newcomen engine was soon recognized throughout the western world, and engineers worked strenuously, both to improve its functioning, and to use it for other tasks than the pumping of water.

Most notable among the improvements was Watt's invention of the *separate condenser* in 1769. Instead of condensing the steam within the cylinder, Watt connected the cylinder, at the appropriate point of the piston movement, to another vessel; this was kept cool the whole time, and held at a low pressure by means of an *air-pump* (Fig. 1.2). This invention brought about a notable saving of fuel, much of which had been wasted by the Newcomen engine in heating the cylinder again after each injection of cold water. Watt and his associates were also responsible for systematizing measurements of the power and fuel consumption of their engines. It was Watt, for example, who invented the *horse-power* unit to express the rate at which work is done.

Manufacturing industry benefited greatly from the existence of the new source of power. By the end of the eighteenth century, steam engines were driving spinning machinery, paper mills, winding engines and much other plant. Experiments in applying steam power to transport were also being pursued.

Power developments in the nineteenth century. Because of the weight and bulk of the machinery, the first successful application of steam power to transport was in the driving of ships. Here the United States of America made notable contributions: the first commercial steam-boat was launched by Fulton on the Hudson River in 1807.

On land, after unsuccessful attempts to make steam engines travel over the inadequate roads of the time, the use of railways made possible the support, not only of the heavy engines themselves, but also of the long trains

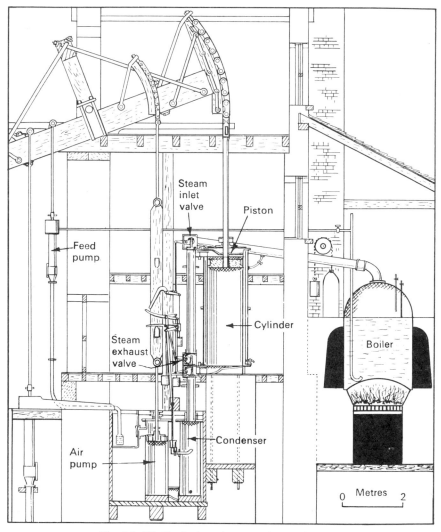

Steam inlet valve

Piston

Feed pump

Cylinder

Steam exhaust valve

Boiler

Air pump

Condenser

0   Metres   2

Fig. 1.2 Watt's single-acting pumping engine, 1788.
(Reproduced by permission of The Science Museum)
Crown copyright reserved

of goods and passengers which they could pull. Throughout the century, networks of railways spread over the civilized regions of the world, transforming both the scenery and the mode of life of millions.

The reciprocating steam engine was not however to remain unchallenged. The improvements in manufacturing technique which it had made possible, together with the great spread of technical knowledge, prompted engineers to look once more at an idea often mooted before, but abandoned as impracticable: the elimination of steam as the intermediary between the fuel and the working piston. Already by 1678 Hooke had proposed the use of gunpowder within a cylinder to operate a piston; a century later, Street patented an internal-combustion engine which was to explode vaporized turpentine and air. It was not however until the second half of the nine-teenth century that commercial success was achieved. The *gas engine* using a fuel gas derived from the partial combustion of coal, and the *gasoline** (petrol) *engine* using a vaporizable fuel, each employing either electrical or external-flame ignition, were both developed in this period; the former was primarily for static operation, the latter for transport. *Heavy-oil engines* followed shortly; in these, an oil-air mixture was ignited in the cylinder, either by contact with a "hot-bulb" or as a result of the temperature rise of the compressed air. French, German and British engineers were prominent in these developments.

So encouraging was the performance of the new prime movers, that it appeared as though steam power would soon lose its importance. Prophecies to this effect were falsified however by the successful combination by Parsons, in 1884, of the use of steam with the principles of the wind-mill. The new engine was the *reaction steam turbine* (Fig. 1.3) in which high-pressure steam from a boiler flowed steadily through rows of fixed vanes, projecting from the inner surface of a casing, alternating with moving blades mounted on a rotating shaft. In this way very great powers were obtained from a small and smoothly running piece of machinery; moreover the fuel consumption of plant employing turbines was significantly lower than when reciprocating engines were employed. The result was a new lease of life for the steam power plant, which, in its new form, retained for many years its dominance in the driving of ships; it is still pre-eminent in the production of electrical power on land.

Power production in the first half of the twentieth century. The beginning of the twentieth century was remarkable more for the intensive development of the prime movers already discussed than for the introduction of new types. In the course of this development, the relative advantages of the various engines were gradually discovered, largely by trial-and-error, and their appropriate areas of employment became delineated. The boundaries of these are however always shifting, as a consequence of minor improvements in performance.

From the start, the gasoline engine has had a clear lead in propelling road vehicles, although the heavy-oil (diesel) engine is often preferred for com-mercial vehicles because of its smaller fuel consumption. Reciprocating steam engines, which once provided the power of the world's railways, are

---

* The term "gasoline" is preferred to "petrol" as the technical term for the fuel of this class of engine.

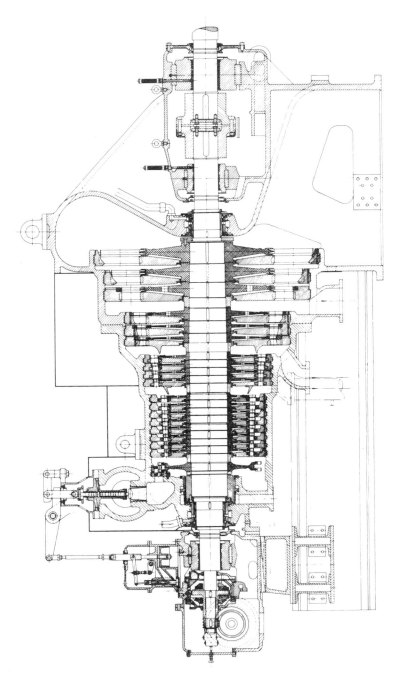

Fig. 1.3 Section through a steam turbine.

now giving way to diesel-powered locomotives, or to electricity derived from central steam-turbine-driven power stations. The heavy-oil engine has made great advances also in the propulsion of ships, both large and small.

The reciprocating gas engine was foiled of its expected future in central power stations by the advent of the steam turbine. Its use is now restricted mainly to regions where a ready supply of fuel gas is obtainable directly from wells in the earth. This field of application also is threatened by a relative newcomer: *the gas-turbine engine.*

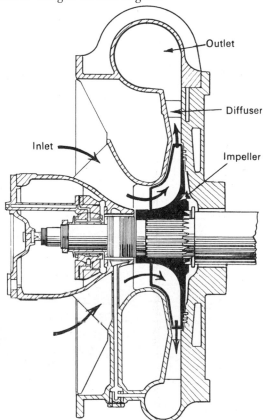

Fig. 1.4 Centrifugal compressor.

The early years of the century saw the development of powered flight. From being a spectacular novelty, the aeroplane has become a major vehicle of transport, both commercial and military. Aero-engines must be both powerful and light, which requirements, until recently, were met only by the gasoline engine. In the 1930's, however, development of the gas-turbine engine in Switzerland, England and Germany provided a new power plant with advantages which have greatly diminished the use of the gasoline engine. Once again, the idea had been proposed in the eighteenth century; but the means for realizing it were not present until 150 years later.

The gas turbine employs the principle which Parsons used to such effect in 1884: the reciprocating piston is dispensed with and the gases flow through alternate rows of moving and stationary blades. Instead of water, however, air has first to be compressed in a rotary compressor (Fig. 1.4); liquid or gaseous fuel is burnt in the compressed-air stream, and the hot gases thereby produced flow out through one or more turbine wheels. The

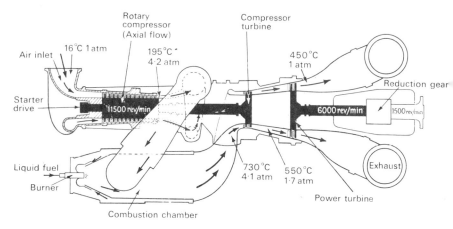

Fig. 1.5 Industrial gas-turbine engine.

power output is effected either through the shaft coupled to the turbine or by way of the propulsive force of a jet (Figs. 1.5 and 1.6). Nowadays the

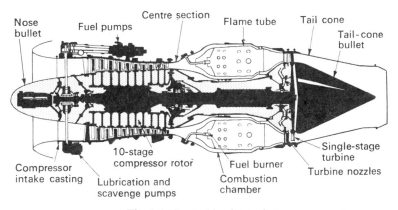

Fig. 1.6 Gas-turbine jet engine.

gas turbine dominates aircraft propulsion; but on land its advance, in competition with other modes of power production, has been much slower. Here fuel consumption is often more important than the lightness and mechanical simplicity which the gas turbine has to offer; in this respect the older-established engines are still supreme.

Fuel consumption is all the more important because coal and oil have to be extracted from the mines and reservoirs beneath the earth. Not only is this extraction difficult and expensive, but the supplies there are not inexhaustible. Already the more far-sighted see serious grounds for anxiety in man's improvident use of the world's fuel resources. It is therefore fortunate that a new type of fuel has been discovered: uranium and its products. Whereas coal and oil are useful because they can undergo *chemical reaction*, uranium carries out a *nuclear reaction*: the nuclei of the atom become split and so release energy. Although still a mineral fuel, and very expensive to extract and process, uranium is distributed in the earth's crust in relative plenty; it may well supply at least part of man's power requirements for many generations; for the work obtainable from a tonne of uranium is equivalent to that from $2\frac{3}{4}$ million tonnes of coal. Already nuclear power is making a significant contribution to the world's electricity supplies.

Despite the differences in its mode of reaction, nuclear fuel is used, like chemical fuels, in conjunction with steam-turbine plants. Although it is likely that gas turbines will also find a part to play, there is no entirely new type of nuclear-driven power plant in prospect at the present time.

The role of electricity. Electrical effects were known in ancient times, but the starting point of modern knowledge and use of electricity may be said to

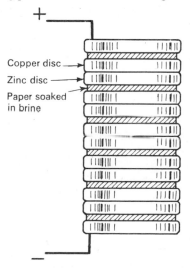

Copper disc

Zinc disc

Paper soaked
in brine

Fig. 1.7 Volta's pile.

be the discovery by Volta in 1799 of the possibility of producing an electrical current from a chemical reaction, namely that between paired discs of copper and zinc in brine (Fig. 1.7). Although developments of it are used for car accumulators and the batteries of flashlamps, Volta's discovery has not yet led to any large-scale production of electrical power; the reason is that, on the one hand, the substances with which it is easy to produce electricity are expensive, while on the other hand the chemical reactions undergone by common fuels are difficult to harness in this way, for technical reasons. Nevertheless the possibility is of great theoretical interest, as will be seen later in this book.

Modern use of the electricity in power production derives from Faraday's discovery, in 1831, of how electricity could be produced by moving an electrical conductor across a magnetic field, thus eventually making possible the convenient *transmission* of power. Dynamos and electric motors should therefore be thought of, not as prime movers, but as modern equivalents of the levers, gears, belts and pulleys which have been used for transmission since engineering began.

Power production in the future. Advances in the techniques of power production have been so swift that speculation as to future developments is hazardous. It can be said with certainty that chemical fuels will become increasingly scarce within relatively few generations unless drastic changes in their use are introduced. One such change would be the application of Volta's discovery to the production of electric power directly from the reaction of coal and oil with air, by the commercial development of a *fuel cell*; in this way about three times as much power as at present could be produced per unit of fuel. Although the difficulties are great, even more formidable ones have been overcome in the development of nuclear power.

A possibility of solving the power difficulties of mankind more permanently is currently engaging serious attention. This is the use of the reaction of nuclear *fusion*, as opposed to *fission*. In fusion, the essential process of the hydrogen bomb, the nuclei of the atoms of light elements combine to form heavier nuclei with a very great release of energy. Since hydrogen is extremely plentiful, no shortage of raw material need be feared. However it has not yet been demonstrated that the power consumed in processing the hydrogen to prepare it for the fusion reaction is less than that which can be released by the reaction. Moreover the reaction is known to proceed only at temperatures which are far above the present range of engineering experience. So far, all that can be said is that the thermo-nuclear reaction is as likely to prove the saviour of mankind as its destroyer.

At high temperatures, gases conduct electricity; and a conducting gas, when flowing through a magnetic field, can act in the same way as the moving solid conductors of a conventional dynamo. Much study is currently being given to the employment of this effect in the production of electrical power, the aims being to decrease fuel consumption and capital cost. The relevant scientific discipline is known as *magneto-hydrodynamics*, commonly abbreviated to MHD.

Summarizing remarks about the means of power production

The choice of power plant. In our brief survey of the history of man's search for power, some important considerations have emerged which will now be recapitulated.

First, the power plant should have sufficient *magnitude*. This require-ment, coupled with the aims which man has set his heart on, has been the major stimulus to development.

Secondly, the *fuel consumption* should be low. This is as true of gasoline engines and nuclear power plants as it was when the food for slaves and animals was the main item of current expenditure.

Often, *concentration* of power is the prime requirement, as in aircraft propulsion. No matter how many oxen are available, and however rich and plentiful their fodder, they are incapable of exerting enough power to lift themselves off the ground.

Finally power sources must be *reliable* and *flexible*. It is for this reason that, though the winds are strong and cost nothing, we prefer to pay for fuel for our ships in order that they should "sail" with certainty, even when the wind is contrary or absent.

The elements of modern power plant. Finally we draw particular attention to some components of engines, which, as has been seen, have long been known to man, and which will greatly occupy our attention in the present book.

The first is the *bladed wheel*. The earliest form was the water-wheel; later it appeared in the wind-mill, and finally as the steam or gas turbine or in the form of the rotary compressor.

The second component is the *piston-cylinder combination*, used first as a pump for water or air and later as the mechanism by which the pressure of steam or hot gases impels the motion of beams, shafts, propellers and motor-cars.

The third is the *Voltaic* cell, still the Cinderella of power sources, waiting for the crystal slipper to be found.

Finally, we note that, whereas solid shafts and gears often transmit the power, it is fluids which make the machines go round. Water and steam, air and combustion products: these are of prime concern to the power plant engineer. They will occupy much of our attention in subsequent chapters.

The Nature of Thermodynamics

The role of science in the development of power plants

Observation, speculation, experiment; abstraction of essentials, predic-tion: these have always been man's most characteristic activities. Nowhere are they more noticeable than in the way he has striven for mastery of his environment. To most men, the first three come more easily; yet the last two characteristics must be superadded before man's activity can be called truly scientific. The history of engineering provides many illustrations.

The systematic *observation* of nature in the seventeenth century, parti-cularly the study of the properties of fluids, led to *speculation* about the production of power from steam and fire. After many failures, *experiment* showed that such engines could indeed be made and used. The pragmatic approach was amply justified. Continued experiment produced continued improvement in performance, particularly in the amount of water which could be pumped with the consumption of a fixed amount of fuel.

It is questionable however whether *science* made much contribution before Watt, recognizing what was "really" going on in the Newcomen cylinder, and interpreting the processes in the light of Black's discovery of the "latent heat" of steam, *predicted* that the provision of a separate condenser would reduce the fuel consumption. This step required a high degree of abstract thought.

Yet deeper questions remained: could the fuel consumption be reduced indefinitely; and if not, why not? In the search for the answers, the science of *thermodynamics* was born.

Nowadays these questions and many others can be answered by engineering students before the end of their first year of study. Attention is paid to them, not primarily because of the intellectual exercise which they provide, but because *experiments are expensive*. "Bright ideas" about new engineering devices are easy to conceive; yet, if they are to be tested in practice, raw material must be bought, machines diverted from other production, and the testing personnel must be paid. No one is likely to authorize this expenditure until the proposals have been examined as to their concordance with known scientific principles, and until the best possible prediction of their outcome has been made. Experiment and trial-and-error development can seldom be dispensed with entirely; but a case for the new ideas must be made out "on paper" before the first order is placed on the workshop.

The discovery of the laws of thermodynamics

At the beginning of the nineteenth century, the questions about the fuel consumption of engines were being thought of in terms of two abstractions: *work*, the thing that the engineer wanted to get; and *heat*, the thing that, via the burning of fuel, he had to employ to get it. What were the laws governing the conversion of heat into work?

A partial answer was provided by Carnot in France in 1824. By a brilliant feat of abstract thought, he perceived that *temperature* provided the key. He used the fruitful method of arguing by analogy. It was known that water, flowing from a high level to a low one, could do an amount of work which increased in proportion to the differences of level. Could not heat, then, in the steam plant, be like water in the hydraulic plant? Could not the temperature at which the heat was transferred to the steam in the boiler, and the temperature at which it was transferred from the steam in the condenser, be the counterparts to the levels of the water? These thoughts, and their development in the hands of Carnot himself and of others, are embodied in what is now known as the *Second Law of Thermodynamics*.

In one respect Carnot was wrong; moreover it was the argument by analogy which misled him. For whereas the same amount of water flows away from a water-wheel as flows to it, the heat transferred from the thermal plant is *less* than that transferred to it. Carnot can hardly be blamed for not noticing this: the difference amounted to less than ten per cent even in the best engines of his time.

Carnot's work attracted little attention until more than twenty years later. Meanwhile the Englishman Joule was following a different track. In a long series of careful experiments he showed what had happened to the "lost" heat: it had "turned into" work. This was not, of course, quite the way in which the matter appeared to him; his experiments were not made on engines but on specially designed apparatus; but his discovery, published in 1850, has this import among many others. It is now known as the *First Law of Thermodynamics.*

The reconciliation of Carnot's and Joule's principles soon followed, and in a few years the main structure of classical thermodynamics was established. Kelvin in Scotland and Clausius in Germany were outstanding in this work, which has proved to be significant far beyond the boundaries of power-plant engineering: physicists, chemists, biologists and even philosophers must nowadays be familiar with the First and Second Laws.

The establishment of the laws of thermodynamics did not of course render further observation and experiment unnecessary. Rather it stimulated them; for now a theoretical framework was available into which the new experimental data could be fitted; and by showing what was possible and what was not, the efforts of the experimenters could be directed towards the most fruitful ends. This book will be largely concerned with the results of their experiments and with an account of the two laws which give them meaning.

The content of thermodynamics

Definition. Thermodynamics is the science of the relations between heat, work and the properties of systems.

The words *heat, work, property* and *system* will all be the subject of precise definition below. For the moment we can substitute *matter* for *system* and attribute to the other terms their meanings in ordinary speech. Thermodynamics then relates the changes which matter undergoes to the influences to which it is subjected: for example, it tells how the pressure and temperature of the steam in an engine depend on the extent to which the steam is heated and the amount of work which is produced.

A characteristic feature of thermodynamics is that these relations can often be stated in complete ignorance of the *details* of the process in question. The following parlour puzzle, which appears in many versions, will serve as illustration.

Problem: Two equal tumblers are each half full, one with white wine, the other with red. A teaspoonful of the white wine is taken from the first tumbler and transferred to the second, the contents of which are then thoroughly mixed; then a teaspoonful of the mixture in the second tumbler is transferred back to the first tumbler, which is similarly stirred. *Question:* Which of the two resultant mixtures is the more pure?

Solution: (This should not be read until attempts have been made to imagine the process in detail, noting that pure white wine is transferred to the second tumbler but that some of it returns in the second transference. Since, further, the first teaspoonful of white wine is mixed with half a tumbler of red wine, whereas the second teaspoonful is mixed with a smaller quantity, algebra arises of sufficient complexity to make mistakes and conflicting answers abound.)

The thermodynamicist's approach is as follows:—

1. Draw a *diagram* in which each tumbler is surrounded by an imaginary boundary, Fig. 1.8. (Later these boundaries will be called *control surfaces*.)

Fig. 1.8

2. Consider the *net* transfer of wine across one of the boundaries. Since one teaspoonful of wine (regardless of colour) travels in each direction, the net transfer of wine is zero.

3. Consider the *net* transfer of *white* wine *out of* the first tumbler. Let this quantity have the volume x. Then, since there is zero net transfer of wine, an equal volume x of *red* wine must have been transferred *into* the first tumbler, and *out of* the second.

4. *Conclusion.* The net process is symmetrical: *the two final mixtures are equally pure (or impure).*

This economical proof, though hardly belonging to engineering thermodynamics, has features which are typical of many which will be found in this book.

Thermodynamics as the scourge of inventors

Another important feature of thermodynamics is that it furnishes simple tests of whether an inventor's proposal for a new power plant is sound. Usually the inventions which are submitted for development are so complicated that to attempt to follow the proposals in detail is to become lost in a maze in which fact and supposition are impossible to disentangle. Often however the thermodynamicist, by proofs as elegant as that above, can demonstrate conclusively that a fallacy must exist. This rarely satisfies the inventor, it must be admitted. The following examples will indicate the sort of problems which arise, although we cannot follow out their solutions here.

Invention 1. Most schoolboys destined to be engineers, soon after learning the function of (*a*) windmills, and (*b*) propellers for aircraft, hit on the idea of combining them in the way shown in Fig. 1.9. A windmill, mounted near

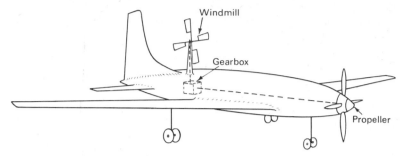

Fig. 1.9 Invention 1. Proposal for aircraft propulsion.

the tail, is rotated by the air rushing past; this windmill is connected by shafts to the propeller and so serves to drive the aircraft forward without the need of an engine. The more sophisticated young inventors recognize that friction in the bearings may be a draw-back; they therefore insert step-up gearing in the transmission. Launching by catapult is usually recommended in order that the propulsion system should get under way.

Discussion: Unfortunately it has not been found possible to make this scheme work in practice, even with step-up gearing! Most readers will accept this without recourse to thermodynamics, but a strict proof is possible only with the help of its laws.

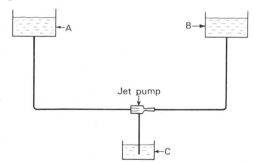

Fig. 1.10 Invention 2. A proposal for pumping water.

Invention 2. Acquaintance with the workings of the filter pump (jet pump) used in school chemistry laboratories sometimes inspires the proposal for pumping water shown in Fig. 1.10. Water from reservoir A flows through the jet pump into reservoir B where the water is *at the same level* (A and B could in fact be the same reservoir. Friction has to be ignored). The mixing region of the jet pump is connected to a third reservoir C at a lower level. Water from C is drawn into the pump and so flows up to B with the main stream.

Discussion: The formal proof that this proposal will not work rests on the laws of thermodynamics (see Problem 10.11, page 216). The invention has been introduced only because of its similarity to a rather more subtle proposal, which now follows.

Invention 3. The arrangement is the same as that of *invention 2*, except that the water flowing from A to the jet-pump first passes through a coiled pipe exposed to a flame (Fig. 1.11); there it is vaporized so that it enters the pump in the form of steam. Water from C is drawn into the pump and flows up to B as before. Will this system work?

Discussion: The answer is "Yes", provided that the temperature of the water in C is below that of the steam; indeed the device will even work if B is higher than *both* the other reservoirs. The reasons are too complicated to discuss here, but it is interesting to note that it was invented by Giffard in 1858, and has been widely used for pumping fresh water into the boilers

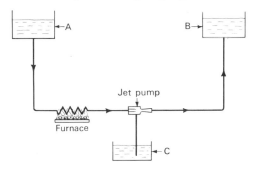

Fig. 1.11 Invention 3. Another proposal for pumping water.

of railway locomotives. It has recently been re-invented, in another form, under the name *aerothermopressor*, as a device for driving supersonic wind-tunnels.

Concluding remarks. The fallacies of the first two inventions were easy to discern, because only mechanical considerations were involved. Some readers may have supposed that *invention 3* was likewise a vain attempt to "lift oneself up by the bootstraps". Evidently however, when heat is involved, a correct decision is harder to reach. It is in these cases that the laws of thermodynamics require specific application.

What thermodynamics leaves out

Rate processes. We have already implied that thermodynamics is concerned with the "before" and "after" of a process and often ignores the "in-between". It should not be inferred that the "in-between" is of no interest to the engineer. Indeed the designer is often as much concerned with the *rate* at which a process advances as with its end-result.

It must therefore be emphasized that thermodynamics has no time scale; to determine the rate of a process, on which often depends the size of the apparatus, the engineer must turn to neighbouring subjects. The one bulking largest in engineering curricula is *heat transfer*, which is concerned with conduction, insulation, radiation and convection. Allied to it is *mass transfer*, which deals with mixing processes, for example between air and fuel in a furnace. Finally there is the subject of *chemical kinetics*, which is the study of how rapidly chemical reactions proceed. Engineers need some knowledge of all three processes in their work on power production; however, only the purely thermodynamic aspects of the subjects are dealt with in the present book.

The microscopic structure of matter. Another body of knowledge excluded from thermodynamics is that concerned with the ultimate constitution of matter, its construction from atoms and molecules and from the still smaller elementary particles now known to physics. Although understanding of the *microscopic* properties of matter is helpful in interpreting some of the *macroscopic** properties which thermodynamics deals with, the thermodynamicist treats matter as a continuum.

This is not the result merely of a lack of interest in small-scale phenomena, but of the fact that the laws of thermodynamics do not apply when only a few particles are in question; for example the distinction between heat and work then becomes blurred. Although we shall have no occasion to discuss the matter below, the laws of thermodynamics are essentially *statistical* in nature, and have meaning only when large numbers of particles are considered at once.

The Scope of the Book

Aim

The book is intended to provide the knowledge of thermodynamics required by every engineer, whether his career is to be spent in mechanical, chemical, civil, electrical or aeronautical engineering. The emphasis is on the fundamental theory, and applications are primarily dealt with as illustrations. The reader must look elsewhere for detailed discussion of how particular plants are designed and operated. The coverage of the fundamentals is fairly complete, although chemical engineers will need knowledge of the theory of chemical equilibrium, which we have had no space for.

Content

The subject matter falls into four main parts.

Chapters 2 to 5 are devoted to treating, with the extra care which thermodynamics demands, concepts which will already be familiar to the reader from his study of mechanics and physics. The main new concept introduced is the *system*, which assumes great importance throughout the book.

* Large-scale.

The second part (Chapters 6 to 9) begins with Joule's discovery: the First Law of Thermodynamics. Here *energy* is introduced. The field of attention is then narrowed to the sort of materials with which the power engineer is chiefly concerned, namely the *pure substance*, and its properties. In this part we discuss a particular form of the First Law, the *Steady-Flow Energy Equation*, with the aid of which a wide variety of engineering processes can be analysed.

Chapters 10 to 13 discuss engines and their *efficiency*. Here we introduce Carnot's Principle: the Second Law of Thermodynamics. The subsequent deductions constitute a body of theory which is both elegant and practically important. In order to bring the theory to bear on engineering analysis, Clausius' concept of *entropy* is defined and studied. This completes the main theoretical apparatus which the thermodynamicist needs.

In the final part we turn once more to the properties of materials. In Chapter 14, Ideal Gases are discussed; the reader will already have some knowledge of these. Then, in Chapter 15, we deal with gaseous mixtures, concentrating particular attention on mixtures of air and water vapour because of their familiarity and importance, although the principles are of wider application. Finally, in Chapter 16, fuels and combustion are treated.

Final remarks

At the end of each chapter, problems are provided. The reader is strongly advised to solve at least some of these before proceeding; for each chapter pre-supposes a knowledge of what has gone before, and it is only by practice that the significance of the text, packed as it will sometimes seem to be with pedantic hedgings and qualifications, can be rightly appreciated. The only justification for this book is that it may help engineers to solve practical problems of design and analysis with certainty and ease; we hope it does so.

BIBLIOGRAPHY

ANON. *Classified Lists of Historical Events: Mechanical and Electrical Engineering.* Science Museum: H.M.S.O., London, 1955.

BILLETT, H., DUNN, P. D., WHITBY, H. C. Generation of electricity by unconventional methods. *Proc.I.Mech.E.*, 1962, **176**, No. 18.

BUCHANAN, R. A. The History of Technology. *The Technologist*, **2**, No. 2, 1965–66.

BURSTALL, A. F. *A History of Mechanical Engineering.* Faber & Faber, London, 1963.

DICKINSON, H. W. *A Short History of the Steam Engine.* Cambridge University Press, 1938.

KAYE, J. and WELSH, J. A. *Direct Conversion of Heat to Electricity.* Wiley & Sons, New York, 1960.

LENARD, P., *Great Men of Science.* G. Bell & Sons, London, 1933.

SOO, S. L. *Direct Energy Conversion.* Prentice Hall, Inc., New Jersey, 1968.

UBBELOHDE, A. R. *Man and Energy.* Penguin Books Ltd., London, 1954.

CHAPTER 1—PROBLEMS

1.1 The watt is a unit of power equivalent to a rate of working of one newton metre per second. Another unit of power, the horse-power is equivalent to a rate of working of 550 foot pounds-force per second. Also one newton metre equals 0·7376 foot pound-force.

Express the following work quantities in newton metres:

(a) One kilowatt hour.

(b) One horsepower hour.

1.2 The performance of a power plant is often expressed in terms of its specific fuel consumption, namely in pounds-mass of fuel consumed per horsepower hour or in kilograms per kilowatt hour.

Evaluate and compare the specific fuel consumptions of the following methods of power production:

(a) A man working a treadmill can produce an average of 103 newton metres of work per second for 8 hours out of 24. The mass of his daily food intake is 2 kilograms. (One pound-mass ≡ 453·6 grams; one newton = 0·2248 lbf; one inch = 2·54 centimetres.)

(b) The average performance of fifteen Newcomen steam engines in England in 1769 was to raise 5·59 million pounds-mass of water one foot in height (i.e. $2·54 \times 10^6$ kilograms, 30·48 centimetres) while consuming one bushel (84 pounds-mass or 38·1 kilograms) of coal. (In a standard gravitational field one pound-mass experiences a force of one pound-force and one kilogram experiences a force of 9·81 newtons.)

(c) The guaranteed performance of Watt steam engines in 1800 was to raise 35 million pounds-mass of water one foot in height (i.e. $15·9 \times 10^6$ kilograms, 30·48 centimetres) while consuming one bushel of coal.

(d) A Worthington steam-driven pumping plant (early twentieth century) pumped 13 400 gallons of water per minute through a vertical height of 54 feet (i.e. 16·45 metres) while consuming coal at the rate of 456 pounds-mass per hour. (One gallon of water has a mass of 10 pounds-mass; one pound-mass = 0·4536 kilograms.)

1.3 Evaluate the specific fuel consumption of each of the following modern power-producing installations:

(a) An oil engine for a submarine develops 1200 kilowatts. Its rate of fuel consumption is 5 kilograms per minute.

(b) A gas engine drives a dynamo which delivers 120 amperes DC at 110 volts. The gas is obtained from a gas-producing plant which is fed with coke at the rate of 150 kilograms in 8 hours.

(c) A gasoline engine drives a motor car at an average speed of 40 miles per hour (64·37 kilometres per hour) against a total resisting force of 275 pounds-force (1223 newtons). Its gasoline consumption is 20 miles per gallon (7·11 kilometres per litre). The specific gravity of the gasoline is 0·73. (1 gallon of water has a mass of 10 pounds-mass; 1 litre has a mass of 1 kilogram.)

(d) A base-load power-station incorporates a steam turbine plant driving AC generators. The hourly coal consumption of the boilers is 178·5 tonne and the electrical power delivered is 500 000 kilowatts. (1 tonne = 1000 kilograms.)

(e) A gas-turbine engine is used as a standby for peak-load power production in a generating station. The shaft power delivered by the engine is 105 kilowatts per kilogram of air flowing through the engine per second. The rate of air flow through the engine is 64 kilograms per second and the oil-fuel flow rate is 48 kilograms per minute.

(f) Each of the two reactors of the nuclear central generating station at Berkeley, Gloucestershire consumes 0·051 tonnes per day uranium-235. The *gross* power produced by each of the 4 generators is 85 megawatts. The power required to recirculate the coolant (CO_2 gas) through the reactors and the heat exchangers and to drive auxiliaries is 17 megawatts per reactor. (1 tonne ≡ 1000 kilograms.)

1.4 The performance of aircraft-propulsion devices is often expressed in terms of the number of kilograms of fuel consumed per hour per newton of thrust developed. Compare the following on this basis:

(a) A rocket carries a total fuel load of 60 tonnes. The rocket engine operates steadily for 2·5 minutes and develops a thrust of 1·5 million newtons.

(b) An aircraft powered by four gas-turbine jet engines consumes 135 000 litres of kerosine on a flight of 5600 kilometres when flying at a steady speed of 880 kilometres per hour. Each engine develops a thrust of 45 000 newtons. The specific gravity of kerosine is 0·81. (1 litre of water has a mass of 1 kilogram.)

(c) An aircraft is powered by a gas-turbine propeller engine (turbo-prop). In level flight at constant speed the engine consumes 7·2 kilograms of kerosine per minute and the drag of the aircraft is 7820 newtons.

1.5 An electricity supply for use in isolated districts incorporates a lead–acid secondary cell which may be charged periodically by means of a DC generator driven by a gasoline engine.

The average load on the accumulator during discharge at 100 volts is 40 amperes for 6 hours per day.

During charging the generator output is 33 amperes at 115 volts. The charging period is 8 hours per day and the weekly gasoline consumption of the engine is 100 litres. The specific gravity of the gasoline is 0·73. (1 litre of water has a mass of 1 kilogram.)

Determine (a) the power delivered by the accumulator during discharge, in kilowatts;

(b) the power delivered to the accumulator during charging, in kilowatts;

(c) the specific fuel consumption of the accumulator plus charging plant in kilograms of fuel per kilowatt hour.

2 Mechanical Quantities and their Units

Introduction

Chapter 1 has shown the engineer to be concerned with the production of power. Thermodynamics helps him to measure the work done by power-producing devices and to compare it with what might reasonably be expected of them. It is thus an aid to improvement and innovation.

Physical quantities, like power and work, require for their meaningful handling the establishment of systems of units. In this chapter, we set down the main facts about the quantities and units relevant to the subject of Mechanics, a subject which, in the present context, can be regarded as forming a part of Thermodynamics.

The first section of the chapter, entitled "General Remarks about the Science of Measurement", contains material with which all readers will be partly familiar; it may be skipped on first reading. Some readers will thereafter wish to return to this section to see what is meant by "the Common Convention of Consistency of Units" and "special conventions" (pages 25 and 26); further, if doubts arise about the status of the constant in Newton's Second Law, for example, or the distinction between force and mass, the discussion in this section may help to dispel them.

Symbols

a	Acceleration.	s	Distance traversed.
D	Diameter.	t	Time.
F	Force.	u	Velocity. Speed.
F_g	Force due to gravity (weight-force).	V	Volume.
		v	Specific volume.
G	Constant in Newton's Law of gravitation.	z	Vertical height above an arbitrary datum level.
g	Acceleration due to gravity.	ρ	Density.
g_c	Constant in Newton's Second Law of Motion.		
m	Mass.	*Subscripts*	
N	Speed of rotation.	atm	Atmospheric.
p	Pressure.	g	Gravitational.
r	Distance separating centres of gravity.	1, 2	Particular bodies of matter.
		12	Interrelation between 1 and 2.

General Remarks about the Science of Measurement

Measurement as an operation, real or imaginary

The abstractness of physical quantities. Even infants experience the different sizes of objects, learn what it is to wait or hasten, and test their strength against restraint or obstacle. Among the first words learned are: "big" and "small", "soon", "long" (for both distance and duration) and "strong"; and who can remember at what age he first spoke of "length" and "time"? Indeed, such ideas are acquired so early that only with effort can we recognize that they are *abstractions*, concepts in which man recognizes and mentally manipulates the measurable features of the distances he has travelled, the succession of events endured, and the speed or tardiness of action.

In this chapter we shall handle many such concepts, particularly: length, time, acceleration, mass, density, force and pressure. The reader may feel that he understands some of these better than others; it is therefore helpful to remark that they are *all* abstractions.

The essence of the measuring operation. The quantities in question are measurable; this is to say that their magnitudes can be expressed in terms of *numbers* and of *units*. The measurement of length forms the simplest example: if we adopt a straight stick as the unit, we can measure the length of a street by laying the stick down in successive steps. The street length will be given as a certain number of stick lengths.

Areas can be similarly measured. For example, we could adopt a certain postage stamp as our unit and establish how many would be needed completely to cover, say, a wall. The wall area would then be reported as the resulting multiple of the stamp area.

The measurement of time requires, even in principle, somewhat more complex apparatus than the measurement of length or area, whether the apparatus be an hour-glass, a candle with markings on its side, or a ship's chronometer. The hour-glass will serve to illustrate the essential feature; for the duration of a process could be measured by repeatedly inverting the glass when the sand ran out, and counting the number of inversions. What is observed however is not, let it be remarked, time itself; for abstractions will ever escape our gaze. The eye merely takes note of happenings, in this case the emptying of sand from the upper container; how much time has passed is then a matter for calculation.

The techniques of measuring physical quantities comprise the branch of science known as *metrology*. Here we need not, except occasionally, look in detail at how measurements are made in practice; it will suffice to understand the essence of the measurement operation, namely: *the comparison of the quantity measured with a counted number of standard units.* The number is called the *measure* of the quantity.

The use of more than one unit for a quantity of a given kind

Every school-child learns perforce that the entity *length* can be measured in metres, or in centimetres, or in a miscellany of other units, including miles, inches and feet. Similarly, the entity *time* can be measured in seconds, minutes or hours; and *area* in square metres, square millimetres or acres. The relations between two units for the same entity may be of two kinds, namely, those which are *defined* and those which arise from different *independent* standards. We now deal with each type in turn.

Definitional relations. The metre is a length unit which is one hundred times as big as the centimetre; the centimetre is ten times as big as the millimetre; and the kilometre is 1000 times as big as the metre. These relations result from the definitions of units in terms of other units; for example a centimetre is *defined* so that there are one hundred centimetres in a metre. In definitional relations, the multiplying factor needed to convert a measurement from one unit to another is usually a whole number (or unity divided by a whole number). In the International System of units (Système Internationale d'Unités, abbreviated to S.I.), the multiplying factors are powers of ten. Further, prefixes to the names of the basic units are used to identify the magnitudes of the factors; Table A. 2, Appendix A, contains the names and symbols of the prefixes which have been adopted.

Relations arising from the independence of standards. The Imperial system of units and the Metric system were established independently; for each, therefore, there was created an independent set of standards; thus for length, a standard yard was specified for the Imperial system and a standard metre for the Metric system. These standards were special metal bars carefully isolated from disturbing influences. When the two standards were compared experimentally, it was found that the ratio of their lengths was not a whole number; indeed only chance could have made it so. This is why, in converting a length measured in inches to the centimetre unit, one must multiply by a number which is not an integer, namely 2·54.

As metrology advances, the number of independent standards tends to be reduced; thus, both the standard-yard bar and the standard-metre bar have been abandoned. When this is done, previously experimental relations between units become definitional ones (see page 27 for examples).

Practice in technical work. In the interests of clarity, engineers and scientists try not to use more than one unit for a given entity in a single set of calculations.* Thus, either seconds *or* hours may be used, but seldom

---

* When a single unit is adopted, for example the foot or the centimetre, there is, incidentally, no special merit in either the Imperial or the Metric system; for to express, say, the speed of sound as $1·09 \times 10^3$ feet per second is neither more nor less clumsy than to call it $3·32 \times 10^4$ centimetres per second. There *is* a case for abandoning the Imperial in favour of the Metric system; but this is based more on the advantages of having a single system, whatever its basic unit, than on the facts that, in the metric system, the *decimal* multiples of the basic unit are used and *all the multiples have names.*

both. Departures from the practice are sometimes found however, usually as a concession to tradition; some examples will be found in the present book.

Elementary and composite units

Combination of units. So various are the quantities occurring in technical work that, were a special unit required for each, the metrologist's task would be almost impossible. Fortunately, however, the possibility exists of measuring many types of quantities by reference to a few elementary units; for example, speed may be measured in kilometres per hour, and acceleration in metres per second per second; there is no need, therefore to set up a "standard speed" and a "standard acceleration". Units for speed and acceleration can be regarded as *composite*, in contrast to those for length and time, which are *elementary*. Other composite units are: square millimetres, for area; cubic centimetres, for volume; and cubic feet per hour, for the volume flow rate of a fluid.

Composite units can be employed, and usually are so, whenever the entity in question is related by definitions to the entities chosen as elementary. Such are the relations between length, area and volume, or between time, length, velocity and acceleration.

The arbitrariness of the choice of elementary units. Lest it be supposed that entities measured in elementary units differ *inherently* from those measured in composite units, it should be stated that the distinction is arbitrary; it is entirely possible to choose, for example, the time unit and the velocity unit as elementary; then distance is measured by a composite unit. Thus the time unit might be the *year*, and the velocity unit the speed of light, called, shall we say: the *light*. With these elementary units, distances would be expressed in multiples of the composite unit, the *light-year*; this practice is indeed in use among astronomers.

Further, anyone may refrain, if he wishes, from employing a particular composite unit; for example, it is possible to use in a single calculation both a volume unit *and* a length unit, say the litre and the centimetre, or the pint and the foot. Units, and practices relating to them, have been invented by man for his convenience; and what is convenient in one context may be troublesome in another.

Abbreviations. The magnitude of a physical quantity is usually represented by a number followed by an abbreviation for the relevant unit. Thus a speed of 4·52 metres per second is written: 4·52 m/s. For many years, unit abbreviations were written in a variety of ways (e.g. for velocity: m./s, m/sec, m./secs, m./sec., m. sec.$^{-1}$). Now however, as a result of the work of the British Standards Institution (BSI) and similar bodies in other countries there is a wide measure of agreement; to use non-standard abbreviations without special (explained) reasons makes the interpretation of one's work by others difficult, and sometimes impossible.

Appendix A, Table A.1 contains some of the recommendations of BSI which are particularly relevant to the thermodynamicist; they should be adopted whenever there is no valid reason for a departure.

Single names for composite units. Some composite units have been given special names; for example, the unit cm²/s, when used for kinematic viscosity, is called a *stoke*; and a unit one hundredth of its size is called a *centistoke*. This practice burdens the memory, without relieving the tongue or pen. Other normally composite quantities are sometimes measured in units having single-name labels; for example, a volume may be measured in *litres*; and the speed of ships may be expressed in *knots*. Fortunately, such quantities find little favour among engineers; nevertheless, a few examples will be found in the present book (pages 34, 41).

Algebraic representation of physical quantities

The "consistency" of equations. Consider the well-known equation from elementary kinematics:

$$s = ut + \tfrac{1}{2} at^2 \tag{2.1}$$

Here s, u and a respectively "stand for" the distance moved by, the initial velocity of, and the acceleration of a particle, while t "stands for" the time which has elapsed; this means that the numerical measure of the relevant quantity may be substituted for each letter; and that then the equation will be true.

Now it is easy to understand that, if distance were measured in metres, time in seconds and velocity in kilometres per hour, the equation would *not* be true. Therefore a convention is implicit in the use of the equation: *only one elementary unit is used for each entity.* Thus, if distance is measured in metres and time in seconds, then velocity *must* be measured in m/s and acceleration in m/s². This implied practice may be called the *Common Convention of Consistency of Units.* It can be assumed to apply when no statement is made to the contrary. It can also be assumed, in the absence of contrary statements, that entities related by definition to those which have elementary units will be expressed in composite units formed from those elementary units.

The incorporation of conversion factors. Although a Consistency Convention is almost always employed during the course of algebraic manipulations, another convention may be preferred for the presentation of a formula which is frequently to be used for numerical evaluations. Consider, for example, the following formula *and its accompanying statements:*

$$u = 0 \cdot 0524 \, DN \tag{2.2}$$

where $u \equiv$ speed of wheel rim, in m/s,
  $D \equiv$ diameter of wheel, in m,
and $N \equiv$ speed of rotation, in rev/min.

Here two different units are used for the entity time (s and min), in departure from the convention of consistency; consequently, the units of *all* the quantities require explicit specification. The numerical constant, equal to $\pi \div 60$, incorporates the "conversion factor" from minutes to seconds.

Special conventions. Equations like (2.1) are superior to those like (2.2) in that they hold for any system of units, provided only that the Common Consistency Convention is obeyed; the acceptance of this one restriction brings freedom from the necessity to state in any further detail *what* unit system is to be used. Further freedoms can be bought at the price of further restrictions, as will now be illustrated.

Consider the equation:

$$m = \rho V \qquad (2.3)$$

where $m \equiv$ the mass of a liquid,
$\rho \equiv$ the density of the liquid, and
$V \equiv$ the volume of the liquid.

Since no statements are made about the units of m, ρ and V, it can be assumed that the consistency convention is obeyed; thus the units of the three terms might be: kg, kg/m³ and m³; or g, g/m³ and m³; or g, g/cm³ and cm³.

Now suppose that the liquid is water and that the last set of units is used; and recall that the density of water is, very nearly, 1 g/cm³. It follows that, so long as these units are used, the equation can be written more simply, namely as:

$$m = V \qquad (2.4$$

Indeed, this equation holds for a more general set of units: it is true *whenever the units of mass and length are such that the numerical value of the density of the liquid is unity.* The italicized clause describes a *special convention;* if it is accepted, the symbol ρ can be dropped from the equation. In the example, the saving in space seems trivial; but in a complex algebraic derivation it may be worthwhile.

Mechanical Units

The measurement of length

Relations between common units. The following Table 2.1* contains conversion factors connecting length units in common use in engineering. The numbers in bold type are *exact*, to distinguish them from the others which are rounded.

The currently accepted standard. It may be surprising, in view of the separate origins of the Imperial and Metric systems of measurement, that

---

* Appendix A, page 409, contains this table and many others relating to conversion factors for various quantities.

TABLE 2.1 *Units of Length*

	metre (m)	inch (in)	foot (ft)	kilometre (km)	mile
1 metre = (m)	1	39·370 1	3·280 84	0·001	6·213 71 × 10⁻⁴
1 inch = (in)	0·025 4	1	0·083 333 3	25·4 × 10⁻⁶	1·578 28 × 10⁻⁵
1 foot = (ft)	0·304 8	12	1	304·8 × 10⁻³	1·893 94 × 10⁻⁴
. kilometre = (km)	1000	3.937 01 × 10⁴	3.280 84 × 10³	1	0·621 371
1 mile =	1609·344	63 360	5280	1·609 344	1

exact relations should exist between units of the two different families. It was not always so; recently however *both* of the metal-bar standards have been abandoned, in favour of a standard length which can be reproduced in any standards laboratory in the world, namely the wave-length of the orange light emitted by the gas krypton under special experimental conditions. Both the yard and the metre are now defined in terms of the krypton wave-length and bear the exact relation to each other given in the table. At the time of the adoption of the new standard, it was found convenient to adjust the values of the length units slightly; the adjustments were so small that few practising engineers were even aware that they had been made.

Related composite units. Implied by the relations in Table 2.1 are further relations between the composite units for area and volume. Some of these are presented in Appendix A, Tables A.4 and A.5.

The measurement of time

General remarks. The time standard is based on the duration of the periods of the radiation of the atom of caesium-133 : 1 second = **9 192 631 770** periods. The second, minute, hour, day and year are used throughout the world; their interrelations require no description here.

Composite units. Quantities involving both length and time units are: velocity, and acceleration. Conversion factors for the former are presented in Table A.6 of Appendix A.

The measurement of mass

The concept of mass. The mass of a body is a measure of the amount of matter in the body; it is a property most commonly measured by weighing, i.e. by counting the number of standard masses which, when placed on the scale-pan opposite to the body in question, make the balance arm horizontal. Although the force of gravity thus plays a part in the measuring process, the mass of the body is no more exclusively connected with gravity than time is with the sand in an hour-glass; for the mass measured by weighing is independent of the actual value of the gravitational force; the weighing operation on a given body would give the same result for the mass, whether conducted on the earth or on the moon.*

The mass of a given collection of matter is found to be a constant; it cannot be altered by a change of location, by a deformation of shape, by a change of temperature, by chemical reactions which the matter undergoes, or by any other processes which it is normally possible to conduct.† These facts express the content of the Law of Conservation of Mass; it is the validity of this law which renders mass such a useful concept.

Relations between common units of mass. The Imperial and Metric systems originally possessed their own separate standards of mass, the pound and the kilogram respectively. Now only one standard is in use : the mass of the international prototype of the kilogram (a platinum block in France); the pound is *defined* by reference to this. Useful conversion factors are contained in Table 2.2.

TABLE 2.2 *Units of Mass*

	kilogram (kg)	pound (lb_m)	slug
1 kilogram (kg) =	1	2·204 62	0·068 521 8
1 pound (lb_m) =	0·453 592 37	1	0·031 081 0
1 slug =	14·593 9	32·174 0	1

Note also that the tonne is equivalent to 1000 kg; the UK ton ≡ 2240 lb_m ≡ 984·207 kg; the US ton ≡ 2000 lb_m ≡ 907·184 74 kg.

The abbreviation for pound. The international standard abbreviation for the unit of mass known as the pound is: lb. In the present book, we use the

* The inessentiality of gravity can be understood, as suggested by P. Melchior ("Masse und Kraft. Kilogramm und Kilopond." *Wäger und Wägung*, 1956, pages 133–135, 140–143, 148–150), by analogy with the Wheatstone Bridge; this measures an electrical resistance by comparison with standard resistances; it does not measure the voltage which is applied to the Bridge; indeed this voltage could have any value. In weighing, the gravitational field plays the part of voltage; it too could have any value, without affecting the result.

This reference appears in the textbook *Thermodynamik* by H. D. Baehr (Springer, Berlin/Göttingen/Heidelberg, 1962), which contains an excellent account of units and their meanings.

† Processes which *do* alter the mass of a given collection of matter are: a nuclear reaction; and acceleration to velocities which are not small compared with the speed of light.

deviant: lb_m; the reason is that we want to make quite sure that the *pound-mass* is distinguished, on every occasion that it is encountered, from the *pound-force* (see next section, p. 30), for which the abbreviation is: lbf. The use of the subscript m is our one departure from the recommended standard abbreviations.

Specific volume. A quantity of importance in thermodynamics is the specific volume of a body; this is defined as the volume of the body divided by its mass. The symbol adopted in the present book is v. A consequence of the definition is that the specific volume of a body is equal to the reciprocal of its density, thus:

$$v = \frac{V}{m} = \frac{1}{\rho} \tag{2.5}$$

Specific volume is measured in composite units: volume/mass. Conversion factors for various of these units are given in Table A.8 of Appendix A.

The measurement of force

The concept of force. The interactions between bodies which are described in common speech by words such as "push" and "pull" are epitomized by the abstraction: force. The concept would present little difficulty were it not for the unfortunate accident that some of the units, by reference to which force is measured, have nearly the same names as are used for mass units. For example, in the Imperial system pound-force and pound-mass are the terms which are used; but, in essence, however, the pound-force is as different from the pound-mass as both of them are from the pound sterling. In SI, however, this difficulty does not arise because the units for force and mass have different names: newton and kilogram.

In principle, a force can be measured by comparison (with the aid of a balance arm for example) with a counted number of standard forces; the latter can be, for example, standard springs compressed to a specified extent.

Units of force. Conversion factors connecting some common units of force are given in Table 2.3.

TABLE 2.3 *Units of Force*

	N	lbf	pdl	kgf
1 newton (N) =	1	0·224 809	7·233 01	0·101 972
1 pound-force (lbf) =	4·448 22	1	32·1740	0·453 592
1 poundal (pdl) =	0·138 255	0·031 081 0	1	0·014 098 1
1 kilogram-force (kgf) =	9·806 65	2·204 62	70·931 6	1

Currently accepted standard. In practice, springs cannot be manufactured with the accuracy, or maintained with the constancy, required of standards. Therefore, as in the case of length, a material standard of force is dispensed with; just as the metre is *defined* so as to equal a prescribed number of wavelengths of the orange light of krypton, so is the force unit *defined* to give a prescribed value to the constant which appears in Newton's Second Law of Motion. This important law, the foundation of the science of dynamics and an important element in thermodynamics is dealt with in the next section. Anticipating that section, we here state the definition of the unit of force which will be mainly used in the present book, the *newton*, with abbreviation: N.

One newton is the magnitude of the force which, acting on a body of mass equal to one kilogram, causes it to accelerate at the rate of 1 m/s² in the direction of the force.*

This definition is adopted because it is convenient to metrologists; its adoption is rendered possible by the universal validity of Newton's Second Law of Motion. However, this detail of metrological practice does not diminish the status of force as an independent entity, any more than the status of length is affected by the employment of the orange light of krypton. Forces could be measured even if Newton's law were not valid, just as lengths were measured for thousands of years before the gas krypton was discovered.

Newton's Laws of Motion and Gravitation

Newton's Second Law of Motion

Statement. When a force F acts on a body of mass m, the body experiences an acceleration a, in the direction of the force, in accordance with the relation:

$$F = \left(\frac{1}{g_c}\right)ma \tag{2.6}$$

where g_c is a universal constant, i.e. one which has the same value whatever the material of which the body consists.

Eq. (2.6) is a generalization of the results of innumerable experiments; no exception to its validity has ever been discovered.

---

* The corresponding definition for the pound-force unit is: one pound-force is the magnitude of the force which, acting on a body of mass equal to one pound-mass, causes it to accelerate at the rate of 32·1740 ft/s² in the direction of the force. The number 32·1740 is a rounded one; the exact value is 9·806 65 ÷ 0·3048. The figure 0·3048 is the number of metres in a foot, and 9·806 65 m/s² is the internationally accepted standard value of the acceleration of any body falling freely, in a vacuum, near the earth's surface

The magnitude of the constant g_c depends, of course on the units which are employed for the measurement of F, m and a; this means that g_c itself is "dimensional", and has composite units. Thus we can write:

$$g_c = 1 \text{ kg m/N s}^2 \tag{2.7}$$

Here the mass, length, force and time units are those used for the measurement of F, m and a in eq. (2.6).*

The dimensional nature of g_c arises out of the use of a four mechanical-unit set: mass, length, time and force.

An alternative procedure is to use mass, length, and time only in a three-unit set; in this case g_c would have the same numerical values as in the corresponding four-unit set, but would be dimensionless. Thus if SI units are used, g_c equals unity, and the unit of force, N, would be regarded as equivalent to the unit combination kg m/s². Such a practice corresponds to writing Newton's Second Law as $F = ma$.

The treatment of g_c in equations. Mathematical equations expressing the consequences of Newton's Second Law of Motion, if they obey the Common Consistency Convention (page 25), will contain the symbol g_c; computations based on them will be valid whatever unit system is selected for the calculations. If one of the four systems of units is adopted which gives g_c the numerical value of unity, this symbol can be omitted from equations; the omission may be a welcome relief when a large amount of algebraic manipulation has to be undertaken. The adoption of such a system can be regarded as a *special convention* (page 26); it occurs frequently in textbooks on the dynamics of solids and fluids.

In the present book, the amount of algebraic work is small; moreover Newton's Second Law of Motion is not of central interest. Consequently we shall include the symbol g_c in all relevant equations, which will thus conform to the Common Consistency Convention, but not to a special one. The presence of g_c in an equation is a sign that Newton's Second Law of Motion is operative in the underlying phenomenon. When specific units are required, we shall mainly use those of the International System: kg, N, m, s, for which g_c has the value 1 kg m/N s².

It should not however be presumed that the present practice is obligatory, or that special conventions are to be deprecated. We shall indeed, later in the book (page 99), make use of a different special convention, which sets equal to unity the value of a different universal constant: the mechanical equivalent of heat.

* The values of g_c for other common unit systems are as follows: 32·174 lb$_m$ ft/lbf s²; 4·1698 × 10⁸ lb$_m$ft/lbf h²; 1 g cm/dyne s²; 1 lb$_m$ ft/pdl s²; 9·806 65 kg m/kgf s²; 1 slug ft/lbf s². The numbers 32·174 and 9·806 65 have appeared in the footnote on page 30; 4·1698 × 10⁸ is (3600)² × 32·174. The last four values of g_c can be regarded as providing definitions of the respective units: the dyne, the pdl and kgf, for force; and the slug for mass.

Newton's Law of Gravitation

Statement. Two bodies 1 and 2, having masses m_1 and m_2 with centres a distance r_{12} apart, exert on each other, along the line joining the centres attractive forces of magnitude F_g, given by:

$$F_g = G\,m_1\,m_2/(r_{12})^2, \tag{2.8}$$

where G is a universal constant.

The magnitude of the quantity G depends, like that of g_c, on the units in which the other quantities are measured. We shall here quote only one example, namely:

$$G = (6{\cdot}670 \pm 0{\cdot}015) \times 10^{-11}\,\text{N m}^2/\text{kg}^2 \tag{2.9}$$

Here the $\pm 0{\cdot}015$ denotes the limits on the precision with which it is possible to measure G experimentally; for with the newton, the kilogram and the metre all defined already, it is not possible* to *define* G, in the manner adopted for g_c.

Gravitational force per unit mass on earth. Almost all human interests are confined to the surface of the Earth, the radius of which is approximately constant and equal to $6{\cdot}37 \times 10^6$m, and the mass of which is approximately equal to $5{\cdot}97 \times 10^{24}$ kg. It follows that, when the body "2" is the Earth the gravitational force per unit mass experienced by a body of mass m is from eq. (2.8), given by:

$$\frac{F_g}{m} \approx \frac{(6{\cdot}67 \times 10^{-11}\,\text{N m}^2/\text{kg}^2) \times (5{\cdot}97 \times 10^{24}\,\text{kg})}{(6{\cdot}37 \times 10^6\,\text{m})^2}$$

$$\approx 9.81\,\text{N/kg}. \tag{2.10}$$

Expressed in other units, the force per unit mass is, approximately 1 lbf/lb_m; 981 dyne/g; 32·17 pdl/lb_m; 1 kgf/kg; 32·17 lbf/slug.

It is of course the proportionality of the gravitational force to the mass of the body which permits masses to be measured (i.e. compared with counted number of standard masses) by means of weighing operations.

The "weight" of a body. The word "weight" is used in common speech in a number of different ways: for "body in a gravitational field", for the result of a weighing operation, i.e. for the mass, and for the gravitational force exerted on the body by the Earth. Lest this ambiguity lead to confusion it is wise to use the term "mass" exclusively, when that quantity is meant; the term "weight-force" can be used as an acceptable synonym for the gravitational force exerted by the attraction of the Earth.

---

* It *would* of course be possible to define G and so fix the force unit in terms of the mass and length units; but then g_c would have to be determined experimentally. The choice of practice is an arbitrary one, settled by reference to metrological convenience.

Whereas the mass of a body is a constant, the weight-force of the body varies (by a few tenths of a per cent) with position on the Earth's surface. Of course it decreases greatly in outer space, soon becoming negligible compared with the gravitational attraction exerted by other heavenly bodies.

Since, as stated above, for several systems of units, the gravitational force per unit mass, F_g/m, is closely equal to unity, the weight-force of a body and the mass of the body have measures which are almost equal; the unit systems in question are those for which the force unit and the mass unit have nearly the same name. This is no accident; it is a consequence of the difficulty that man has experienced, historically, in disentangling the concept of mass from the concept of force.

The interaction of Newton's Laws of Motion and Gravitation

The gravitational acceleration g. When the only force on a body is the weight-force F_g, the acceleration which it experiences, g, is known as the gravitational acceleration; this quantity is deducible from eq. (2.6), which can be put in the form:

$$g = \frac{F_g}{m} \cdot g_c \qquad (2.11)$$

The magnitude of the quantity g varies slightly with the position on the earth's surface, because F_g/m varies; it also depends on the system of units in use. Approximate values, accurate enough for almost all engineering work on the Earth's surface, are:

$$\left.\begin{aligned} g &\approx 9 \cdot 81 \text{ m/s}^2, \\ &\approx 32 \cdot 17 \text{ ft/s}^2, \\ &\approx 4 \cdot 17 \times 10^8 \text{ ft/h}^2. \end{aligned}\right\} \qquad (2.12)$$

A major implication of eq. (2.11) is that all bodies would fall, in a vacuum, at precisely the same rate, irrespective of their masses or constitutions. This is the famous hypothesis of Galileo, boldly based on observations in which the influence of air resistance was not completely excluded; it was advanced about a century before the formulation of Newton's laws.

Eq. (2.11) allows *the strength of the local gravitational field* to be specified by a statement of the local value of g; it is indeed more common to specify g than to state the value of F_g/m.

The relation between g and g_c. It will have been noticed that the numerical values of g are approximately equal, in some systems of units, to the numerical values of g_c for those unit systems. Thus, in the Imperial system, g_c may be expressed as $32 \cdot 1740 \text{ lb}_m\text{ft/lbf s}^2$; and g is approximately equal to $32 \cdot 17 \text{ ft/s}^2$.

These coincidences are not accidental; they derive from the historically explicable practice of adopting force and mass units such that F_g/m is close to unity on the Earth's surface. However, they should not be held to support in any way the fallacy that g_c is simply a standard value of the gravitational acceleration. It is not; for Newton's Second Law of Motion could prevail, as far as is known, in a universe from which gravitation was altogether absent; and certainly there are some unit systems (for example SI), in which the two quantities have different numerical measures.

Pressure and its Measurement

We conclude this chapter with a brief discussion of the pressure of a fluid, and of some of the means which are adopted in practice for its measurement: the *manometer*, the *barometer* and the *Bourdon gauge*. An elementary knowledge of hydrostatics is presumed.

Pressure. A fluid exerts forces on its boundaries; these forces are not concentrated at particular points, but are distributed. It is therefore useful to define the quantity *pressure* as the force exerted on a surface, normal to the surface, divided by the area of the surface. The units of pressure are correspondingly composite, for example: N/m^2. Tables 2.4 and 2.5 contain some conversion factors relevant to various units used for pressure.

TABLE 2.4 *Units of Pressure*

		N/m^2	lbf/in^2	lbf/ft^2	atm	kgf/cm^2
1 N/m^2	=	1	$1.450\ 38$ $\times\ 10^{-4}$	$2.088\ 54$ $\times\ 10^{-2}$	$9.869\ 23$ $\times\ 10^{-6}$	$1.019\ 72$ $\times\ 10^{-5}$
1 lbf/in^2	=	$6.894\ 76 \times 10^3$	1	144	$6.804\ 60$ $\times\ 10^{-2}$	$7.030\ 70$ $\times\ 10^{-2}$
1 lbf/ft^2	=	47.8807	6.944 $\times\ 10^{-3}$	1	4.725 $\times\ 10^{-4}$	4.882 $\times\ 10^{-4}$
1 atm	=	$1.013\ 250$ $\times\ 10^5$	14.6959	$2.116\ 215$ $\times\ 10^3$	1	$1.033\ 23$
1 kgf/cm^2	=	$9.806\ 65$ $\times\ 10^4$	14.2233	$2.048\ 160$ $\times\ 10^3$	$0.967\ 841$	1

TABLE 2.5 *Additional Conversion Factors for Units of Pressure*

1 bar (bar) = 10^5 N/m^2
1 pascal (Pa) = 1 N/m^2
1 standard atmosphere (atm) = 760 barometric millimetre of mercury, mmHg
= 29.9213 barometric inch of mercury, inHg

The manometer. Fig. 2.1 illustrates the principle of the manometer: a U-tube, the two arms of which need not be of equal, or even uniform, cross-sectional area, is connected to two gas containers, in which the pressures are

p_1 and p_2 respectively; the purpose of the manometer is to provide a measure of the difference between p_1 and p_2. The lower part of the U-tube is filled with a liquid having a density ρ which greatly exceeds the densities of both the gases. The vertical distance z, separating the two free liquid surfaces, is measured by a scale laid alongside.

According to the principles of hydrostatics, the pressures, the density, and the height difference are connected by:

$$p_1 - p_2 = \rho z \frac{g}{g_c}$$

$$= \rho z \frac{F_g}{m} \qquad (2.13)$$

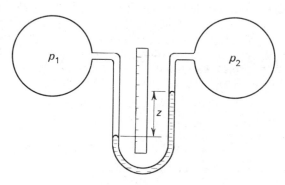

Fig. 2.1 Manometer.

Since ρ and F_g/m are normally known, measurement of z permits the *pressure difference, $(p_1 - p_2)$, to be calculated.* Then if p_2, say, is also known, p_1 can be determined.

EXAMPLE 2.1

Problem. What pressure difference, in N/m^2, corresponds to one centimetre difference in liquid-surface level in a manometer when the liquid is: (i) water, (ii) mercury?

Solution. The densities of the two liquids can be taken as 1000 kg/m^3 and $13\,568 \text{ kg/m}^2$ respectively. (The densities vary somewhat with temperature, and even with pressure, but these values are accurate enough for most purposes.) Also, in the absence of further information, it will be supposed that the strength of the Earth's gravitational field is 9.81 N/kg.

(i) Hence for water, from eq. (2.13),

$$p_1 - p_2 = 1000 \frac{\text{kg}}{\text{m}^3} \times 0.01 \text{ m} \times 9.81 \frac{\text{N}}{\text{kg}}$$

$$= 98.1 \text{ N/m}^2 \qquad\qquad \textit{Answer (i)}$$

(ii) Similarly, for mercury,

$$p_1 - p_2 = 13\ 568\ \frac{kg}{m^3} \times 0{\cdot}01\ m \times 9{\cdot}81\ \frac{N}{kg}$$

$$= 1331\ N/m^2 \qquad\qquad \textit{Answer}\ (ii)$$

Gauge pressure. Often one of the arms of a U-tube is open to the atmosphere, the pressure of which may be denoted by p_{atm}. Then the manometer height z gives a measure of the difference between the pressure in the other arm, say p_1, and the atmospheric pressure, p_{atm}; thus:

$$p_1 - p_{atm} = pz\frac{g}{g_c}$$

$$= \rho z\frac{F_g}{m} \qquad\qquad (2.14)$$

This difference is often called "the gauge pressure"; thus one might write: "the pressure of the gas in the container was $0.3 \times 10^5\ N/m^2$ gauge", meaning "the pressure in the container was $0.3 \times 10^5\ N/m^2$ in excess of the pressure of the atmosphere". So:

$$\text{pressure} = \text{gauge pressure} + \text{atmospheric pressure.} \qquad (2.15)$$

Of course, from the knowledge of the gauge pressure one can deduce the pressure (sometimes termed *absolute* pressure, to give emphasis to the contrast with gauge pressure), when the local value of p_{atm} is known. The latter quantity varies with altitude and with weather; near sea-level, it is usually within $\pm 5\%$ of 1 standard atmosphere, i.e. of $1{\cdot}013\ 250 \times 10^5\ N/m^2$.

"Vacuum." In steam-plant practice, negative gauge pressures may be expressed by use of the word "vacuum"; thus, "the pressure in the condenser was 700 millimetres of mercury, vacuum" is a way of saying that the gauge pressure was -700 mmHg gauge. Like "gauge", "vacuum" is a piece of jargon which needs to be understood, but which is of merely local significance. "Vacuum" is related to the pressure and the atmosphere pressure by way of eq. (2.15).

The barometer. A barometer is a form of U-tube; one arm is closed at the top, and the space above the liquid in that arm is evacuated* (Fig. 2.2). If the pressure in the open-ended arm is p and the difference of height of the two liquids is z, we have, since the pressure in the closed end is zero:

$$p = \rho z\frac{g}{g_c}$$

$$= \rho z\frac{F_g}{m} \qquad\qquad (2.16)$$

* In practice, complete evacuation is prevented by evaporation from the liquid surface. However the pressure exerted by the vapour is usually small and can be allowed for.

Thus, the barometer is an instrument which measures absolute pressures (i.e. simply pressures), not pressure differences (e.g. gauge pressures).

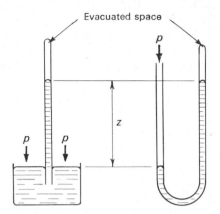

Fig. 2.2 Barometer.

EXAMPLE 2.2

Problem. A mercury manometer, measuring the pressure in a steam-plant condenser, reads 63 centimetres, vacuum. The barometer in the plant room reads 76 centimetres of mercury. What is the absolute pressure in the condenser in N/m²?

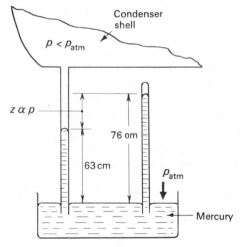

Fig. 2.3.

Solution. Fig. 2.3 illustrates the situation.

From the data: $\qquad\qquad p_{\text{atm}} = 76 \text{ cmHg}$

From eq (2.15): $\qquad\qquad p - p_{\text{atm}} = -63 \text{ cmHg}$

Hence, $\qquad\qquad\qquad P = (76 - 73) \text{ cmHg}$

From example 2.1: $\qquad 1 \text{ cmHg} = 1331 \text{ N/m}^2$

So $\qquad\qquad\qquad\qquad p = 13 \times 1331 \text{ N/m}^2$

$\qquad\qquad\qquad\qquad\quad = 17\,303 \text{ N/m}^2$ $\qquad\qquad\qquad$ *Answer.*

The Bourdon gauge. This instrument consists of an oval-sectioned metal tube, shaped as shown in Fig. 2.4. One end of the tube is open; it is fixed rigidly to the casing, and connected by further tubing to the vessel of which the pressure p is to be measured. The other end of the tube is closed; it is connected, through a gear-and-lever mechanism, to a pointer. A change in p tends to deform the section of the tube and so to deflect the free end; thus the location of the pointer on a suitably inscribed scale gives a measure of the pressure p.

Since it is the difference of pressure, $p - p_{atm}$, which deforms the tube section, the Bourdon gauge measures *gauge* pressure, not absolute pressure.

The angular movement of the pointer resulting from a unit change in p depends on the kinematics of the gear-and-lever mechanism, and on the

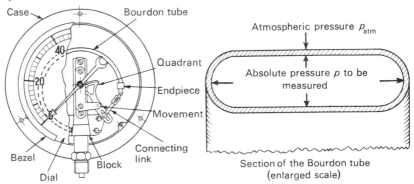

Fig. 2.4 Bourdon pressure gauge.

thickness and shape of the tube. Normally the instrument requires calibration, i.e. the marks on the scale must be numbered to agree with a series of experiments in which known pressure differences are applied to the Bourdon tube.

BIBLIOGRAPHY

BS 1991 : 1961. *Recommendations for letter symbols, signs and abbreviations. Part 2. Chemical engineering, nuclear science and applied chemistry.* British Standards Institution.

ANON. *Symbols, signs, and abbreviations recommended for British scientific publications.* The Royal Society, London, 1969.

BS 350 : 1959. *Conversion factors and tables. Part 1: Basis of tables; conversion factors.* British Standards Institution.

BS 1780 : 1960. *Pressure gauges.* British Standards Institution.

COHEN, E. R., CROWE, K. M. and DUMOND, J. W. M. *Fundamental Constants of Physics.* Interscience Publishers, New York, 1957.

JERRARD, H. G. and McNEILL, D. B. *A Dictionary of Scientific Units, including Dimensionless Numbers and Scales.* Chapman & Hall, London, 1963.

CHAPTER 2—PROBLEMS

Note: Unit-conversion tables are given in Appendix A, p. 409.

2.1 The mass of one of the pistons of a gasoline engine is 0·4 kg. At a particular point in its motion the piston has an acceleration of 6000 m/s². Calculate the net force on the piston in newtons.

2.2 A force of 8 N acts on a mass of (a) 100 g (b) 0·1 lb_m. Find the acceleration of each mass in cm/s².

2.3 A material unit of force is to be set up. The unit force, the *cant* (c), is specified as that force which, when acting on the end of a standard cantilever beam, will cause the end of the beam to deflect 0·100 cm.

It is found that 1 c when acting on 1 kg causes it to accelerate at 100 m/s².

(a) Evaluate g_c for the c, kg, m, s, set of units.

(b) Determine the relation between the N and the c.

(c) Compare the numerical values of the mass and the weight-force of a body in the c, kg, m, s, unit system at a place where the local gravitational acceleration is 9·81 m/s.

2.4 (a) The first artificial earth satellite was reported to have encircled the earth at a speed of 18 000 miles/h and its maximum height above the Earth's surface was stated to be 560 miles. Taking the mean diameter of the earth to be 7920 miles, and assuming the orbit to be circular, evaluate the value of the gravitational acceleration at this height. (The acceleration of a body moving with velocity V in a circular path of radius r is V^2/r towards the centre of rotation.)

(b) The mass of the satellite is reported to have been 86 kg. Taking the sea-level value of g as 9·81 m/s² estimate the gravitational force acting on the satellite at the operational altitude.

2.5 A mercury barometer reads 30·1 in. Express the reading in lbf/in². (Take $g =$ 32·2 ft/s² and the density of mercury $= 850\ lb_m/ft^3$.)

2.6 The air supply to an internal-combustion engine is metered by observing the pressure drop across an orifice in the air supply line to the engine. The pressure drop is measured by means of a manometer containing kerosine having a specific gravity of 0·81

Express a difference of level of 25 cm in the manometer in (a) N/m² and (b) metres of air. (Take the density of water as 1000 kg/m³ and of air as 1·2 kg/m³; take g as 9·81 m/s².)

2.7 A turbine is supplied with steam at a gauge pressure of 14 bar. After expansion in the turbine the steam passes to a condenser which is maintained at a vacuum of 710 mmHg by means of pumps (the wet and dry air pumps). The barometric pressure is 772 mmHg. Express the inlet and exhaust steam (absolute) pressures in N/m². Take the density of mercury as 13·6 × 10³ kg/m³.

2.8 A British manufacturer of internal-combustion engines is to supply a supercharged gasoline engine to a customer in Johannesburg.

The engine specification calls for a guaranteed absolute boost pressure (i.e. the pressure at which the petrol-air mixture enters the engine cylinders) of 18 lbf/in².

What reading should the customer obtain on a mercury manometer used to measure the boost pressure, and by how much (per cent) will it differ from that obtained during the manufacturer's test in Britain.

Use the following data:

	Altitude, feet	Atmospheric pressure, lbf/in²	Atmospheric temperature, °F	g ft/s²
Britain	180	14·6	60	32·2
Johannesburg	5740	12·0	79	32·1

The variation of the density of mercury, ρ, with temperature, is given by $\rho =$ 850[1 − 1·01 × 10⁻⁴(t − 60)] when ρ is measured in lb_m/ft^3 and t in °F.

3 Work

Introduction

In Chapter 1 it was stated that a main pre-occupation of the mechanical engineer is the production of *work*, whether for pumping water, for driving machine tools or for transportation. In the present chapter therefore a precise definition of work is given, in order that its amount can be measured in the various circumstances arising in engineering.

It will be assumed that the reader is familiar with the concept of work used in mechanics; however, this concept is too restricted to suffice for the more complex problems of thermodynamics. A new definition therefore has to be introduced which covers mechanical work but which includes other forms also. As a preliminary, one of the central ideas of thermodynamics, the *system*, has to be defined.

The remainder of the chapter will be taken up by discussions of the various forms of work which are of importance in engineering, and of the ways in which work can be evaluated.

Symbols

A	Area of piston face. Area of part of a system boundary.	P_s	Shaft power.
		p	System pressure.
a	Area of indicator diagram.	p_atm	Atmospheric pressure.
F	Force.	p_b	Pressure at a system boundary.
I	Current.		
k	Constant in $pV = k$ and $pV^n = k$.	p_1	Pressure on lower surface, eq. (3.7).
		p_m	Mean effective pressure.
L	Distance. Displacement. Length of piston stroke.	p_u	Pressure on upper surface, eq. (3.7).
		$\mathcal{Q}$	Quantity of electrical charge.
l	Length of indicator diagram.	r	Radius.
m	Mass of a system.	S	Indicator spring number.
N	Rotational speed.	T	Torque.
n	Exponent in $pV^n = k$.	V	System volume. Velocity.
P	Power.		
P_i	Indicated power.	V_c	Clearance volume in engine cylinder.

V_{sw} Volume swept by part of a system boundary. Swept volume of an engine cylinder.

$\overset{\cdot}{V}$ Potential difference.

v Specific volume of a fluid system.

W Work done by a system.

W_d Displacement work, "$p\,dV$" work.

W_s Shear work.

θ Angular displacement.

σ Shear stress.

τ Time.

Definition and Measurement of Work

Work as defined in mechanics

Most readers will be familiar with the following definition:

Work is done when the point of application of a force moves in the direction of the force. The amount of the work is equal to the product of the force and the distance moved in the direction of the force.

This definition is expressed symbolically by the equation

$$W = F \times L \tag{3.1}$$

where W is the work, F is the force, and L is the distance (see Fig. 3.1).

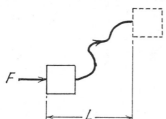

Fig. 3.1 Illustrating the definition of work used in mechanics.

Units of work. The definition implies that the units of work are composite. In the international system (SI), the unit of work is the newton metre, corresponding to the units used for F and L. Other units can of course also be used, e.g. the foot pound-force (ft lbf); in some cases, special names are used. thus 1 dyne centimetre is called an *erg*, while 10^7 ergs is called 1 *joule*; the latter is identical with one newton metre.

Although in thermodynamics a different definition of work is used, the units are the same. Table 3.1* contains some conversion factors for changing from one unit system to another.

Power is the rate at which work is done. It may be measured, for example, in N m/s when SI units are used; 1 N m/s is identical with one *watt* (W). Often a larger unit of power is used, the *kilowatt* (kW); it corresponds to 1000 W. In the Imperial system, the unit of power is the ft lbf/s; the larger unit, the horsepower (hp), (introduced by Watt in 1783), is equivalent to 550 ft lbf/s. Table 3.2 contains factors for converting some units of power.

* Appendix A, p. 409, contains this table and many others relating to conversion factors for various quantities.

TABLE 3.1 Conversion Factors for Units of Work

	N m	ft lbf	kWh	hp h	m kgf
1 N m (≡ 1 joule (J)) =	1	0·737 562	$0·277\ 778 \times 10^{-6}$	$0·372\ 506 \times 10^{-6}$	0·101 972
1 ft lbf =	1·355 82	1	$0·376\ 616 \times 10^{-6}$	$0·505\ 051 \times 10^{-6}$	0·138 255
1 kWh =	$3·6 \times 10^{6}$	$2·655\ 22 \times 10^{6}$	1	1·341 02	$0·367\ 098 \times 10^{6}$
1 hp h =	$2·684\ 52 \times 10^{6}$	$1·98 \times 10^{6}$	0·745 700	1	$0·273\ 745 \times 10^{6}$
1 m kgf =	9·806 65	7·233 01	$2·724\ 07 \times 10^{-6}$	$3·653\ 04 \times 10^{-6}$	1

TABLE 3.2 Conversion Factors for Units of Power

	N m/s	ft lbf/s	metric hp	hp	kgf m/s
1 N m/s (≡ 1 watt (W)) =	1	0·737 562	$1·359\ 62 \times 10^{-3}$	$1·341\ 02 \times 10^{-3}$	0·101 972
1 ft lbf/s =	1·355 82	1	$1·843 \times 10^{-3}$	$1·818\ 18 \times 10^{-3}$	0·138 255
1 metric hp =	735·499	542·476	1	0·986 320	75
1 hp =	745·7	550	1·013 87	1	76·040 2
1 kgf m/s =	9·806 65	7·233 01	0·013 333 3	0·013 150 9	1

Positive and negative work. The length L in eq. (3.1) is measured in the direction of F. In Fig. 3.1, L is a positive quantity so the product $F \times L$ is positive: in this case we say that *positive work* has been done.

However, the point of application of the force might have been "pushed backwards", making L, and the product $F \times L$, negative quantities: in that case we would say that *negative work* had been done by F.

A similar distinction of *sign* is made in the thermodynamic definition.

The system

Before proceeding further, we introduce a concept which, despite its simple appearance, is of profound importance in thermodynamics: the *system*. Its definition is as follows:

A system is any prescribed and identifiable collection of matter.

Nearly all the statements and laws of thermodynamics relate to a system. Fig. 3.2 shows the usual representation of one. The important part of this

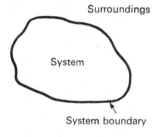

Fig. 3.2 Conventional representation of a system.

amoeba-like sketch is the *boundary*, which completely encloses the system and separates it from the *surroundings*. Of course the surroundings can also be regarded as a system.

The boundary will usually not be a material envelope but an imaginary closed surface which contains (prescribes and identifies) the collection of material we are talking about. Consequently no material crosses the boundary of a system, by definition. Often the system will change *position* and *shape* while we are watching it, but the boundary then moves so as always to hold the same collection of matter. This means that a system has *constant mass*.

Thermodynamics is concerned with *interactions* between systems, i.e. the "influences" of a system on its surroundings and vice versa. *These inter-actions are of only two kinds*: the first is work, the subject of the present chapter; the second is heat, which is discussed in Chapter 5. Thereafter we shall be concerned with the changes within the system brought about by these interactions at its boundaries.

The thermodynamic definition of work

We are now in a position to give the definition of work which will be used throughout this book. It is

Positive work is done by a system, during a given operation, when the *sole* **effect** *external* **to the system** *could be reduced to* **the rise of a weight.** *

This definition may appear arbitrary; yet some such form is forced on us by the need to make a distinction between work and heat which the Second Law of Thermodynamics states as existing in the physical world. This necessity will not be fully apparent until Chapter 10 is reached. Below, each of the italicized phrases will be discussed. First however, we hasten to show that the definition covers the restricted kind of work treated in mechanics.

Relation between "mechanical" and "thermodynamic" work. Fig. 3.3a shows a system S exerting a force, *F*, on its surroundings at one point of its

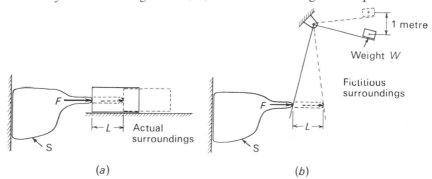

Fig. 3.3 Illustrating the relation between "mechanical" and "thermodynamic" work.

boundary. The surroundings comprise a block on a rough surface, which is pushed along as the system changes shape. Therefore positive work is done, according to the "mechanics" definition.

Now in this process no weights have been raised in the surroundings. Yet by carrying out the *same operation* (i.e. the same change of shape of S, with the same forces at the boundaries) in *changed surroundings*, a weight *could* have been raised, as is shown by Fig. 3.3b; and this could have been the *sole external effect*. Therefore the system does positive work in the operation according to the thermodynamic definition.

Measurement of work. The magnitude of the positive-work interaction is accomplished by counting the standard weights which could have been raised through a standard vertical height in a standard gravitational field. In the international system of units, the weights are each taken to comprise an amount of material on which the gravitational force is 1 N; the standard height is 1 m, and the gravitational field has the standard value of 9·806 65 N/kg. The work unit is the N m as already stated.

Application of the principle of the lever to the operation in the fictitious surroundings of Fig. 3.3b shows that the number of standard weights raised

---

* Here and throughout this book the word "weight" means a body in a gravitational field.

through 1 m would be $F \times L$. The work done by the system is therefore $F \times L$, which is in accordance with the mechanics definition.

Width of application of the thermodynamic definition of work. The example just considered does not explain our discontent with the mechanics definition. An example will now be given to show that the thermodynamic definition covers a wider field, and is therefore superior.

Consider the system S shown in Fig. 3.4*a*, comprising an electrical storage battery. External to the system boundary, the terminals are

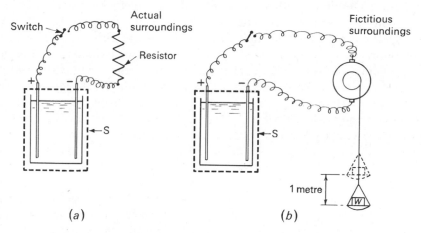

(*a*)   (*b*)

Fig. 3.4 Illustrating the application of the thermodynamic definition of work to an electric storage battery discharging through a resistor.

connected to an electrical resistance coil through a switch. We suppose that the switch is closed for a period, during which current flows through the battery and the resistance; the latter becomes warmer as a result. The system has therefore interacted with its surroundings. Can this interaction be classified as work?

According to the mechanics definition the answer is "No", for no force has moved its point of application. Yet, according to the thermodynamic definition, the answer is "Yes", as will now be demonstrated.

Imagine the surroundings of the system to be altered by the replacement of the resistance by an electric motor which winds up a string on which a weight is suspended (Fig. 3.4*b*). This operation can be arranged so that, when the switch is closed, exactly the same current flows as when the resistance was there; as far as events within the battery (system) and at the terminals are concerned, there has been no change: the battery has performed the same operation at the boundary. Yet this time, provided care has been taken to eliminate friction from the motor bearings and to use thick wire so that the motor does not become warmer, the sole effect external to the system has been the rise of a weight.

Therefore according to the thermodynamic definition the system does positive work in the operation, whether its surroundings comprise the resistance or the motor.

Remarks on the thermodynamic definition of work. We now consider the reasons for the wording used in the definition.

(*a*) *"Sole effect"*. The qualification "sole" is necessary because there is another kind of interaction between a system and its surroundings which can have the rise of a weight as *part* of its effect: this we shall later learn to call *heat*. Thus a hot body (system) placed in contact with water can cause it to boil; the steam can then drive an engine which raises weights. It will be found however that there are always other effects; either the water does not return to its original state after passing through the engine, or the surroundings become warmer. The weight-raising cannot be the sole effect.

(*b*) *"External"*. This word emphasizes that work is defined only with reference to a system boundary, separating the system from its surroundings. Work is an interaction at the boundary. If the boundary is drawn differently, i.e. if a different system is chosen, the magnitude of the work may be different. Thus, if in Fig. 3.3*a* the boundary were to enclose the block and the rough surface as well, the work would be zero; for there are *no* effects external to this new boundary.

(*c*) *"Could be reduced to"*. The examples already given show that, for work to be done by the system, it is not required that weights actually *are* raised. It suffices for us to imagine a means by which this could be the sole effect. Imagination must be kept within the bounds of what is physically possible however. It is permissible to imagine the friction in the imagined mechanism to be zero, because we know that, by taking sufficient trouble, friction can always be made negligibly small. On the other hand, merely to postulate a "black box", which will convert *any* interaction completely into weight-raising, is not imagination but fantasy; it is not permissible.

The imaginary reconstruction of the surroundings must not involve any alteration to the operation performed by the system. If it does, we find ourselves evaluating the work in a different operation; there is no point in that.

(*d*) It will have been noted that the definition is restricted to *positive* work only. Negative work will now be discussed.

Thermodynamic definition of negative work

When a system does positive work, its surroundings do an equal quantity of negative work. Conversely, when the surroundings do positive work, the system does an equal quantity of negative work.

In symbols:

$$W_{system} + W_{surroundings} = 0 \qquad (3.2)$$

This definition enables us to evaluate the work when, for example, the system of Fig. 3.3 is compressed rather than extended, or when the battery of Fig. 3.4 is charged instead of discharged. We do it by turning our attention to the surroundings, treating *them* as the system, and applying the definition of positive work to the operation performed by them.

It may be thought that negative work could have been defined more simply as occurring when the sole external effect could be the *fall* of a weight. This is not so. Once again the apparent circumlocution results from the need to distinguish work from the interaction heat in accordance with the Second Law; for there exist interactions *not* involving negative work according to the above definition which, as far as the system is concerned, could be caused solely by the fall of an external weight. (See for example p. 96.)

Sign convention. When positive work appears in equations, it is given the symbol W. If, on evaluation, W turns out to stand for a negative number,

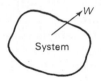

Fig. 3.5 Conventional representation of a system doing work.

the work done by the system is negative. This is the usual algebraic sign convention used for other physical quantities such as length.

A diagram of the system is an essential part of every thermodynamic calculation. Work is indicated on the diagram by an outwardly directed arrow as shown in Fig. 3.5.

In speech, it is often convenient to describe the direction of work by emphatic use of the prepositions "*by*" and "*on*". We say, "work is done *by* the system", when W stands for a positive number; and "work is done *on* the system", when W stands for a negative number. It follows that if work is done *by* the system it is done *on* the surroundings, and vice versa.

Occasionally, the work interaction is called a work *transfer*; specifically, if work is done on a system, the transfer is said to be "to" the system. It must be understood, however, that the idea of transference is at best a metaphor.

Concluding remarks on the definition of work

Four points should be noted before leaving this subject. The first is that the measurement procedure for work involves the arbitrary specification of a standard environment (surroundings) for the system (the standard weights) and a standard effect in the surroundings (the increase of height in a standard gravitational field). This pattern will be repeated when we come to define heat in Chapter 5.

Secondly, the measurement procedure implies an observer; for an increase of elevation means motion relative to some point taken as fixed. Usually we think of the observer as standing on the earth counting the weights rising past his head, but this is not necessary or even always convenient. For example, the observer *could* sit on the weight in Fig. 3.3*b*, in which case he would say that no weight had been raised past *him*, and so would assert that the system had done no work. From his point of view he would be right. That no inconsistency is involved will not be apparent until energy is discussed in Chapter 6.

Thirdly, work is *transient*. It is present during the interaction but does not exist either before or after the interaction. It is something which happens to a system but it is not a characteristic of the system. It therefore differs from, for example, pressure or force, which can have finite values in a system at rest.

Fourthly, work can be associated with a particular part of the boundary of a system (that place where the force *F* is exerted in Fig. 3.3; the terminals in Fig. 3.4). It will often be convenient to consider individually the work interactions associated with events occurring at particular parts of the boundary. To compute the magnitudes of these individual work interactions, we have to imagine that the individual parts of the surroundings are separately replaced by weight-raising mechanisms. The net work done by the system will then be given by the sum of individual work interactions (see also page 72).

Displacement Work

The remainder of the chapter will be taken up by the derivation of relations between the work done by a system and measurements made at the system boundary. To derive these relations we have to consider various ways in which work can be done. Since in engineering thermodynamics the system often comprises a fluid, rather than a rigid structure, special attention is given to this case in the present section.

It will be observed that, in the examples dealt with first, the mechanics definition of work is just as good as the thermodynamic one. In the later examples the greater scope and power of the thermodynamic definition will become evident.

Stresses at the boundary of a system

The force exerted by a system on its surroundings across an element of area of the system boundary can be split into two components: one normal to the area element; the other in the plane of the element. Since the magnitudes of these components vary, in general, with position, it is useful to consider the magnitudes of the forces *per unit area*; these are known as the *stresses*.

The stress acting normally to the area in the direction away from the system is known as the *pressure*, p_b.* The component in the plane of the area element is known as the *shear stress*, σ. Both these stresses can result in work being done when the system boundary moves. The normal stress is usually the more important in thermodynamics; we consider it first. The mode of work associated with the normal stresses is known as *displacement work*, or, for reasons which will appear, "*p* d*V*" *work*.

Displacement work done at part of a system boundary

Work done on a piston. Fig. 3.6 illustrates a mechanism comprising a piston and a cylinder, for example for a steam engine. The cylinder is filled

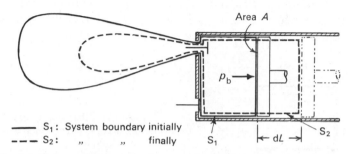

Area *A*

$p_b \longrightarrow$

—— S_1: System boundary initially
– – – S_2: „   „   finally S_1 |← d*L* →| S_2

Fig. 3.6 Piston-cylinder mechanism used in the derivation of eq. (3.6).

by a working fluid, for example steam, which enters and leaves through valves at the end remote from the piston.

The mass of working fluid will be considered which is enclosed by the system boundary S. Initially this boundary is in the position marked S_1 in Fig. 3.6; finally it occupies that marked S_2. Attention will be restricted to that part of S which is in contact with the face of the piston.

The pressure at all points of the piston face will be taken as having the uniform value, p_b. If the piston-face area is A, the force F on it is given by

$$F = p_b A \qquad (3.3)$$

Suppose that the piston moves an infinitesimal distance, d*L*, to the right. Then the displacement work done by the system as a result of this movement, dW_d, is given by eq. (3.1) as

$$dW_d = p_b A \, dL \qquad (3.4)$$

An alternative expression is derived by noting that $A \, dL$ represents the infinitesimal volume *swept out* by the part of the system boundary in question. Calling this volume increment dV_{sw}, we can write

$$dW_d = p_b \, dV_{sw} \qquad (3.5)$$

whence the name "*p* d*V* work" for displacement work.

* The subscript b here acts as a reminder that the pressure at the boundary is in question. There may be other pressures prevailing *within* the system.

Eq. (3.5) is a differential equation. If the piston moves through a finite distance, the work done on the piston face must be evaluated by integrating eq. (3.5). If 1 and 2 denote the initial and final piston positions, and W_d is the displacement work done in the process, we have

$$W_d = \int_1^2 p_b \, dV_{sw} \qquad (3.6)$$

If this is to be evaluated, the value of the boundary pressure at each increment of volume must be given; for in general p_b varies with piston position.

The interpretations of p_b and dV_{sw} in the integral are important. Particularly to be noted is that dV_{sw} is *not* in general the differential of the system volume. For example when a rigid mass is suddenly placed on a

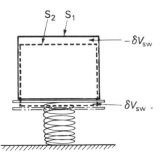

Fig. 3.7 Illustrating that swept volume is not the same as the change in system volume.

spring-supported platform (Fig. 3.7), it does displacement work in its subsequent motion; for, although the volume of the mass does not change, the pressure on the lower surface exceeds that on the upper surface. Let these two surface pressures be p_1 and p_u respectively. As the lower surface sweeps out a small volume, δV_{sw} say, the upper one sweeps out the volume $-\delta V_{sw}$. The net displacement is therefore

$$\delta W_d = p_1 \delta V_{sw} + p_u(-\delta V_{sw})$$
$$= (p_1 - p_u)\delta V_{sw} \qquad (3.7)$$

which is positive.

It should also be noted that, in general, the total work done by a system is not equal to $\int p_b \, dV_{sw}$, because other sorts of work, e.g. electrical, may be done simultaneously.

Resisted expansion; engine indicators

Engineers need to measure the work done on the piston of an engine for comparison with the work which is actually delivered at the output shaft (these quantities usually differ because of bearing friction). Eq. (3.4) shows that this may be done by measuring the pressure exerted by the fluid on the piston face throughout its movement. However this is not easy to do directly because of the rapid movement and the inaccessibility to observation of the piston.

Fortunately most piston-cylinder machines operate at a sufficiently low speed for the fluid pressure at any instant to be very nearly *uniform* throughout the cylinder. This situation is known as *fully-resisted expansion* (or compression) of the fluid and arises when all the effort of the fluid goes into moving the piston, and none into moving itself. More precisely, it occurs when the speed of piston motion is much less than the velocity of sound in the fluid. The latter is usually some hundreds of metres per second; the

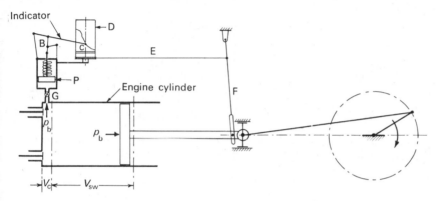

Fig. 3.8 The essential features and the general arrangement of an engine indicator.

former rarely exceeds 20 m/s even in the engines of racing cars. Sometimes a fully-resisted situation is called *quasi-static*.

A consequence of the uniformity of pressure in the cylinder is that the piston-face pressure, p_b, can be inferred from a measurement of the pressure at a part of the cylinder wall where a pressure gauge can be more easily fixed.

The engine indicator. In order to evaluate the work done on the piston, the piston-face pressure needs to be recorded at each piston position. This is accomplished for slow-speed engines by a device known as the spring-and-piston engine indicator, invented by J. Southern in 1796.

Figs. 3.8 and 3.9 show the essential features. The cylinder pressure is measured by the motion of a small piston, P, moving in a small cylinder communicating with the main one. The motion of this piston is opposed by a spring, the compression of which is directly proportional to the pressure in the cylinder. The motion of the indicator piston is transmitted through a mechanical linkage B to a stylus C; this may be pressed manually against a piece of paper. A rise of cylinder pressure causes the stylus to move vertically over the paper.

The cylinder pressure has to be plotted against piston displacement. Instead of making the stylus move horizontally in response to piston position, it is found easier to transmit this second motion to the paper, which is therefore mounted on a drum D rotating about a vertical axis. The

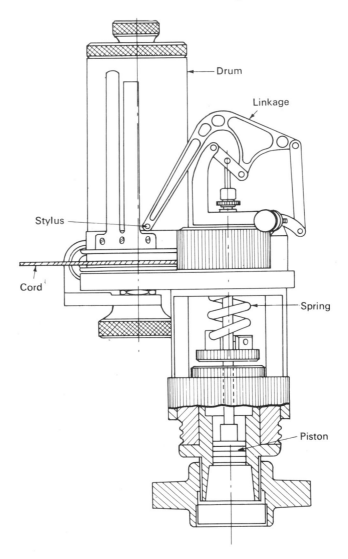

Fig. 3.9 Spring-and-piston indicator.

rotation is made proportional to engine-piston movement by a system of strings and levers (E, F).

The indicator diagram. When the engine is running steadily, the stylus is pressed against the paper for a revolution or two. Since events in the cylinder repeat themselves, the stylus traces out a closed curve on the paper;

this curve is called the *indicator diagram* or *indicator card*. Fig. 3.10a shows such a diagram, typical of a steam engine. It discloses the following events:—

When the piston is in its extreme left-hand position, the cylinder pressure has its highest value (at point a). The cylinder is now in communication with the steam-supply main through the inlet valve. The pressure remains high at first as the piston moves to the right, but begins to fall when the inlet-valve is closed (the *cut-off* point, b). The stylus then traces out a roughly hyperbolic path which steepens as the piston approaches the extreme right-hand position. This is due to the opening of the exhaust valve (at the *release* point, c), which causes the cylinder to communicate with the

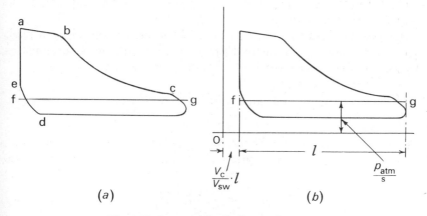

(a)   (b)

Fig. 3.10 Steam-engine indicator diagrams.

condenser which is at a low pressure. The cylinder pressure remains low throughout almost the whole of the return path of the piston, but rises again towards the end of the piston travel, as a consequence of the closure of the exhaust valve (at the *compression* point, d). Thus the steam pressure is already high when the inlet valve opens once more (at the *admission* point, e).

In addition to the closed curve, an *atmospheric line*, fg, is recorded. This is usually achieved by setting the indicator cock, G (Fig. 3.8), so that the indicator is isolated from the engine cylinder and is placed in communication with the atmosphere. The atmospheric line is required if the absolute pressure of the fluid in the cylinder has to be measured.

Evaluation of the work done per revolution: indicated work. We have seen that the displacement work done at a piston face may be determined from eq. (3.6),

$$W_d = \int p_b \, dV_{sw} \qquad (3.8)$$

provided that the pressure at the piston face, p_b, and the corresponding volume swept by the piston, dV_{sw}, are known. The indicator diagram provides this information since it is a plot, to scale, of the pressure on the piston face for each position (displacement) of the piston. The area of the diagram is therefore proportional to the displacement work done at the piston face; for vertical height represents p_b, while horizontal movement represents dV_{sw}; these are the two quantities appearing in eq. (3.8).

The evaluation of the work done on the piston face usually proceeds in the following manner. It is necessary to know the vertical and horizontal scales of the diagram. The first scale is known as the *spring number*, S; this is measured in N/m^3 and represents the change in steam pressure in N/m^2 which causes the stylus to rise one metre on the paper. The second is equal to V_{sw}/l, the volume swept out by the engine piston divided by the horizontal width of the indicator diagram.

The area enclosed by the stylus trace is measured by means of a planimeter. If this area a is in square metres, if V_{sw} is in m^3, and if l is in metres, the *indicated work* per revolution, in N m/rev, is given by:

$$\text{indicated work} = aS \times \frac{V_{sw}}{l}$$

Sometimes it is thought preferable to calculate in terms of the *mean effective pressure* p_m. This is given by S times the average height of the diagram:

$$p_m = \frac{aS}{l} \tag{3.9}$$

Then:

$$\text{indicated work} = p_m V_{sw} \tag{3.10}$$

It is often convenient to express this work on a time basis. If the number of revolutions per second is N, the rate at which work is done on the piston face, usually called the *indicated power*, P_i, in N m/s, (or watts), is given by:

$$P_i = N p_m V_{sw} \tag{3.11}$$

Expressed in kilowatts, and with V_{sw} replaced by the product of the length of piston stroke L and the piston area A, eq. (3.11) becomes

$$P_i = p_m LAN \times 10^{-3} \tag{3.12}$$

a formula which some engineers find easy to remember.

EXAMPLE 3.1

Problem. A single-cylinder steam engine has a cylinder diameter of 0·25 m and a stroke of 0·35 m. The area of the indicator diagram is 880 mm² and its mean length is 69 mm; the indicator spring number is 15×10^6 N/m³. Calculate (a) the indicated work per revolution, and (b) the indicated power, given that the engine speed is 140 rev/min.

Solution

(a)   Swept volume, $V_{sw} = \dfrac{\pi}{4} \times 0.25^2 \times 0.35 = 0.017\ 18\ m^3$

Mean effective pressure, $p_m = \dfrac{880}{69} \times 10^{-3} \times 15 \times 10^6 = 191.5 \times 10^3\ N/m^2$

$\therefore$ From eq. (3.10) we have:

$$\text{Indicated work} = 191.5 \times 10^3 \times 0.017\ 18$$
$$= 3290\ N\ m/rev \qquad\qquad \textit{Answer} (a)$$

(b) From eq (3.11):

$$\text{Indicated power, } P_i = \text{indicated work} \times N$$
$$= 3290 \times \dfrac{140}{60}$$
$$= 7670\ N\ m/s$$
$$= 7.67\ kW \qquad\qquad \textit{Answer} (b)$$

Remark: Note that p_{atm} does not affect the work done on the piston; since the atmosphere does work equal to $p_{atm}\ V_{sw}$ as the piston moves into the cylinder and to $-p_{atm}\ V_{sw}$ as it moves out, the net work done by the atmosphere in a revolution is zero.

Two-stroke and four-stroke engines. The above analysis applies to engines in which the sequence of events occurring in the engine cylinder repeats itself once for every revolution of the crankshaft. Such engines are referred to as *two-stroke* engines. A *four-stroke* engine on the other hand completes the sequence of events in *two* revolutions of the crankshaft. In this case the indicated work, given by eq. (3.10), is the work done per two revolutions; so, to obtain the indicated power, the indicated work has to be multiplied by $N/2$; correspondingly the indicated power will be *one half* that given by eq. (3.12). Reciprocating internal-combustion engines operate either the two-stroke or the four-stroke cycle, whereas steam engines invariably use the two-stroke cycle.*

Single-acting and double-acting engines. A feature of the engine considered above is that only one face of the piston is exposed to the working fluid; for this reason it is termed a *single-acting* engine. In *double-acting* engines, however, the working fluid is admitted to both ends of the cylinder with the result that work is done on each face of the piston. In this case there is an indicator diagram for each end of the cylinder and consequently the indicated power will be approximately double† that of the corresponding single-acting engine. Most reciprocating steam engines are double-acting, and so are many marine diesel engines; this arrangement gives the larger

* The use of the term "cycle" to describe the complete sequence of events occurring in the engine cylinder is common practice. Later, however, we use the term to describe a special sort of thermodynamic process.

† It is only approximately double for two reasons: (i) the indicator diagrams for each end of the cylinder are never exactly similar and, (ii) the piston rod reduces the effective area of the piston on one side, so that the swept volumes are not equal for both ends of the cylinder.

power per unit mass of engine. Internal-combustion engines for road transport are always single-acting; such engines are too small to warrant the complication of the piston-cooling devices which would be needed if hot gases were in contact with both sides of the piston.

Further remarks about indicator diagrams. The spring number, S, fixes the pressure scale of the diagram since 1 metre vertical movement of the stylus corresponds to S N/m² pressure change in the engine cylinder. It follows that the line of zero pressure is p_{atm}/S metres below the atmospheric line (Fig. 3.10b). The swept volume, V_{sw}, fixes the horizontal scale of the diagram, since l metres corresponds to V_{sw}, Fig. 3.10b; correspondingly the line of zero volume is (lV_c/V_{sw}) metres from the left-hand end of the diagram, where V_c is the *clearance volume*, i.e. the "dead" space at the end of the cylinder which is not traversed by the piston (Fig. 3.8).

With the axes of the indicator diagram so fixed, the (absolute) pressure and the volume of the *fluid in the cylinder* may be determined for any piston position.

Although an indicator diagram has pressure as ordinate and volume as abscissa, it must not be confused with the pressure-volume relation of a system, discussed below (p. 57); for even during one revolution, the quantity of material within the cylinder changes as a result of the valve openings and the piston motion.

Displacement work done at the whole system boundary

Having discussed the displacement work done at *part* of a system boundary, we now consider the system *as a whole*. To evaluate the displacement work for the whole system boundary, we take the summation of the values of displacement work for each part of the boundary.

The case most frequently met in practice is that of a system the boundary of which may be divided into sections which have uniform, but different, pressures acting on them. For such a system we may, from eq. (3.6), write the summation as

$$W_d = \int_A p_b \, dV_{sw} + \int_B p_b \, dV_{sw} + \dots \qquad (3.13)$$

in which the terms on the right-hand side represent the displacement work done at sections A, B, ... of the boundary, on each of which there is a different uniform pressure.

To illustrate this procedure we consider two cases. The first is the specially simple case in which the pressure is uniform throughout the system; in the second the system can be sub-divided into uniform pressure regions.

Displacement work when the system pressure is uniform. Consider, for simplicity, the case in which a fluid system S is enclosed within a cylinder by a leakproof piston (Fig. 3.11). The cylinder has no valves, so the same collection of matter is present at all times. If only slow motions of the piston occur, the expansions and compressions can be regarded as "fully-resisted", so the pressure is uniform at all points, as has been seen*; a single pressure

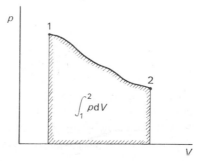

Fig. 3.11 Cylinder and piston enclosing a fluid system.

gauge will therefore indicate the system pressure p, which also prevails over the whole system boundary.

Since both pressure and volume have definite values at any instant, we can represent the history of the system on a pressure~volume diagram. Fig. 3.12 illustrates this, the system volume V being employed as abscissa. Such a plot is called a *state diagram*.

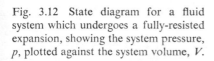

Fig. 3.12 State diagram for a fluid system which undergoes a fully-resisted expansion, showing the system pressure, p, plotted against the system volume, V.

$$\int_1^2 p\,dV$$

At this point it is convenient to introduce some definitions for the sake of precision. Some of the terms have already been used.

A *property* is any (macroscopic) observable characteristic of a system. For example, pressure, volume and density are properties.

The *state* of a system is the totality of the properties of the system, i.e. it is the complete description of the system.

A *state-point* is a point on a diagram representing the properties of a system at any moment.

A change of state of the system is the consequence of a *process;* the latter comprises the succession of states passed through, known as the *path* of the process, and the succession of interactions at the boundary.

* Differences associated with gravity (hydrostatic pressure) will be neglected.

On Fig. 3.12 the points marked 1 and 2 are state-points representing two states of the system. The line joining them represents a possible path of a process in which the system changes from state 1 to state 2; it shows how the system pressure and system volume change during the process.

On page 50 it was stated that, in the $\int p_b \, dV_{sw}$ expression for displacement work, dV_{sw} cannot be interpreted in general as the change in system volume. In the present case however the pressure at the boundary, p_b, is the same at all points and is equal to the system pressure, p; this interpretation therefore *is* permissible here. For, if p_b equals p, we have from eq. (3.13)

$$W_d = \int_A p_b \, dV_{sw} + \int_B p_b \, dV_{sw} + \ldots$$

$$= \int p[(dV_{sw})_A + (dV_{sw})_B + \ldots]$$

Now $[(dV_{sw})_A + (dV_{sw})_B + \ldots]$ equals dV, the incremental change in the system volume, V. We can therefore write our expression for the displacement work done by a system, on changing from state 1 to state 2, as

$$W_d = \int_1^2 p \, dV$$

The integral has an obvious geometrical interpretation: it is proportional to the area beneath the curve of Fig. 3.12, shown shaded. Inspection of the diagram shows that merely specifying the points 1 and 2 does not determine the area: the run of the connecting curve needs to be known also; it may be arched upwards or it may sag downwards; and the area will vary according-ly. The mathematical significance of this is that we need to know p as a function of V if we are to evaluate the integral $\int_1^2 p \, dV$. Some examples are dealt with below. Before discussing these however we note that, if the system has mass m, the expression for the displacement work becomes

$$W_d = \int_1^2 p \, dV$$

$$= m \int_1^2 p \, dv \tag{3.14}$$

since the system volume V equals mv, from the definition of v, and m is constant.

In most cases of fully-resisted expansion, the displacement work is the whole work done by the system; other forms are usually absent.

Displacement work in processes undergone by systems of uniform pressure

(a) *The constant-pressure process.* Fig. 3.13a shows a vertical cylinder fitted with a frictionless leakproof piston enclosing a fluid system S_1. A weight rests on the piston, and adjusts its position so as to keep the system pressure constant in accordance with the principles of hydrostatics. If the system is heated, for example by an external flame, it will expand, causing the piston to rise slowly; S_1 becomes S_2.

The corresponding $p \sim V$ diagram is shown in Fig. 3.13b: the path of the process is a horizontal line, because p is constant.

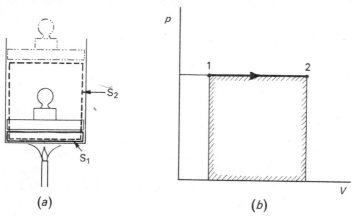

Fig. 3.13 Illustrating a constant-pressure process.

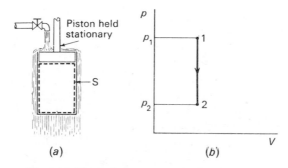

Fig. 3.14 Illustrating a constant-volume process.

In this case the evaluation of the displacement work done by the system is easy. It is, from eq. (3.14), simply

$$W_d = p_1 \int_1^2 dV$$

$$= p_1(V_2 - V_1)$$

$$= mp_1(v_2 - v_1) \qquad (3.15)$$

Although seldom occurring in cylinders, constant-pressure processes are encountered in many engineering situations; for example they occur in boilers and condensers.

(b) *The constant-volume process.* A process permitting a still simpler evaluation of W_d is that in which the piston is prevented from moving, so that the system volume remains unchanged. An example is found in primitive steam engines in which condensation occurred by a sudden chilling of the steam by cold water before the piston had time to move from its uppermost position.

Fig. 3.14a illustrates the arrangement schematically. Fig. 3.14b is the corresponding state diagram. The path of the process is a vertical straight line. In this case

$$W_{\mathrm{d}} = \int_1^2 p \, \mathrm{d}V$$

$$= 0 \qquad (3.16)$$

for the volume does not change.

(c) *The process: pV = constant.* We now consider a process represented by a simple algebraic formula. Later (Chapter 14) it will be shown that many (but not most) substances carry out such processes when expanding in a fully-resisted manner at constant temperature. The algebraic relation is

$$pV = constant$$

$$= k, \text{ say}; \qquad (3.17)$$

this represents a rectangular hyperbola on the state diagram (Fig. 3.15).

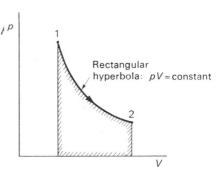

Fig. 3.15 State diagram for a system which undergoes the process $pV = constant$.

To evaluate the work we write

$$W_{\mathrm{d}} = \int_1^2 p \, \mathrm{d}V$$

$$= k \int_1^2 \frac{\mathrm{d}V}{V}$$

$$= k \ln \left(\frac{V_2}{V_1} \right)$$

$$= p_1 V_1 \ln \left(\frac{V_2}{V_1} \right) = m p_1 v_1 \ln \left(\frac{v_2}{v_1} \right) \qquad (3.18)$$

Alternative formulae (e.g. containing the logarithm of p_1/p_2) can also be derived. We see, as is to be expected, that the work increases as the volume increases; not proportionately however, for the pressure is falling.

(*d*) *The process:* $pV^n = constant$. Finally, a family of processes will be discussed which often provide useful approximations to real processes. This family is characterized by the formula

$$pV^n = constant$$

$$= k, \text{ say} \tag{3.19}$$

where n is some number. Fig. 3.16 contains sketches of some members of

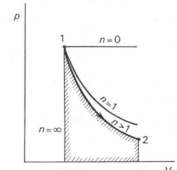

Fig. 3.16 State diagram for a system undergoing processes characterized by $pV^n = constant$.

the family. Such processes assume some importance in Chapter 14.

Evaluation of the displacement work is carried out as follows:

$$W_d = \int_1^2 p \, dV$$

$$= k \int_1^2 \frac{dV}{V^n}$$

$$= \frac{k}{1-n} (V_2^{1-n} - V_1^{1-n})$$

$$= \frac{p_2 V_2^n \cdot V_2^{1-n} - p_1 V_1^n \cdot V_1^{1-n}}{1-n}$$

$$= \frac{p_2 V_2 - p_1 V_1}{1-n} = m \left(\frac{p_2 v_2 - p_1 v_1}{1-n} \right) \tag{3.20}$$

together with alternative formulae.

It will be noted that, although the process $pV = constant$ is a member of the family, having n equal to 1, formula (3.20) fails in this case, becoming indeterminate ($W_d = 0/0$). This is why it was treated separately above.

Displacement work when the system pressure is not uniform

We now evaluate the displacement work in three important special cases of systems of non-uniform pressure. In the first two, the dV_{sw} in $\int p_b \, dV_{sw}$

can again, as it happens, be interpreted as the differential of system volume dV, because the only *moving* part of the system boundary has uniform pressure. In the third example, two parts of the boundary move and the pressures acting on them differ; dV_{sw} can no longer be regarded as the change in system volume.

EXAMPLE 3.2.   Filling a balloon from a bottle.

Problem. A balloon of flexible material is to be filled with air from a storage bottle until it has a volume of 0.9 m³. The atmospheric pressure is 1.013×10^5 N/m². Determine the work done by the system comprising the air initially in the bottle, given that the balloon is light and requires no stretching.

Solution. The first step in any thermodynamic analysis is to sketch the apparatus, indicating the system boundary. Fig. 3.17 shows this. Initially the system boundary

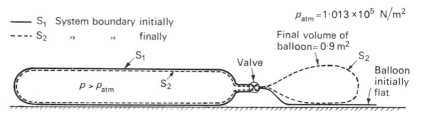

Fig. 3.17   Filling a balloon from a bottle. Example 3.2.

coincides with the inner surface of the storage bottle: boundary S_1 in the sketch. At the end of the process the boundary also encloses the 0.9 m³ content of the balloon: S_2 in the sketch.

The displacement work, the only work in the process, is obtained by taking the summation of the values of $\int p_b \, dV_{sw}$ for each part of the boundary, according to eq. (3.13). dV_{sw} is zero for the part of the boundary in contact with the bottle surface, for the bottle can be taken as rigid; we therefore do not need to know the pressure of the air in the bottle.

The moving part of the boundary sweeps out a volume of 0.9 m³. The pressure on this part is uniform and constant at 1.013×10^5 N/m². (Since the balloon is light and requires no stretching, the pressures on both sides of the fabric are equal.) From eq. (3.13) the displacement work is therefore

$$W_d = \underbrace{\int p_b \, dV_{sw}}_{\text{Balloon}} + \underbrace{\int p_b \, dV_{sw}}_{\text{Bottle}}$$

$$= p_{atm} \int dV_{sw} + 0$$

$$= 1.013 \times 10^5 \times 0.9$$

$$= 91\ 170 \text{ N m} \qquad\qquad \textit{Answer}$$

Remarks.

1. From the definition of negative work it follows that the work done by the atmosphere is $-91\ 170$ N m.

2. The solution of the problem rests on the assumption that the pressure in the balloon is equal to p_{atm} at all times. It is valid if the balloon fabric is light, inelastic and unstressed

If, however, the fabric were elastic, and stretched during the filling process, the pressure within the balloon could exceed p_{atm} and vary with time; the pressure on the outside surface of the balloon is, of course, equal to p_{atm} at all times. It follows therefore that, while the displacement work done by the atmosphere would be $-91\ 170$ N m, as before, the W_d for the gas system would be greater than $91\ 170$ N m by an amount equal to the work done in stretching the balloon. However, for the system comprising the gas *and* the balloon, the displacement work would be $91\ 170$ N m as calculated above.

3. It should be noted that the balloon forms a convenient way of marking part of the system boundary. In examples 3 3 and 3.4 which follow, this *material* system boundary is absent.

EXAMPLE 3.3. Filling an evacuated bottle from the atmosphere.

Problem. Determine the work done by the air which enters an evacuated bottle from the atmosphere when the cock is opened. The atmospheric pressure is $1 \cdot 013 \times 10^5$ N/m^2 and $0 \cdot 2$ m^3 of air (measured at atmospheric conditions) enter.

Solution. Fig. 3.18 shows a sketch with initial and final system boundaries. Once again no work is done by the part of the boundary in contract with the bottle; only the moving

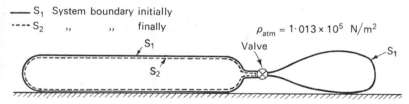

Fig. 3.18 Filling a bottle from the atmosphere. Example 3.3.

part external to the bottle need be observed. Over this part the pressure is uniform at $1 \cdot 013 \times 10^5$ N/m^2. The displacement work done by the system is therefore evaluated from eq. (3.13) as follows:

$$W_d = \int p_b \, dV_{sw} + \int p_b \, dV_{sw}$$

Free-air   Bottle
boundary

$$= p_{atm} \int dV_{sw} + 0$$

$$= 1 \cdot 013 \times 10^5 \times (-0 \cdot 2)$$

$$= -20\ 260 \text{ N m} \qquad\qquad\qquad Answer$$

It should be noted that this time the work is *negative*; $\int dV_{sw}$ for the free-air boundary s negative because the boundary is contracting; thus the surroundings do positive work at the boundary, and the work done by the system is negative.

EXAMPLE 3.4 The induction process of an internal-combustion engine.

The piston of an oil engine, of area 45 cm^2, moves downwards $7 \cdot 5$ cm, drawing in ˙90 cm^3 of fresh air from the atmosphere. The pressure in the cylinder is uniform during he process at $7 \cdot 7 \times 10^4$ N/m^2, while the atmospheric pressure is $1 \cdot 013 \times 10^5$ N/m^2, the difference being accounted for by flow resistance in the induction pipe and inlet valve. Determine the displacement work done by the air finally in the cylinder.

Solution. Fig. 3.19 sketches the apparatus with the initial and final system boundaries S_1 and S_2. Two parts of the system boundary perform work; this is evaluated from eq. (3.13) as

$$= \int p_b \, dV_{sw} + \int p_b \, dV_{sw} + \int p_b \, dV_{sw}$$

| Piston | Free-air boundary | Stationary boundaries |

$$= \left(7 \cdot 7 \times 10^4 \times \frac{45}{(100)^2} \times \frac{7 \cdot 5}{100}\right) - \left(1 \cdot 013 \times 10^5 \times \frac{290}{(100)^3}\right) + 0$$

$$= 26 \cdot 0 - 29 \cdot 4$$

$$= -3 \cdot 4 \, \text{N m}$$

Answer

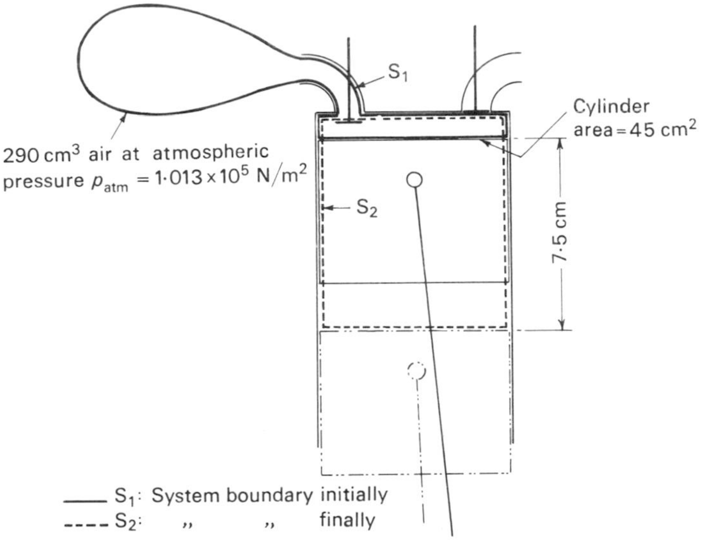

Fig. 3.19 Induction process of an internal-combustion engine. Example 3.4.

In this case the first two terms have opposite signs since the piston moves in the direction of the pressure on it while the free-air boundary moves against the pressure exerted on it by the system. The negative work term happens to be the larger. Such calculations have to be made as part of the determination of the state, and so for example of the temperature, of the air finally contained within the cylinder.

It should be noted that, during the process, very complicated events are occurring *within* the system, and individual parts of the system do work on other parts. All these can however be ignored in the above calculation; we need consider the system boundary alone.

General case: displacement work done by the whole system when the pressure varies continuously along the boundary

In the above examples, the moving parts of the system boundary have different uniform pressures acting on them. We now consider the genera case in which the pressure varies from point to point along the system boundary. To do this, we refer once more to eq. (3.5) which, although it was developed for the piston-cylinder mechanism of Fig. 3.6, is of general applicability; for it also expresses the displacement work done by an element of the system boundary which sweeps out the local volume element dV_{sw} where the local pressure is p_b. In general p_b varies with position *on* the boundary and with position *of* the boundary; so, to obtain the displacement work for the whole boundary, eq. (3.5) has to be integrated. There results

$$W_d = \int_S p_b \, dV_{sw}$$

in which the symbol $\int_S$ is used to emphasize that the summation has to be carried out over the whole boundary *and* its movement.

In practice, for example in fluid systems where the hydrostatic pressure at the boundary varies from point to point, it may be convenient to give W_d the form of a double integral.

Displacement work in unresisted expansion

In the final example of the present section, a fluid expands in an unresisted manner, i.e. the relative velocities of the parts of the system boundary are not small compared with the velocity of propagation of a sound wave through the system. This example will demonstrate the power of the thermodynamic technique of referring work to a system boundary.

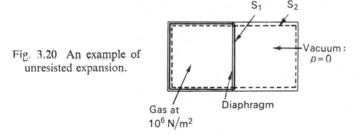

Fig. 3.20 An example of unresisted expansion.

S_1 S_2

Vacuum: $p = 0$

Gas at 10^6 N/m^2

Diaphragm

Collapse of a diaphragm. Consider the rigid vessel shown in Fig. 3.20, divided by a light diaphragm. Initially, the left-hand half of the vessel is filled with gas, while the right-hand half is evacuated. The problem is to evaluate the work done by the gas when the diaphragm gives way under the pressure on it and the gas expands to fill the whole vessel. For definiteness, each half of the vessel will be supposed to have a volume of 1 m^3, and the gas in the left-hand will be supposed to have the pressure 10^6 N/m^2 initially. The initial pressure in the right-hand half is of course zero.

The problem will be solved in two ways. The first follows the treatment given earlier, but requires some physical insight into the nature of the expansion process. The second is simpler and shows what can be done by judicious choice of the system boundary.

First method. Consider a system boundary S_1 enclosing the left-hand half of the vessel. As a result of the process, this boundary finally takes up the position S_2 in Fig. 3.20. From eq. (3.13) we have to evaluate

$$W_d = \int p_b \, dV_{sw} + \int p_b \, dV_{sw} \qquad (3.21)$$
$$\underset{\substack{\text{Moving} \\ \text{part of S}}}{} \qquad \underset{\substack{\text{Stationary} \\ \text{part of S}}}{}$$

Clearly the volume swept out by the boundary is finite. However, reflection reveals that the pressure on the moving part of the boundary, i.e. the face of the diaphragm exposed to the vacuum, is always zero while that part of the boundary is in motion; for, during the motion, there is no material to the right of the moving boundary to exert a pressure. We conclude

$$W_d = \int_1^2 0 \, dV_{sw} + 0 = 0 \qquad (3.22)$$

There is *no* displacement work done by the gas in expanding.

Second method. The same result can be achieved more directly by considering the system boundary to occupy position S_2 initially as well as finally. The system is still the same, because the boundary still encloses nothing but the gas and the diaphragm.

When we now come to evaluate the displacement work, we see that dV_{sw} equals 0 at *all* points of the boundary. This leads to $W_d = 0$ immediately.

Examples of unresisted expansion occur in practice, though seldom in quite the simple form just discussed; an example is the discharge of gas from a diesel-engine cylinder when the exhaust valve is opened. Unresisted expansion will be discussed again later in this book (Chapter 11, page 225).

Work Done as a Result of Shear Forces

We now turn to *shear work*, which is associated with the force exerted by a system on its surroundings, in a direction lying in the plane of the local system boundary. It is necessary to distinguish several cases. The first is that in which the system boundary is an imaginary one cutting a rigid solid; the most important example in engineering thermodynamics is *shaft work*. Then two sorts of friction will be discussed: *fluid friction* and *solid friction*. Finally we deal with *stirring work*.

Shear work at part of a system boundary in a solid

Consider a block being pushed slowly along a rough horizontal surface by an agent which applies a force F (Fig. 3.21). We suppose that it is desired to determine the work done on the lower part of the block by the system comprising the agent and the upper part of the block; accordingly we consider a system boundary S around the agent and the upper half. A shear stress exists along that part of S *in* the block *tending* (of course in vain) to make the upper half of the block slide relative to the lower half. If this shear stress has the uniform value σ, if the area over which it operates is A, and if the force applied to the block is F, the principles of statics lead to the relation

$$F = \sigma A \qquad (3.23)$$

By use of the "mechanics" definition of work, which suffices here, it is easily seen that, when the block, and with it the part of the system boundary

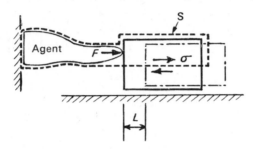

Fig. 3.21 Illustrating shear work at a system boundary in a solid.

in the block, has moved a distance L in the direction of the shear stresses, the shear work by the system W_S is

$$W_\mathrm{S} = \sigma A L \qquad (3.24)$$

Generalizing this result to the case in which the shear stress is not uniform or constant, we write

$$W_\mathrm{S} = \int\!\!\int \sigma \, \mathrm{d}A \, \mathrm{d}L \qquad (3.25)$$

which simply means that we must consider each element, $\mathrm{d}A$, separately, and multiply it by the value of σ prevailing there at the moment in question and by the elementary distance, $\mathrm{d}L$; this is the distance moved, by the material on both sides of that part of the system boundary within the block, in the direction of the stress. The total shear work is evaluated by adding up all these elementary contributions. A particular example of this integration now follows.

Shaft work. Consider a shaft penetrating the system boundary shown in Fig. 3.22a. The system comprises a flat-coiled circular spring, which is unwinding and so raising weights by means of a winch. Fig. 3.22b shows a section through the shaft at the system boundary. At each element of area, dA, there is a shear stress σ, in a tangential direction, tending to cause relative rotation of the two parts of the shaft on either side of S.

The shear stress depends on the radius, r, at which the element is situated; it is related, according to the principles of statics, to the torque T exerted by the shaft across the boundary by

$$dT = (\sigma\, dA)\, r$$

or
$$T = \int \sigma r \, dA \tag{3.26}$$

in which the integral represents a summation over all the elements of area.

When the shaft turns through a small angle dθ, each element (on both sides of S) moves a distance r dθ in the direction of the stress. The shear

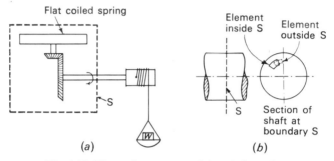

(a)

(b)

Fig. 3.22 Illustrating a system doing shaft work.

work done by the element dA on its surroundings during the elementary rotation is given, according to eq. (3.24), by

$$dW_s = (\sigma\, dA)(r\, d\theta) \tag{3.27}$$

It follows from eq. (3.27) that the shear work done by the system on its surroundings is given by

$$W_s = \int\!\!\int \sigma r \, dA \, d\theta \tag{3.28}$$

and hence, from eq. (3.26), we obtain

$$W_s = \int T \, d\theta \tag{3.29}$$

This form of shear work is known as *shaft work*. Its engineering importance lies in the fact that most power plants deliver their output in the form of a shaft rotation at some stage, because of the mechanical ease with which it can be transmitted in this form.

Eq. (3.29) enables the shaft work to be evaluated when T varies with θ. Often, however, steady conditions exist, in which case T is constant and independent of θ. Eq. (3.29) then becomes

$$W_s = T \times \theta$$

Expressing θ in terms of the speed of rotation of the shaft, N, measured in rev/min, and with T in N m, we obtain the rate at which shaft work is done, i.e. the shaft power, P_s, expressed in N m/s, as

$$P_s = T \times \frac{2\pi N}{60}$$

wherein the factor $\dfrac{2\pi}{60}$ converts the shaft speed to radians/s.

It follows that the shaft power, in terms of kilowatts, is given by

$$P_s = T \times \frac{2\pi N}{60} \times 10^{-3}$$

Fluid friction

When a fluid flows past a solid surface, for example the inner wall of a pipe, its velocity is never uniform. Fig. 3.23 illustrates the distribution of

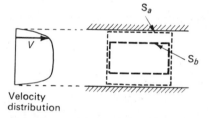

Fig. 3.23 Fluid flow in a pipe.

Velocity
distribution

longitudinal velocity which is found in practice. Its most striking feature is that discontinuities of velocity are absent, even at the interface between fluid and pipe wall.* Thus, although there is relative motion between the pipe and the fluid, treated as a whole, it is not possible to find any surface for which a finite difference of velocity exists between the materials *immediately* on either side of it.

In this respect therefore the situation is similar to that of a solid body. An important consequence is that the shear work at a system boundary in a fluid exerting shear stresses can be evaluated by means of the same formula, namely eq. (3.25). Two cases will be considered.

Shear work done by a fluid system on a solid surface at rest. Consider the system with the boundary S_a in Fig. 3.23. If the shear work done by the fluid on the wall at the boundary is evaluated, we find, from eq. (3.25)

$$W_s = \int\int \sigma \, dA \cdot dL$$

$$= 0, \text{ because } dL = 0; \qquad (3.30)$$

* This ceases to be true only for very rarefied gases, which will not be dealt with in this book.

for the material at the part of the boundary in question does not move. We conclude that a fluid system does no shear work on a surface at rest, even though a shear stress is exerted.

Shear work done by a fluid system on adjacent fluid. To illustrate the next point, a smaller system is chosen, S_b in Fig. 3.23. A glance at the velocity diagram shows that a finite velocity exists at the upper and lower system boundaries. Suppose that these velocities both have the value V. Then the boundaries at which the shear stress σ is exerted move a distance, dL, in an element of time, dτ, given by

$$dL = V \, d\tau \tag{3.31}$$

If we take for simplicity the case in which the stress σ is uniform over the relevant system boundary area A, the shear work done by the system is given by

$$W_s = \int \sigma A V \, d\tau \tag{3.32}$$

Correspondingly, the *rate* of doing work, i.e. the power P, is given by

$$P = \frac{dW_s}{d\tau} = \sigma A V \tag{3.33}$$

Such calculations are important in the theory of lubrication, where the shear work causes the oil to become hot, and in supersonic aerodynamics, where very high temperatures result from the intense shear work adjacent to surfaces.

Stirring work

Allied both to shaft work and to fluid friction is *stirring work*. This occurs when a fluid is stirred in a more or less random manner, usually by a paddle wheel on the end of the shaft. Fig. 3.24 illustrates the apparatus. If

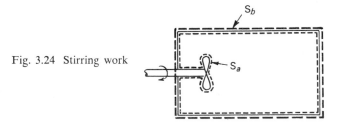

Fig. 3.24 Stirring work

the system boundary S_a is considered, enclosing the fluid but excluding the wheel and shaft, it is found that both normal forces and shear forces are exerted on the wheel surface. In principle the work associated with these forces can be evaluated in the manner indicated above.

Usually however the force distribution over the wheel surface, and its variation with time, are not known with sufficient precision. It is therefore simpler to re-draw the system boundary in the position S_b. Now we have merely shaft work at the boundary; it can be determined by means of the eq. (3.29).

It should be noted that stirring work is *negative*, for the surroundings do work on the system.

Solid friction

The case of solid friction has been left to the end because it exhibits a special difficulty. Suppose that we wish to determine the shear work done by the whole block of Fig. 3.21 on the plane beneath. An attempt to use the formulae already given, eq. (3.24), leads to the question: What is the value of L? For, though the material above the system boundary moves, that below does not. There does not seem to be any reason to give either the priority.

The answer to the question cannot be given here, for, surprisingly enough, the value of the shear work depends on the relative *thermal conductivities* of the two materials. This topic will be dealt with, after thermal quantities have been precisely introduced, on p. 107; although of minor practical importance, this matter throws further light on the thermodynamic definition of work.

Electrical Work

There are as many sorts of work as there are forces to cause change— capillary, magnetic, electric, electromagnetic, and so on. Here, however, we shall discuss electrical work only.

Already on page 45 it was shown that the thermodynamic definition of work covers electrical effects. It remains to relate the units of electrical work to those used for the mechanical variety. The important statement is:

When $\mathcal{Q}$ coulombs of positive electricity cross a system boundary at a point "A" and re-enter the system at a point "B", the electric potential of "A" being $\mathscr{V}$ volts above the potential of "B", the system does positive work on its surroundings equal to $\mathcal{Q}\mathscr{V}$ joules.

This statement is the definition of the volt in a form acceptable to thermodynamicists. The coulomb is a unit of electric charge, which is defined in texts on electricity. The joule has already been introduced on page 41.

Further, one coulomb per second is called a current of one ampere. Consequently when a steady current of I amperes flows between the points A and B of the above definition the system does work at the rate of $I\mathscr{V}$ joules per second.

Other factors relating units of power are given in Table 3.2, p. 42.

Net Work done by a System

In the present chapter we have considered the various ways in which work may be done by a system. Often several of the different sorts of work occur simultaneously during a process executed by the system. When all the sorts of work done by a system have been evaluated, the *total* or *net* work done by the system, *W*, can be computed by simple *algebraic* addition of the separate work terms. Note that the algebraic sum must be taken and therefore due attention must be paid to the sign of each work term. Expressed symbolically, our conclusion is:

$$W = W_{\text{displacement}} + W_{\text{shear}} + W_{\text{electrical}} + W_{\text{stirring}} + \cdots \quad (3.34)$$

CHAPTER 3—PROBLEMS

3.1 A rigid body of mass 10 kg rests on a rough horizontal table. An agent applies a horizontal force of 30 N to the body and so slowly pulls the latter across the table. How much work is done by the agent in moving the body a distance of 1·5 m?

3.2 A rigid body rests on a rough horizontal plane. A coiled compression spring is attached to the body with its axis horizontal; the spring is such that a force of 4 N causes it to compress 10 mm.

(*a*) An agent applies a horizontal force to the free end of the spring causing the block to move slowly along the table. When the block has moved a distance of 150 mm it comes into contact with a rigid stop fixed to the surface of the table. During the motion the resisting force exerted by the plane on the block is 20 N.

How much work is done by the agent?

(*b*) If after the block has contacted the stop, the agent slowly relaxes the applied force, how much work will it have done altogether?

(*c*) How much work is done by the spring in (*a*) and in (*a*) + (*b*)?

3.3 A steel structure is to be tested by measuring the deflections caused by slowly-applied external loads.

In a particular test, a load of 10^5 N causes a deflection of 0·2 mm at its point of application. What is the work done (in N m) by the applied load? (Assume that the deflection increases in direct proportion to the force.)

3.4 Evaluate the work done in the following processes. The systems to be considered are printed in italics.

(*a*) An *agent* slowly raises a *body* of mass 5 kg through a vertical distance of 10 m in a gravitational field. The local gravitational acceleration is 9·81 m/s². Neglect air resistance

(*b*) A *body* of mass 10 kg is lowered by a *crane* slowly through a vertical distance of 30 m in a gravitational field for which $g = 6·0$ m/s². Neglect air resistance.

(*c*) A *body* of mass 10 kg falls freely through a vertical distance of 30 m in a gravitational field for which $g = 6·0$ m/s². The drag force of the *atmosphere* on the body is 4 N.

(*d*) A *body* of mass 10 kg falls freely (in a vacuum) through a vertical distance of 30 m. The gravitational acceleration is 6·0 m/s².

(*e*) A *rat*, weighing 4 N, climbs a stair 0·3 m in height.

3.5 The *electric motor* of a crane is supplied with a current of 14 amperes from a 200 volt DC supply for a period of 15 seconds while the crane lifts a *machine* of mass 2 tonne through a vertical distance of 2 m. The weight of the inextensible *cable* and the resistance of the *pulleys* over which the cable passes are negligible; $g = 9·81$ m/s².

Evaluate the work done by the systems printed in italics.

3.6 A vertical cylinder, closed by a frictionless leakproof piston of area 20 cm², contains a quantity of air at a pressure of 70×10^3 N/m²; the lower face of the piston is exposed to the atmospheric pressure, 100×10^3 N/m² (see figure). The cylinder is connected, by way of a pipe and valve, to a rigid vessel containing nitrogen at a pressure of 20×10^3 N/m²; initially the valve is shut. The valve is opened, so causing the piston to move *slowly* a distance of 10 cm from its initial position. Calculate the work done by each of the following systems:

(a) the piston; (b) the atmosphere; (c) the air plus the nitrogen.

Problem 3.6

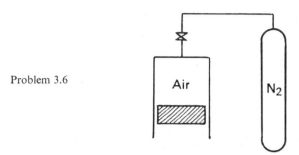

3.7 A cylinder closed by a frictionless leakproof piston contains a gas at a pressure of 150×10^3 N/m². The cylinder area is 220 cm². As the piston is pushed inwards, some of the gas is forced out of the cylinder through a spring-loaded valve in the cylinder cover. An indicator connected to the cylinder shows that during this process the pressure of the gas in the cylinder increases linearly with the piston position to 300×10^3 N/m².

Evaluate the work done at the piston face as the piston is pushed inwards a distance of 200 mm, when the shape of the piston crown (i.e. the surface exposed to the gas in the cylinder) is (a) flat, and (b) hemispherical.

3.8 A single-cylinder, double-acting, 2-stroke oil engine has a cylinder diameter ("bore") of 200 mm and a stroke of 250 mm. When the engine is run at a speed of 300 rev/min, the mean effective pressure is 900×10^3 N/m². Evaluate the work done per "cycle", and the indicated power.

3.9 An indicator spring is found to require an axial force of 60 N to shorten it by $1\cdot0$ mm. The spring is used in an indicator having a piston area of 4 cm² and a pencil mechanism which magnifies the motion of the indicator piston six-fold.

(a) Calculate the spring number in N/m³.

(b) A single-cylinder, single-acting, 4-stroke gas engine of 130-mm bore develops an indicated power of $4\cdot5$ kW when running at 216 rev/min. Calculate the area of the indicator diagram that would be obtained using the above indicator, given that the length of the diagram is $0\cdot1$ times the length of the stroke of the engine.

3.10 A six-cylinder, 4-stroke gasoline engine is run at a speed of 2520 rev/min. The area of the indicator card of one cylinder is $2\cdot33 \times 10^3$ mm² and its length is $62\cdot1$ mm. The spring number is 20×10^6 N/m³. The bore of the cylinders is 150 mm and the piston stroke is 160 mm

Evaluate the indicated power, assuming that all cylinders contribute equal powers.

3.11 The following description is an idealization of the sequence of events which occur in the cylinder of a reciprocating air compressor:

(a) Initially, with the piston at its innermost position (inner dead-centre) and with the inlet and delivery valves shut, the clearance volume contains air at the delivery pressure p_2.

(b) The piston moves slowly outwards causing the clearance air to expand to the inlet pressure p_1 according to $pV^n = constant$; p and V are respectively the pressure and volume of the clearance air at any instant during this process and n is a constant.

(c) When the pressure of the clearance air has fallen to p_1, the inlet valve opens, and, as the piston completes its outward stroke, atmospheric air is drawn into the cylinder. The pressure on the piston face is p_1 throughout this induction process.

(*d*) With the piston in its outermost position (outer dead-centre), the inlet valve closes. The piston now commences its inward stroke and compresses the air in the cylinder (i.e. the clearance air plus the induced air) according to $pV^n = constant$ until its pressure rises to p_2; p and V are respectively the pressure and volume of the air in the cylinder during the compression process.

(*e*) When the pressure has risen to p_2, the delivery valve opens and air is discharged at pressure p_2 as the piston completes its inward stroke.

(*f*) Finally with the piston at the inner dead-centre the delivery valve closes leaving the clearance volume filled with air at pressure p_2. The sequence is then repeated.

All motions of the piston are assumed to be fully resisted. The clearance volume is c per cent of the swept volume V_{sw}.

(i) Sketch the indicator diagram.

(ii) Derive an expression for the net work done at the piston face per "cycle" in terms of p_1, p_2, V_{sw}, c and n.

(iii) Evaluate the net work per "cycle" when $p_2 = 600 \times 10^3$ N/m², $p_1 = 100 \times 10^3$ N/m², $V_{sw} = 2 \cdot 4$ dm³, $c = 5$ per cent and $n = 1 \cdot 2$.

(iv) Evaluate the indicated power when the compressor speed is 390 rev/min assuming single-acting operation. (Note that one complete sequence of events occurs during one revolution of the crankshaft.)

3.12 The figure shows a piston-cylinder mechanism employed to pump a corrosive fluid. The 250 mm-diameter cylinder is closed by a light flexible diaphragm attached to the 150 mm-diameter piston. The stroke of the piston is 250 mm and the length of the cylinder is 125 mm. The shape of the diaphragm in its extreme positions is the frustum of a right cone.

During the delivery stroke an agent pushes the thin piston rod slowly inwards against the constant fluid pressure of 180×10^3 N/m² gauge. The atmospheric pressure acts on the outer side of the diaphragm and piston.

Evaluate the work done by the agent.

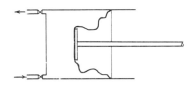

Problem 3.12

3.13 The following description is an idealization of the sequence of events which occur in the cylinder of a reciprocating steam engine.

(*a*) Initially, with the piston at its innermost position (inner dead-centre) and with the inlet and exhaust valves shut, the clearance volume contains steam at the exhaust pressure p_e.

(*b*) The inlet valve now opens, causing the steam pressure to rise rapidly to the supply pressure p_1.

(*c*) The piston commences its outward stroke, so admitting steam to the cylinder. When the piston has moved a per cent of the swept volume, the inlet valve closes ("cut-off"). The pressure on the piston face is p_1 throughout the admission process.

(*d*) As the piston continues its outward stroke, the steam in the cylinder (i.e. clearance steam plus admitted steam) expands according to $pV = constant$ ("hyperbolic expansion"), where p and V are respectively the pressure and volume of the steam in the cylinder during this process.

(*e*) When the piston has reached the end of its stroke (outer dead-centre) the exhaust valve opens, so causing the steam pressure to fall rapidly to the exhaust pressure p_e ("the back-pressure").

(*f*) The piston now performs its return stroke, so discharging the steam from the cylinder at the exhaust pressure p_e.

(*g*) Finally, with the piston at the inner dead-centre, the exhaust valve closes leaving the clearance volume filled with steam at the exhaust pressure p_e. The sequence is then repeated.

All motions of the piston are assumed to be fully-resisted. The clearance volume is c per cent of the swept volume V_{sw}.

(i) Sketch the indicator diagram.

(ii) Derive expressions for the net work done at the piston face per "cycle" and for the mean effective pressure in terms of p_1, p_e, V_{sw}, a and c.

(iii) Evaluate the mean effective pressure when $p_1 = 550 \times 10^3$ N/m², $p_e = 100 \times 10^3$ N/m², $a = 30$ per cent, $c = 5$ per cent. Hence evaluate the net work done per "cycle" given that the bore and stroke of the engine are 230 mm and 280 mm respectively.

(iv) Evaluate the indicated power when the engine speed is 132 rev/min, assuming double-acting operation. Neglect the *area* of the piston rod.

3.14 A fluid system at a pressure of 600×10^3 N/m² occupies a volume of 15 dm³. The system undergoes a fully-resisted process to a final pressure of 200×10^3 N/m², and during the process the system pressure and volume are related as follows:

p	kN/m²	600	500	400	300	200
V	dm³	15	26	34	40	45

Sketch the process on a pressure~volume state diagram and evaluate the work done by the fluid.

3.15 A horizontal cylinder, area 150 cm², is fitted with a freely-floating, frictionless leak-proof piston, i.e. there is no piston rod.

The left-hand end of the cylinder, which contains 0·0045 m³ of air at atmospheric pressure (100×10^3 N/m²), is connected to a light inextensible balloon; the balloon is flat initially. The right-hand end of the cylinder is connected, via a valve, to a rigid vessel containing nitrogen at high pressure.

The balloon is inflated by slowly admitting nitrogen to the right-hand end of the cylinder. During the inflating process the piston moves a distance of 250 mm and when fully inflated the volume of the balloon is 0·003 m³. Any changes in the air pressure p are related to the corresponding air volume V by $pV = constant$.

(*a*) Sketch the trace that would be obtained on an indicator connected to the left-hand end of the cylinder.

(*b*) Evaluate the work done by each of the following systems.

(i) the atmosphere, (ii) the balloon, (iii) the air, (iv) the piston, (v) the nitrogen.

3.16 A steam main carries steam at a pressure of $1·5 \times 10^6$ N/m² and a temperature of 250 °C. A rigid vessel containing steam at a pressure of $0·4 \times 10^6$ N/m² is connected, via a valve, to the main. Over a period of time, 0·1 kg of steam from the main leaks into the vessel thereby raising the pressure of the contents of the vessel to $0·5 \times 10^6$ N/m².

For the system comprising the steam finally in the vessel, find the magnitude and sign of the work. (The specific volume of steam at a pressure of $1·5 \times 10^6$ N/m² and a temperature of 250 °C is 0·1520 kg/m³.)

3.17 An open tank is filled to the brim with a liquid of density 1100 kg/m³. A flexible spherical balloon 0·5 m in diameter is immersed in the liquid with its centre 2·5 m below the free liquid surface. Gas from a storage vessel is used to inflate the balloon thereby causing the tank to overflow. The atmospheric pressure is 100×10^3 N/m².

Assuming that the centre of the balloon remains fixed and neglecting the work done in stretching the balloon fabric, evaluate the work done by the gas in the balloon and the storage vessel as the balloon diameter increases to 1 m. (The gravitational acceleration is 9·81 m/s².)

3.18 A motor-car gasoline engine running at a speed of 3000 rev/min develops a torque of 260 N m. Calculate the shaft power of the engine.

3.19 A ship's propeller is driven by a steam turbine through an 8 : 1 reduction gear. The mean resisting torque imposed by the water on the propeller is 770×10^3 N m and the shaft power delivered by the turbine to the reduction gear is 15 000 kW; the turbine speed is 1440 rev/min.

Evaluate (a) the torque developed by the turbine;

(b) the power delivered to the propeller shaft;

(c) the net rate of working of the reduction gear in N m/s.

3.20 A fluid, contained in a horizontal cylinder fitted with a frictionless leakproof piston, is continuously agitated by means of a stirrer passing through the cylinder cover. The cylinder diameter is 400 mm.

During a stirring process occupying 10 minutes the piston moves slowly outwards a distance of 485 mm against the atmosphere ($p_{\text{atm}} = 101 \cdot 3 \times 10^3$ N/m²). The net work done by the fluid during the process is 2000 N m.

Given that the speed of the electric motor driving the stirrer is 840 rev/min estimate the torque in the driving shaft and the shaft power output of the motor.

3.21 The electric motor in example 3.20 is supplied with current from a 24-volt accumulator. The current taken is 0·35 amperes. Evaluate the net work done by (a) the accumulator and (b) the motor.

4 Temperature

Introduction

In Chapter 3 we discussed the interaction known as work. In thermo-dynamics, only one other type of interaction between systems is considered; this is heat. Heat may be defined by reference to temperature difference; we need to know, therefore, what is meant by temperature.

The sensations of relative hotness and coldness are familiar; temperature is defined in such a way as to make these sensations quantitative. This is important because, as will be seen later, without temperature differences many types of engine would not run. Usually they run "better" the bigger is the temperature difference available; we need to know what "bigger" means.

We begin by discussing temperature equality and the Zero'th Law of Thermodynamics.† Inequality of temperature, temperature measurement and temperature scales are then considered, and the dependence of tempera-ture scales on the nature of the thermometric substance used is emphasized. We then introduce the International Scale of Temperature and make preliminary mention of the Absolute Thermodynamic Temperature Scale. Finally the gas thermometer and the Ideal-Gas temperature scale are discussed.

Symbols

A,B Constants.
p Pressure.
p_i Pressure at the ice-point in a gas thermometer.
p_s Pressure at the steam-point in a gas thermometer.
x "Error" in alcohol-in-glass thermometer reading.
y "Error" in mercury-in-glass thermometer reading.

R_G A constant (eq. (4.3)).
R_t Resistance of a standard platinum-resistance thermometer (eq. (4.3)).
t Temperature.
t_C Celsius (Centigrade) temperature.
t_F Fahrenheit temperature.
t_a Alcohol-in-glass temperature.
t_m Mercury-in-glass temperature.
α A constant.

† So called because, though logically prior to the First and Second Laws, it was not explicitly stated until later.

Equality of Temperature

Definition of temperature equality

Let two bodies, one hot and one cold, be placed in contact; then, after a time, the one feels less hot and the other less cold. In addition, changes in their physical properties (e.g. length, electrical resistance) occur. After contact for a long time, however, no further change takes place: equilibrium is established. The bodies are then said to have the *same temperature*.

Definition. Two systems are equal in temperature if no change in any property occurs when they are brought into communication.

The definition must not be reversed; it does *not* imply that "when two systems are equal in temperature, no changes result from their communication". For example, if an electric light bulb is caused to communicate in the appropriate way with a battery, the bulb "lights up"; that, of course, is the result of a work interaction.

Zero'th Law of Thermodynamics

If we apply the concept of temperature equality to an experiment involving three systems, S_1, S_2, and S_3 (Fig. 4.1), an experimental law regarding temperature can be established.

Suppose that systems S_1 and S_3 are equal in temperature, so that no change occurs in the physical properties of either when they are brought into

Fig. 4.1 Systems which can be brought into contact.

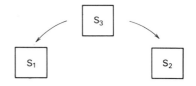

contact. Similarly let S_2 and S_3 be equal in temperature so that no change occurs when they are brought into contact. Systems S_1 and S_2 are supposed not to react with each other chemically or electrically. If now S_1 and S_2 are placed in contact, it is an *observed fact* that no change occurs in their physical properties. Therefore S_1 and S_2 must be equal in temperature.

This experimental fact is embodied in the *Zero'th Law of Thermodynamics* which may be stated as follows: **Two systems which are equal in temperature to a third system are equal in temperature to each other.**

This law provides the basis for temperature measurement.

Inequality of Temperature

We now consider a different case. Suppose that, when system S_3 has contacted S_1 and reached equilibrium with it, and then is subsequently removed and brought into contact with system S_2, observable changes do take place in S_3. Suppose further that work interactions are absent. Then the temperatures of S_1 and S_2 are said to be *unequal*.

To facilitate the recognition of temperature inequality, it is usual to select system S_3 so that changes in its physical characteristics are easily observed. A common choice for S_3 is mercury in a glass capillary tube; changes in the volume of the mercury relative to the glass are readily seen.

In the above experiment therefore, a change in the length of the mercury column, when S_3 contacts S_1 and S_2 in turn, means that S_1 and S_2 have unequal temperatures. Correspondingly an absence of change in the length means that S_1 and S_2 have equal temperatures.

Measurement of Temperature

Temperature scales

Thermometers and fixed points. In order to measure the magnitude of the inequality, a *thermometer* is constructed. For example, the glass tube containing mercury, which was selected for S_3 above, becomes a mercury-in-glass thermometer if the glass tube is marked so that the position of the mercury meniscus can be noted. To construct such a thermometer, the following procedure may be adopted:

Two standard systems are established, being so chosen that their conditions are easily reproducible. The first, S_{ice}, consists of a mixture of ice and water at a pressure of one atmosphere; the second, S_{steam}, consists of water boiling at a pressure of one atmosphere. The position of the mercury meniscus of the thermometer S_3, when it is in contact with S_{ice}, is marked by a scratch on the glass and is called the *ice-point*; the position when S_{steam} is contacted is marked and is termed the *steam-point*. The length of glass tube between these two marks, which are usually called the *fixed points*, is then divided into a number of *equal* intervals marked by further scratches.

Celsius and Fahrenheit temperature scales. Temperature scales are defined by assigning numbers to the ice-point, to the steam-point and to the equally-spaced marks between those points. A mercury-in-glass Celsius (or Centigrade) temperature scale has the ice-point marked as 0, and the steam-point marked as 100, with 100 equal sub-divisions. Then if the thermometer S_3, graduated in this way, is brought into communication with system S_1, whereupon the mercury rises to the sub-division marked 50, we say that the system S_1 has a temperature of 50 degrees Celsius, written 50 °C. The size of each of the 100 subdivisions defines the unit of temperature, the *kelvin* (K); in terms of this unit, we say that for the system S_1, its Celsius temperature is 50 K.

Another common procedure is to mark the ice-point and the steam-point 32 and 212 respectively, with 180 equal intermediate intervals; this procedure defines the mercury-in-glass Fahrenheit scale of temperature. The unit of temperature for this scale is the *rankine* (R).

The two sets of graduations could be marked on the same thermometer.

Examination of Fig. 4.2 shows that there is a simple relation between the Celsius and Fahrenheit temperatures of a given body, t_C and t_F, namely

$$t_C = \tfrac{5}{9}(t_F - 32) \tag{4.1}$$

$$t_F = \tfrac{9}{5}t_C + 32 \tag{4.2}$$

The fraction $\tfrac{5}{9}$ of course is $100/180$.

Thus if S_1 is at 50 °C, its temperature on the Fahrenheit scale is $\tfrac{9}{5} \times 50 + 32$ degrees, i.e. $t_F = 122$ °F.

Extrapolation beyond the fixed points. So far only those temperatures between the ice- and steam-points have been defined. However, the uniform graduation of the glass tube can obviously be continued above and below the fixed points: if the mercury rises 50 divisions above the steam-point mark of a Celsius thermometer, its temperature is then defined as 150 °C.

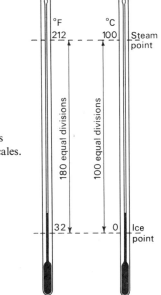

Fig. 4.2 Mercury-in-glass Celsius and Fahrenheit temperature scales.

This extrapolation is limited in practice by the fact that, if the temperature falls too low, the mercury freezes (at −38·9 °C), while at high temperatures glass becomes an unsuitable casing material. This difficulty is met by adopting different fluids and casing materials, or by using an altogether different property from thermal expansion. Thus, ethyl alcohol in glass can be used down to temperatures of −110 °C; while the electrical resistance of platinum wire can be used as a temperature indicator up to 1000 °C. Other examples of properties used to indicate temperature are: the potential difference between the junctions of dissimilar metals (the thermocouple); and the change in the colour of materials.

However, the change to new thermometric substances brings with it a difficulty: if an alcohol-in-glass thermometer and a mercury-in-glass thermometer measure the temperature of the same system, they give, in general, different readings, even though they were graduated with reference to the same standard systems. Fig. 4.3 illustrates this by a plot of alcohol-in-glass temperature t_a versus mercury-in-glass temperature t_m (not to scale).

The alcohol-in-glass Celsius scale and the mercury-in-glass Celsius scale must agree at the fixed points, by definition; i.e. each has the numbers 0 and 100 assigned to the ice- and steam-points respectively; in general these are the *only* points at which the two scales agree. We see, for example, in Fig. 4.3 that, when t_m is 60 °C, t_a is only 59 °C.* Which thermometer is correct?

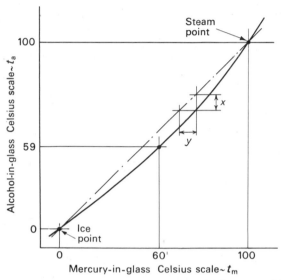

Fig. 4.3 Comparison of temperature scales (not to scale).

Arbitrariness of temperature scales. All temperature scales are arbitrary. Each step, from the selection of the thermometric substance and the casing material to the choice of the number of equal sub-divisions between the fixed points, is unrelated. The answer to the question "Which thermometer is correct?" therefore is that it is up to us which we shall call right and which wrong. All that is necessary is that we should make a decision and stick to it. For example, in Fig. 4.3, if the mercury-in-glass thermometer is taken as standard, then x is the "error" in the alcohol-in-glass reading. Similarly y is the "error" in the mercury-in-glass reading, if the alcohol-in-glass thermometer is taken as standard.

While in principle any thermometer may thus be selected as a standard, it is nevertheless desirable in practice for everyone to use the same standard. International agreement has been obtained for the following arbitrary scale.

The International Scale of Temperature

The International Temperature Scale is an agreed practical standard scale. It is defined by specifying a number of fixed points (standard systems)

* These are not the true figures; the actual differences are smaller.

together with a method of interpolating between them (standard thermo-meters). Table 4.1 gives the basic fixed points* which have been adopted.

TABLE 4.1 *Basic Fixed Points of the International Temperature Scale*

Fixed points	Standard system a pressure of 1 atm	Temperature	
		°C	°F
Oxygen-point	Oxygen boiling	−182·97	−297·35
Ice-point	Ice melting	0·000	32·000
Steam-point	Water boiling	100·000	212·000
Sulphur-point	Sulphur boiling	444·60	832·28
Silver-point	Silver melting	960·8	1 761·4
Gold-point	Gold melting	1 063·0	1 945·4

In order to specify temperatures between the basic fixed points, an inter-polation procedure is prescribed. The range of the scale is divided into four parts and for each part a particular type of thermometer is specified as the indicator of the temperature. For example, between the ice-point and 660 °C, the temperature t is deduced from the resistance R_t of a standard platinum-resistance thermometer, by means of the equation:

$$R_t = R_0 (1 + At + Bt^2) \tag{4.3}$$

The constants R_0, A and B of eq. (4.3) are determined by calibrating the thermometer at the ice, steam and sulphur points. Different means of interpolation are specified for the other three parts of the scale. Full details of the International Scale are given in the references at the end of this chapter.

Absolute Scale of Temperature

The numbers assigned to the various fixed points of the International Scale have not been chosen without reason, but in order that this scale should approximate closely to another temperature scale which is discussed in Chapter 12, subsequent to the introduction of the Second Law of Thermo-dynamics. This, the Absolute (Thermodynamic) Scale of Temperature, has the great merit that it is independent of the properties of particular substances or systems. This matter will be discussed in the appropriate place, but it is convenient here to mention another class of thermometers which give readings which are very close to the absolute scale.

The gas thermometer

We have seen that temperature scales are arbitrary, depending on the properties of the substances from which the thermometer is constructed.

---

* There are secondary fixed points in addition.

However, one group of substances, the so-called "permanent gases" (e.g. oxygen, nitrogen, hydrogen) may be used to define temperature scales which turn out to be almost identical, whichever gas is used; in addition, these scales agree closely with the absolute scale of temperature.

A gas thermometer, Fig. 4.4, consists essentially of a vessel (a glass bulb for example) connected to a U-tube containing a liquid, say mercury. A

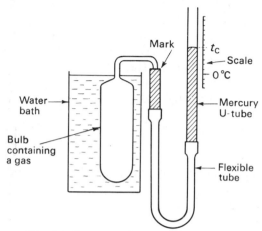

Fig. 4.4 Constant-volume gas thermometer.

permanent gas is enclosed within the bulb and the connecting tube by the mercury, the other limb of the U-tube being open to the atmosphere. During use, with the bulb in contact with systems at different temperatures, the level of the mercury can be adjusted to keep either the volume or the pressure of the gas constant.

Let us consider its use as a constant-volume gas thermometer. Following the above procedure for setting up a temperature scale, we note the difference in the right-hand meniscus level (i.e. the gas pressure) at the two fixed points; the mercury level in the left-hand limb, and so the gas volume, are held the same in each case. Assigning 0 and 100 to the ice- and steam-points respectively, with 100 equal sub-divisions between them, we define a gas-thermometer Celsius scale. This scale, linear in gas pressure p, can be expressed as

$$p = p_i[1 + \alpha t]$$

where $p_i \equiv$ the gas pressure at the ice-point, 0 °C

$\quad t \equiv$ the "gas-thermometer" Celsius temperature

$\quad \alpha \equiv$ a constant, equal to $(p_s - p_i)/100\,p_i$

$\quad p_s \equiv$ the gas pressure at the steam-point, 100 °C.

By repetition of this procedure and the use of equal volumes of the *same gas* at progressively lower gas pressures (i.e. smaller masses of gas), a series of values of α is obtained. By plotting these values of α against the pressure at the ice-point p_i in each determination, and extrapolating the curve to zero pressure, a value of α corresponding to zero pressure is determined. In practice α varies little with pressure.

Similarly, by the use of a number of *different gases*, a value of α corresponding to zero pressure is obtained for each gas; the important experimental result is that these values of α prove to be the same *for every gas*.

The Ideal-Gas Temperature Scale. The behaviour of gases at very low pressures corresponds to that of an Ideal Gas* and the value of α, so obtained by extrapolation, defines the Ideal-Gas temperature scale.

Two Ideal-Gas Scales are commonly used, namely:

The Ideal-Gas Celsius Scale, is given by

$$p = p_i\left[1 + \frac{1}{273\cdot15}(t_C)\right] \tag{4.4}$$

The Ideal-Gas Fahrenheit Scale, is given by

$$p = p_i\left[1 + \frac{1}{491\cdot67}(t_F - 32)\right] \tag{4.5}$$

Corresponding equations, with the same values of α but with volumes replacing pressures, would be obtained using the constant-pressure gas thermometer.

It will be seen in eq. (4.4) and (4.5) that, when the pressure p is zero, $t = -273\cdot15$ °C on the Ideal-Gas Celsius Scale and $t = -459\cdot67$ °F on the Ideal-Gas Fahrenheit Scale. These temperatures will assume particular significance in Chapter 12.

BIBLIOGRAPHY

BS 1041 : 1943. *Temperature Measurement.* British Standards Institution.
DAHL, A. I. *Temperature, its measurement and control in science and industry.* Vol. 3. Reinhold, New York, 1962.

CHAPTER 4—PROBLEMS

4.1 (*a*) The temperature of the coolant (CO_2 gas) at exit from a gas-cooled nuclear reactor is 336 °C. Express the temperature in degrees Fahrenheit.

(*b*) The temperature of the atmosphere at a height of 6000 m above the earth's surface is 5 °F. Express the temperature in degrees Celsius.

(*c*) The correct temperature for the storage of apples during overseas shipment is −1 °C. Express the temperature in degrees Fahrenheit.

(*d*) The temperature of the steam at entry to the turbines in a generating station is 1050 °F. Express the temperature in degrees Celsius.

(*e*) A high-performance gas-turbine jet engine incorporates an axial-flow compressor. During the test-bed running of the engine, atmospheric air enters the compressor at a

* Defined in Chapter 14.

temperature of 68 °F and a pressure 100×10^3 N/m². After compression the air is delivered to the combustion chambers at a temperature of 554 °F and a pressure of 800×10^3 N/m². Express the increase in temperature of the air in degrees Celsius.

4.2 The temperature t on a thermometric scale is defined in terms of a property P by the relation

$$t = a \ln P + b$$

where a and b are constants.

The temperatures of the ice-point and the steam-point are assigned the numbers 0 and 100 respectively. Experiment gives values of P of 1·86 and 6·81 at the ice-point and steam-point respectively.

Evaluate the temperature corresponding to a reading of $P = 2·50$ on the thermometer.

4.3 The readings, t_A and t_B, of two Celsius thermometers, A and B, agree at the ice-point (0 K) and the steam-point (100 K), but elsewhere are related by the equation

$$t_A = l + mt_B + nt_B^2$$

where l, m and n are constants.

When both thermometers are immersed in a well-stirred oil bath, A registers 51 K while B registers 50 K.

(a) Determine the reading on B when A reads 25 K.

(b) Discuss the question: "Which thermometer is correct?"

4.4 A constant-volume gas thermometer containing helium gives readings of gas pressure, p, of 1000 and 1366 mm of mercury at the ice-point and the steam-point respectively.

(a) Express the gas-thermometer Celsius temperature, t_C, in terms of the gas pressure p.

(b) The thermometer, when left standing in the atmosphere, registers 1075 mm. Determine the atmospheric temperature.

5 Heat

Introduction

Equality of temperatures has been defined as the relation between two systems which exists when no change results from their communication. This suggests that, when the temperatures are unequal, some change may result even when work interactions are absent. Examples are: the change in state of the contents of a kettle when a flame plays on the outside; and that of molten iron when poured into a cold casting mould.

In this chapter we shall be concerned with the interaction which invariably results from temperature differences. The interaction is given a name, *heat*; a method of measurement is defined; and some space will be devoted to showing what heat is *not*. Distinctions will be made which at first sight appear pedantic; but they are, in fact, essential if the later chapters are to be understood, particularly those concerned with the Second Law of Thermodynamics.

Symbols

Q Heat, heat transfer.
t, θ Temperature
W Work

Definition and Measurement of Heat

Definition of heat. Heat is the interaction between systems which occurs by virtue of their temperature difference when they communicate.

Comments. This definition describes the procedure by which a heat interaction is *recognized*. It specifies *two* conditions which must be satisfied for a heat interaction between systems to occur: *temperature difference* and *communication*. The two conditions must be fulfilled simultaneously; absence of either requirement means absence of heat. What is meant by communication is explained on p. 89 below.

To be useful, the definition must be combined with a statement about how heat is to be *measured*. Measurement is particularly important because the effect of heat on, say, an engine which it is driving, or on a structure in which it is causing undesirable stresses, is directly dependent on the magnitude of the heat involved.

In defining a measurement procedure for work, it was necessary to define a standard system (a specified amount of any substance) and a standard process which the system was caused to undergo by the interaction (a specified change of elevation in a standard gravitational field). The magnitude of the work interaction causing this process, i.e. *the* work, was then established as the unit of work and given a name, one newton metre.

Corresponding steps are needed in defining a measurement procedure for a heat interaction.

The magnitude of the interaction will be called *the* heat.

A measurement procedure for heat

Standard system:

1 kg of water at a temperature of 14·5 °C and a pressure of 1 atm.

Standard process:

The standard system communicates with a second system at a different temperature. A heat interaction takes place which causes the temperature of the standard system to rise to 15·5 °C.

Definition of the heat unit:

The magnitude of the heat interaction in the above experiment, i.e. *the* heat, is defined as one kilocalorie (1 kcal).

Remarks on the measurement of heat

1. It is found experimentally that the second system mentioned above has to be at a temperature higher than 15·5 °C. This is an instance of the rule that a heat interaction tends to reduce the temperature difference which causes the interaction. Exceptions to this rule exist however (see p. 91).

2. When a system S_a at high temperature communicates with a system S_b at a lower temperature, the heat can be determined as follows. Imagine S_b to be replaced by a number of standard kilograms of water which rise in temperature from 14·5 °C to 15·5 °C as S_a performs the same process as when communicating with S_b. The number of kilograms is the required magnitude, in kcal.

Of course it is not necessary actually to carry out this process (often it would be impossible to perform), any more than we fetch the standard kilogram from Sèvres when we want to measure the mass of fuel used by a boiler. Physicists have devised numerous more practical devices, in effect *sub-standards*, to serve every-day needs; but these are all referred to, i.e. calibrated in terms of, the standard systems and process described

3. Even this is not the whole story. Firstly there are other standard systems and processes in use. For example, when one pound-mass and a temperature rise of one degree Fahrenheit are used, a unit of smaller size is obtained, namely the British Thermal Unit, or Btu (1 Btu $\equiv$ 0·252 kcal). Again, the use of 1 lb$_m$ and 1 K results in the Centigrade Heat Unit,* or Chu (1 Chu = $\frac{9}{5}$ Btu). These variations of practice between English-speaking and European engineers, and between engineers and physicists, were tiresome but not difficult to comprehend.

* Sometimes called the Pound Calorie (lb cal).

4. As a second complication it has been discovered that there is a quantitative relation between heat and work. This will be discussed fully in Chapter 6 on the First Law of Thermodynamics. The relation makes it possible to do away with the heat standard altogether and refer everything to that of work (see p. 99); for example, the rate at which a heat interaction occurs can be expressed in watts.

The Direction of Heat: "Heat Transfer"

In defining a unit of heat a standard process was described involving a temperature rise of 1 kg of water. It is found experimentally that when, as a result of a heat interaction, the temperature of 1 kg of water *falls* from 15·5 °C to 14·5 °C, the magnitude of the heat interaction is again 1 kcal.* In order to distinguish these two cases, a directional language is used with a corresponding algebraic sign convention.

Directional language. If, in a heat interaction between two systems, S_a and S_b, system S_a is at the higher temperature, we say that a "heat transfer" has taken place *from* S_a *to* S_b. Thus the standard process, in which 1 kg of water rises in temperature from 14·5 °C to 15·5 °C, may be described as a heat transfer of 1 kcal *to* the water and *from* the second system which communicates with it; the process in which the 1 kg of water falls from 15·5 °C to 14·5 °C on the other hand is described as a heat transfer *from* the water *to* the second system.

The words "transfer", "to", and "from" are metaphorical. They are universally used because of their convenience, but it should be clearly understood that there is not really any substance, "hotness", which is being transferred in the sense in which water can be poured from a jug into a glass. The "transfer" metaphor derives historically from the long-discredited *caloric theory* (see p. 91 below).

* This may be established by considering three systems:
S_1, the standard system comprising 1 kg of water at atmospheric pressure.
S_2, any system at high temperature θ_1 ($>$15·5 °C) and
S_3, any system at low temperature t_1 ($<$14·5 °C).
We compare the heat transfers when S_1 and S_2, S_2 and S_3 and S_1 and S_3 communicate in turn as follows:
(i) With S_1 at 14·5 °C, raise its temperature to 15·5 °C by heat interaction with S_2, which, as a result, falls in temperature from θ_1 to θ_2. By definition the heat in this process is equal to 1 kcal.
(ii) With S_2 restored to θ_1, reduce its temperature to θ_2 by heat interaction with S_3 which, as a result, rises in temperature from t_1 to t_2. Since S_2 performs the same process as in (i), the heat must also be 1 kcal.
(iii) With S_1 at 15·5 °C, reduce its temperature to 14·5 °C by heat interaction with S_3. If S_3 is initially at t_1 its temperature will be found to have risen to t_2. But from (ii) a heat interaction of magnitude equal to 1 kcal produced this same change of state in S_3. Therefore when the temperature of S_1 is reduced from 15·5 °C to 14·5 °C, the magnitude of the heat interaction must be equal to 1 kcal.

The nature of heat as an interaction would be better conveyed by saying, as for work, that "heat is *done on*" one system and "*done by*" another. Unfortunately, these phrases conflict too much with common usage.

Algebraic sign convention. When the magnitude of a heat interaction appears in an equation, it is given the symbol Q and is associated with one of the two systems taking part in the interaction. Usually the specified system is called "*the* system" while the other system is called "the surroundings". By convention, Q is positive when the surroundings have the higher temperature, and negative when the system temperature is the higher. This

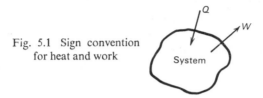

Fig. 5.1 Sign convention
for heat and work

means that heat transfer to the system is positive while heat transfer from the system is negative.

It will be noted that this convention is opposite to that for work, as is shown in Fig. 5.1, where the arrows express symbolically the "transfer" metaphor.

A consequence of the convention is that if, in a given interaction, the heat· for *the* system is positive, that for the surroundings is negative.

Finally it may be mentioned that the directional language and the sign convention may appear simultaneously in phrases such as "the heat transfer to the system was negative", signifying that the system temperature exceeded that of its surroundings and so that the direction of heat transfer was from the system to its surroundings.

How Heat Transfer Occurs

It has been stated above that two systems of different temperature must "communicate" if a heat interaction is to occur. We now briefly consider ways in which the communication may be effected.

Conduction. The simplest form of communication is by contact between the systems, either directly or across a separating layer. Then there occurs the interaction known as *heat conduction* at a rate (e.g. in W/m^2) that is proportional to the existing temperature gradient (e.g. in K/m). The proportionality constant is known as the *thermal conductivity* (e.g. in $W/m\ K$); its magnitude depends on the material, being large for metals, smaller for non-metallic solids and liquids, and smallest of all for gases.

Thermal insulation. In practice, two systems are nearly always in communication by conduction through the intervening material, so that heat transfer occurs constantly. However, the rate of transfer can often be reduced to negligible proportions by separating the systems by thick layers of poorly-conducting materials: the so-called *thermal insulators* or lagging materials. These are usually non-metallic solids containing pores filled with air, e.g. glass wool.

Convection. If the conducting medium is a fluid in motion, the mode of heat transfer is called *convection*.

The mechanism is as follows: heat transfer from a hot body to the particles of fluid in contact with it occurs by *conduction*, causing the fluid temperature to rise. The warmed fluid, being in motion, subsequently comes into contact with a body at a temperature lower than that of the fluid and heat transfer occurs by conduction from the fluid to the cold body.

Radiation. Even when all intervening matter between systems is removed, the systems are still in thermal communication, for electromagnetic radiation occurs between their surfaces. Radiative heat transfer from the sun to the earth is a familiar example. The rate of radiative heat transfer increases rapidly with the temperature of the surfaces. It is often negligibly small at room temperature.

Adiabatic processes. It is clear from the above that systems are always in communication and that no process is entirely without heat interactions. Nevertheless the magnitude of the interaction may be negligibly small if the systems are separated by insulators, if the separation distance is large, or if the period of time under consideration is small. Idealizing this situation, we therefore find it useful to imagine processes in which the heat is zero. These processes are termed *adiabatic*. Similarly, a system which is isolated from its surroundings as regards heat transfer is called an *adiabatic system*.

"Adiabatic" is sometimes used to denote a type of zero-heat process which we shall later learn to call "reversible adiabatic" (Chapter 11). This restriction of meaning is undesirable: in the present book "adiabatic" means neither more nor less than: "$Q = 0$".

"Heat transfer" as a branch of engineering science. The design of equipment for the power-producing, process and other industries is often dominated by the necessity of effecting heat interactions. The design procedures form a separate subject known as "heat transfer", some textbooks on which are mentioned at the end of this Chapter. "Heat transfer" differs from thermodynamics in being concerned with the *rate* at which interactions proceed.

What Heat is Not

By defining heat as above, we exclude a number of meanings attached to the word "heat," in common speech and elsewhere. Some of these will now be brought to light to avoid later confusion.

1. It used to be thought that heat was a fluid which passed from hot bodies to cold. This fluid was given the name "caloric" and attempts were made to weigh and isolate it. Such phenomena as the temperature rise which occurs in metal-turning were explained as the escape of the caloric as the material became sub-divided. The caloric theory was disproved by Rumford in 1798 in a famous experiment: by boring a cannon with a blunt cutting-tool, he showed that the effects of heat, namely the rise in the temperature of the water used to cool the metal, could be produced indefinitely without cutting (sub-dividing) any material at all. Many an apprentice mechanic must wonder why it was so long before this fact was observed.

Heat, therefore, is not a conserved fluid; it is consequently inconsistent with thermodynamic usage to speak of "the heat in the steam", "the heat contained in the exhaust gases", etc. More precisely, *heat is not a property of a system.*

All that remains of the caloric theory is its metaphorical use in the terms "heat transfer to" and "heat transfer from", denoting the direction of a heat interaction.

2. Heat is not "that which inevitably causes a temperature rise". To demonstrate this it suffices to consider a system comprising ice and water. If a heat transfer to the system occurs, for example from a container at higher temperature, it is well known that the temperature of the system does not rise, at least until all the ice has melted. In exceptional circumstances, heat transfer to a system causes a fall in temperature, for example if the system is a gas flowing in a straight pipe at a velocity just below that of sound.

3. Heat is not "that which is always present when a temperature rise occurs". To demonstrate this, it may be noted that work can cause a temperature rise, as in the cannon-boring experiment just mentioned. Since the boring tool and the cannon will, in general, have the same temperature, the definition of heat makes it clear that heat is absent from this process.

As another example, consider a well-lagged rigid vessel containing hydrogen and oxygen, Fig. 5.2. Ignition of the contents of the vessel results in the formation of H_2O accompanied by a temperature rise. However, since the vessel is lagged, there can be no heat transfer from it; for, even though the temperature of the H_2O exceeds that of the surroundings, there is no communication; so Q equals zero. A similar process occurs in the cylinder of a gasoline engine. Ignition of the gasoline-air mixture in the cylinder initiates a reaction in which the fuel and air combine chemically to produce gases of higher temperature. As this occurs very quickly, whereas heat transfer takes time, heat transfer cannot be held responsible for the temperature rise. Indeed if heat transfer did occur, it would be *from* the system comprising the high-temperature gases and *to* the water-cooled combustion-chamber walls. Another example of the lack of an inevitable connexion between a temperature change and a heat interaction, is the

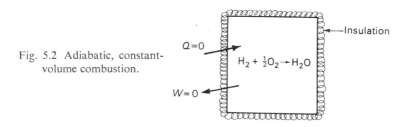

Fig. 5.2 Adiabatic, constant-volume combustion.

$Q=0$

$H_2 + \tfrac{1}{2}O_2 \rightarrow H_2O$

$W=0$

←—Insulation

adiabatic dissolving of common salt in water; this is accompanied by a fall in temperature.

Final remarks about heat. The most important point about heat, as defined above, is that it has meaning only when referred to the boundary of a system; in this it is similar to work. Heat is also similar to work in that it is transient: it exists during the interaction only. Heat is a happening not a substance. It may be likened in this respect to speech between two persons: though the effects may remain for ever, the communication comes to an end with the last word.

BIBLIOGRAPHY

ECKERT, E. R. G. and DRAKE, R. M., *Introduction to the Transfer of Heat and Mass*. McGraw-Hill, New York and Maidenhead, 2nd Edition, 1959.

JAKOB, M. and HAWKINS, G. A., *Elements of Heat Transfer*. Wiley & Sons, New York and London, 3rd Edition, 1957.

McADAMS, W. H., *Heat Transmission*. McGraw-Hill, New York and Maidenhead, 3r Edition, 1954.

FISHENDEN, M. and SAUNDERS, O. A., *Introduction to Heat Transfer*. Clarendon Pres Oxford, 1950.

SCHENK, H., *Heat Transfer Engineering*. Longmans, Green & Co., London, 1960.

CHAPTER 5—PROBLEMS

5.1 One joule is equal to $0 \cdot 238\,8 \times 10^{-3}$ kilocalories. Express one British thermal unit in joules.

5.2 A mass of two kilograms of water at a temperature of 18 °C is poured into an insulated copper vessel which initially is at a temperature of 15 °C. When the temperatures have equalized, the water is at a temperature of 17·4 °C. Determine the magnitude and sign of the heat transfer for each of the following three systems:
(i) the vessel and the insulation; (ii) the water; (iii) the vessel and the insulation plus the water.

5.3 The insulated copper vessel of problem **5.2** contains 10 kg of water at a temperature of 15 °C. A steel bar, of mass 0·3 kg, at a temperature of 740 °C, is suddenly dropped into the water. When the temperatures have equalized, the temperature of the water is 17·4 °C. Determine the magnitude and sign of the heat transfer for each of the following four systems:

(i) the water; (ii) the vessel and the insulation; (iii) the steel bar; (iv) the vessel, the insulation and the contents.

5.4 The insulated copper vessel of problem **5.2** contains a quantity of oil at a temperature of 15 °C. A steel bar, of mass 0·3 kg, at a temperature of 740 °C, is suddenly dropped into the oil. When the temperatures have equalized the temperature of the oil is 17·4 °C. Determine the magnitude and sign of the heat transfer for each of the following four systems:

(i) the vessel and the insulation; (ii) the steel bar; (iii) the oil; (iv) the vessel, the insulation and the contents.

5.5 A copper block, with its top and sides insulated, slides slowly down a rough, non-conducting inclined plane. The mass of the block is 1 kg and it falls a vertical distance of 1 m. Evaluate the heat for the system comprising the block. (The local gravitational acceleration may be taken as 9·81 m/s².)

5.6 State whether the heat, Q, and the work, W, are positive, negative or zero in each of the following processes. The systems to be considered are printed in italics.

(a) A rigid steel vessel containing *steam* at a temperature of 150 °C is left standing in the atmosphere which is at a temperature of 25 °C.

(b) 0·1 kg of *gas* contained in an insulated cylinder expands as the piston moves slowly outwards.

(c) A *mixture of ice and water* is contained in a vertical cylinder closed at the top by a piston; the upper surface of the piston is exposed to the atmosphere. The piston is held stationary while a flame, applied to the base of the cylinder, causes some of the ice to melt.

(d) As under (c), but the piston is allowed to move so as to keep the mixture pressure constant.

(e) A *mixture of ice and water* is contained in an insulated vertical cylinder closed at the top by a non-conducting piston; the upper surface is exposed to the atmosphere. The piston is held stationary while the mixture is stirred by means of a paddle-wheel protruding through the cylinder wall. As a result some of the ice melts.

(f) As under (e), but the piston is allowed to move so as to keep the mixture pressure constant.

(g) A rigid vessel containing *ammonia gas* is connected through a valve to an evacuated rigid vessel. The vessels, the valve and the connecting pipe are well insulated. The valve is opened and after a time, conditions throughout the two vessels become uniform.

(h) *One kilogram of air* flows adiabatically from the atmosphere into a previously evacuated bottle.

5.7 Evaluate the magnitudes and signs of the heat transfer and of the work in the following processes. The system to be considered is printed in italics.

(a) A well-insulated, sealed vessel contains 0·001 kg of fuel-oil and some oxygen gas. The oil ignites, causing a rise in the temperature of the *vessel and its contents*.

(b) A *sealed calorimeter* containing powdered coal and oxygen gas is immersed in a tank containing 2 kg of water. In the first half-minute after ignition of the coal, the water rises in temperature 0·03 K.

5.8 A rigid insulated vessel is divided into two parts by a diaphragm. One part of the vessel contains sulphuric acid at a temperature of 15 °C and the other part contains water at a temperature of 15 °C. The diaphragm is removed so allowing the two fluids to mix; the pressure and temperature of the contents rise and after a time conditions are uniform throughout the vessel.

(a) Considering the vessel and its contents, state whether the heat transfer Q and the work W are positive, negative or zero.

(*b*) The insulation is subsequently removed, allowing the temperature of the contents to fall to that of the atmosphere, 15 °C. State whether Q and W are positive, negative or zero for this process.

(*c*) State whether Q and W are positive, negative or zero for the combined processes (*a*) and (*b*).

6 The First Law of Thermodynamics

Introduction

Now that heat has been defined, it is useful to recall that heat interactions between systems are essential processes in many forms of prime mover. To take the steam power plant as an example, unless the boiler feed-water is brought into communication with a hotter substance (usually the reaction products of some fuel), no steam will be formed and the plant will not run: the production of *work* is conditional on the transfer of *heat* to the steam.

This example has great economic importance, for fuel is often expensive and in short supply, yet the demand for power grows insistently. We must therefore establish what quantitative connexions exist between the "raw material" and the "end-product" of the power-plant engineer, i.e. between heat and work. One connexion is suggested by Rumford's cannon-boring experiment, in which, it will be recalled, work is done continuously on the boring-tool and cannon, while heat is transferred continuously to the cooling water; meanwhile the cannon and boring tool remain unchanged. This is a case of work being "converted into" heat.

In the present chapter, the "rate of exchange" between heat and work will be shown to be a constant and universal one, a fact that will be expressed by the First Law of Thermodynamics. In formulating and discussing this law, the concepts of *cyclic process* and *energy* will be introduced; and new light will be thrown on what is meant by a *property*.

Symbols

E Energy of a system.
e Specific energy of a system.
J Mechanical Equivalent of Heat.
Q Heat transfer to a system.
W Work done by a system.
Δ Increase in ($\equiv$ final value − initial value).

$\underset{\text{cycle}}{\Sigma}$ Algebraic summation around a cycle.

$\oint$ Integral around a cycle, cyclic integral.

Subscripts
1,2 Initial, final state of a system.

The Rate of Exchange Between Heat and Work

There is no *logical* relation between heat and work; so, if a relation does exist, it must be found in an *experimental* investigation. Such a study was carried out between 1840 and 1849 by J. P. Joule, to whom credit for discovering the First Law of Thermodynamics is usually given.* His experiments were of two types, which will now be described.

Experimental evidence for the equivalence of heat and work

Experiments using heat and work to obtain equal effects. We have already seen that a heat interaction can bring about a temperature rise in one of the participating systems. For example, a heat transfer of 1 kcal will cause 1 kg of water to rise from 14·5 °C to 15·5 °C. (Fig. 6.1a)

But a precisely similar change in the water can be brought about without any heat transfer at all: for example, a shaft carrying a paddle-wheel can

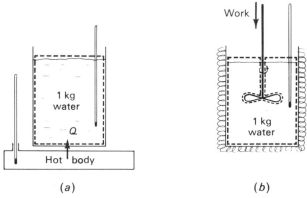

(a)   (b)

Fig. 6.1 Experiments in which the same effect is brought
about (a) by heat and (b) by work.

project into the water; its continued rotation causes the water temperature to rise steadily, even though the container is thermally insulated.

Fig. 6.1b illustrates this case. The shaft may be supposed uniform in temperature so that no heat interaction occurs. The change in the state of the system (the water) has therefore been brought about by the stirring work.

Since heat and work can bring about equal effects, measurements of the magnitudes of the two interactions which separately cause a given change in a system should provide interesting information. Joule carried out many experiments of this type with various systems and various sorts of work. He established that, within the limits of accuracy of his experiments, the number of units of work required to accomplish a given effect, divided by the number of units of heat required to bring about the same effect, was equal to a constant.

Symbolically we write this experimental result as

$$\frac{W}{Q} = a\ constant$$

* R. Mayer deduced a form of the Law in 1840, from a consideration of the properties of the permanent gases.

where W and Q are respectively the work and the heat which cause the same change of state in identical systems.

The constant in this equation is known as "The Mechanical Equivalent of Heat" and is given the symbol J; it has the value $4 \cdot 1868 \times 10^3$ N m/kcal, when the work is measured in N m and the heat in kcal. For SI units, when work is measured in newton metres (N m) and heat in joules (J):

$$J = 1 \text{ N m/J}$$

When Imperial units are used J has the value $778 \cdot 17$ ft lbf/Btu.*

Experiments involving cyclic processes. In experiments of the type just mentioned, the initial and final states of the system are different. These experiments, together with what we know about heat and work, suffice as a basis for the First Law. The modern formulation of this law can be introduced more smoothly however by considering a second type of experiment, also carried out by Joule, in which heat and work are caused to *undo* each other's effects rather than to reproduce them. To make this idea precise, a new definition is required :

A process is cyclic if the initial and final states of the system executing the process are identical, i.e. if all system properties have the same values at the end of the process as at the beginning.

As an example simultaneously of a cyclic process and of Joule's second type of experiment, consider Fig. 6.1 once more. Let the stirring process of Fig. 6.1*b* be carried out first, resulting in a rise in water temperature. Then let the stirring be stopped, and let the water container be brought into contact with another body as shown in Fig. 6.1*a*. This time the body is chosen to be at lower temperature than the water, so that the heat transfer is *from* the water, not *to* it; as a result the water temperature falls.

Suppose that, as soon as the water temperature has fallen to the value obtaining before stirring began, the communication between the container and the cold body is broken. Then the water is in exactly the same state at the end as at the beginning of the process; this may be tested by measuring all conceivable properties of the water in the two cases. The stirring and cooling of the water have combined to form a cyclic process.

Many such cyclic experiments have been carried out, by Joule and later workers. Both the nature of the system and the nature of the process have been varied. Thus the system may comprise a different fluid; chemical reaction may occur; a mechanism may be incorporated; the heat and work interactions may take place in several stages of unequal amount and direction; shear work, displacement work and electrical work may all take part. Yet in each cyclic experiment, when the magnitudes of all the work interactions are added and then, separately, the magnitudes of all the heat

---

* Joule, in his early experiments, decided that $773 \cdot 4$ was the right value. The rounded value of 778 is often used.

interactions, it is found that the first sum is equal to a constant times the second, this constant being J, the Mechanical Equivalent of Heat. The results of these experiments may be written symbolically as

$$\sum_{\text{cycle}} W = J \sum_{\text{cycle}} Q \qquad (6.1)$$

where $\sum_{\text{cycle}} W$ stands for the algebraic summation of all the work interactions in the cycle, i.e. the net work, and $\sum_{\text{cycle}} Q$ the summation of all the heat transfers in the cycle, i.e. the net heat.

J has the same value, as might be expected, in these experiments as in those of the first type, namely $4 \cdot 1868 \times 10^3$ N m/kcal. J does vary, of course with the units used. Thus J is 1 in N m/J and $778 \cdot 17$ ft lbf/Btu.

The First Law of Thermodynamics

Although the number of systems and processes which have been investigated in the way described has obviously not exhausted all possibilities, the uniformity of the outcome of the experiments which have been performed has led to eq. (6.1) being regarded as universally true. We use it as our statement of the *First Law of Thermodynamics* which, for formality's sake, will also be stated in words, as:

When a system executes a cyclic process, the net work is proportional to the net heat.

Eq. (6.1), which is a symbolical expression of the First Law, is applicable to processes in which the heat and the work occur in finite steps. When the interactions occur as a succession of infinite steps, the summation signs are replaced by integrals, to give a second symbolical form of the law:

$$\oint dW = J \oint dQ$$

or

$$J \oint dQ - \oint dW = 0 \qquad (6.2)$$

Here dQ and dW represent infinitesimal elements of the heat and work interactions respectively, with the conventional signs, while $\oint$ stands for adding up all the elements occurring in the cyclic process. $\oint$ is known as the "cyclic integral" or "integral around the cycle".

EXAMPLE 6.1

Problem. In a cyclic process the heat transfers are $+10$ J, -24 J, -3 J and $+31$ J. What is the net work for this cyclic process?

Solution. We apply the First Law in the form of eq. (6.1).

In this example $\sum_{\text{cycle}} Q$ stands for the sum $(10 - 24 - 3 + 31)$ J.

$\therefore$ $\sum_{\text{cycle}} Q = 14 \text{ J}$

Hence $\sum_{\text{cycle}} W = J \times 14 \text{ J}$

$= 1 \text{ N m/J} \times 14 \text{ J}$

$= 14 \text{ N m}$

$\therefore$ Net work $= 14 \text{ N m}$ *Answer*

Discussion of the First Law

Engineering implication. An immediate consequence of the First Law is that a power plant relying on heat transfer, from fuel reaction products for example, cannot produce more than $4 \cdot 1868 \times 10^3$ N m of work for every kcal of heat transferred. As will be seen below (p. 379 *et seq.*), the combustion of a given mass of fuel under specified conditions allows a definite heat transfer to its surroundings. With this quantity known, the lowest fuel consumption per unit of work can be calculated.

That no more than $4 \cdot 1868 \times 10^3$ N m of work can be obtained for each kcal of heat may be disappointing, but the First Law also suggests that this amount of work should at least be forthcoming. In Chapter 12 it will be shown that, for a reason unconnected with the First Law, often no more than 10^3 to $1 \cdot 5 \times 10^3$ N m of work are obtainable for each kcal of heat *transferred from the fuel reaction products.*

The N m of heat and the kcal of work. The fixed rate of exchange between heat and work enables each to be expressed in the units of the other. Thus heat can be measured in N m or ft lbf, while work can be expressed in, for example, kcal or Btu, just as the price of an article can be expressed in dollars or pounds sterling regardless of whether it is a product of the United Kingdom or the United States. This does not mean of course that heat and work are the same thing, any more than the dollar is the same, in all its capabilities, as its nominal equivalent in sterling.

When both quantities, the heat and the work occur in the same calculation, it is convenient to use the same unit for both. This means that J can be dropped from the equation. The First Law of Thermodynamics, for example, now becomes

$$\oint dQ - \oint dW = 0$$

or $$\oint (dQ - dW) = 0 \qquad (6.3)$$

in which it is understood that *both* Q and W are expressed in N m, J, or kcal, or other appropriate unit. Eq. (6.3) can be regarded as embodying a *special convention*, in the sense given in Chapter 2, namely: the unit system shall be such that J equals unity. We shall use this special convention extensively.

Abandoning the "heat standard". In Chapters 3 and 5, independent procedures for measuring work and heat were described. The determination

TABLE 6.1   *Conversion Factors for Units of Heat, Work and Energy*

	newton metre (N m)	foot-pound force (ft lbf)	kilocalorie (kcal)	British thermal unit (Btu)	kilowatt hour (kWh)	horsepower hour (hp h)
1 newton metre (N m) ($\equiv$ 1 Joule (J)) =	1	0.737 562	$0.238\ 846 \times 10^{-3}$	$0.947\ 817 \times 10^{-3}$	$0.277\ 778 \times 10^{-6}$	$0.372\ 506 \times 10^{-6}$
1 foot pound-force (ft lbf) =	1.355 82	1	$0.323\ 823 \times 10^{-3}$	$1.285\ 07 \times 10^{-3}$	$0.376\ 616 \times 10^{-6}$	$0.505\ 051 \times 10^{-6}$
1 kilocalorie (kcal) =	4186.8	3088.03	1	3.968 32	0.001 163	$1.559\ 61 \times 10^{-3}$
1 British thermal unit (Btu) =	1055.06	778.170	0.251 996	1	$0.293\ 071 \times 10^{-3}$	$0.393\ 015 \times 10^{-3}$
1 kilowatt hour (kWh) =	3.6×10^{6}	$2.655\ 22 \times 10^{6}$	859.845	3412.14	1	1.341 02
1 horsepower hour (hp h) =	$2.684\ 52 \times 10^{6}$	1.98×10^{6}	641.186	2544.43	0.745 700	1

of J was then shown to be an experimental matter. As experiments of increasing accuracy have been performed, the actual value ascribed to J has altered continually, the later changes being of course only to the last one or two significant figures. In this situation it is natural to suggest that the constant re-adjustment of J, and of calculations involving it, should be stopped once and for all by abandoning either the heat or the work standard and *defining J* to have a definite value.

This has in fact been done. Since work measurements are easier to make than those of heat, the definition of the kcal in terms of 1 kg of water raised from 14·5 °C to 15·5 °C has been abandoned. Instead we define 1 kcal of heat as the equivalent 4·1868 × 10³ N m of work, where the latter unit is defined in terms of the "mechanical" standards. In the Imperial system 1 Btu ≡ 778·170 ft lbf and in SI, 1 J ≡ 1 N m.

This change in standardization procedure, which is a common device of metrologists, should not be regarded as diminishing the status of the heat unit in any way. For most engineering purposes it suffices to imagine the kcal as still defined in terms of the temperature rise of water.

Units of heat and work. Tables 6.1 and 6.2 contain conversion factors connecting some of the more common units in which heat and work are measured.

Concluding remark on equations (6.1) and (6.2). The First Law of Thermodynamics, as expressed by equations (6.1) and (6.2), interrelates the boundary interactions occurring when a system executes a cycle. It is valid whatever the nature of the system or its interactions with its surroundings,

TABLE 6.2 *Additional Conversion Factors for Units of Heat, Work and Energy*

1 joule = 0·101 972 metre kilogram-force, m kgf

= 9·868 96 × 10⁻³ litre atmosphere

= 23·730 4 foot poundal, ft pdl

= 2·389 20 × 10⁻⁷ thermie, th

but one restriction remains: the process must be cyclic. Most processes that the thermodynamicist must deal with are not cyclic; a way must therefore be found of re-formulating the First Law so that it can be brought to bear on non-cyclic processes also. This is done in the next section by the introduction of a new concept: *energy.* The re-formulation has the effect of relating the interactions *at* the boundary to changes *within* the system.

Energy

Properties of properties

The concept "property" was defined on p. 57 as "any observable char acteristic of a system". It is now desirable to point out three logical con sequences of this definition, namely:

(i) The change in the value of a property of a system depends only on th end-states of the process and not on the path of the process.

(ii) If a magnitude related to a system changes during a process by a amount that depends only on the end-states and not on the path of a process that magnitude is a property of the system.

(iii) In a cyclic process the net change in each property of a system is zero

Explanatory examples. The truth of these statements is best perceived b consideration of examples:

The *latitude* and *longitude* of a ship on the ocean are *properties*, for the may be ascribed definite values as a result of observations of the ship at given moment. When the ship moves to a new position, the changes in th latitude and longitude can be calculated from knowledge of the new and th old positions alone: there is no need to know where the ship went to i between. In particular, if the ship leaves port for a voyage *and then return* to that port, the changes in latitude and longitude are zero: *properties suff no net change in a cyclic process.*

The *distance travelled* by the ship is *not a property*. It can not be deter mined from mere observation of the ship's position at any one time, nor ca it be calculated given only the initial and final position. The distance travelled is not zero merely because the ship returns from a round trip to th same berth from which it set out.

The *balance* in a bank account is a *property* because it has a definite val at any one time. Knowing the values of the balance at the beginning a end of a year, we can calculate the net effect of the deposits and wit drawals. What can not be determined from the two end-states, however, whether a decrease in the balance took place as a result of a single lar withdrawal, as a result of many small withdrawals, or as a result of larg withdrawals inadequately compensated by a few deposits of money into t account. The *deposits* and *withdrawals* therefore are *not properties*. It is t balance, which is the sum of the deposits minus the sum of the withdrawa which is the property. If, by careful budgetting, the balance at the end the year is identical with that at the beginning, this does not mean that deposits or withdrawals have taken place, but merely that the differen between them is zero: a cyclic process has been performed.

In mathematics, a *point function* is one which has a definite value when t finite number of variables forming its argument are specified; it therefe defines a property. A *path function* has a value which depends on t specification of a curve, i.e. of an infinite succession of arguments; it cannot therefore be said to define a property.

Definition of energy

The foregoing discussion paves the way for the demonstration that the First Law implies the existence of a property, which will be called *energy*. Although students of mechanics will already have some notion of energy, thermodynamicists require an extended definition. Moreover the pattern of argument proving the existence of a property from a statement about cyclic processes will be repeated in Chapter 13, in connexion with entropy.

First, energy will be defined:

The increase of energy of a system during a change of state is numerically equal to the net heat during the process minus the net work during the process.

In symbols this definition is written as

$$E_2 - E_1 \equiv Q - W \qquad (6.4)$$

where E_2 and E_1 are the energies of the system in the final state 2 and the initial state 1 respectively.

Remarks on the definition of energy

1. Only *changes* of energy have been defined. This merely means that, like longitude for example, energy has to be measured with respect to some arbitrary base.

2. In order that the numerical difference between the heat and the work should be physically meaningful, Q and W must both be measured in the same units, e.g. N m, or kcal, or J, i.e. the "special convention" must be employed. These same units must correspondingly be attributed to E.

3. Often the increase in a magnitude, i.e. its final value minus its initial value, is symbolized by a delta, Δ, followed by the symbol for the magnitude. In this notation, eq. (6.4) is written

$$\Delta E \equiv Q - W \qquad (6.5)$$

4. The energy of a system of unit mass, sometimes called the *specific energy*, will be given the lower-case symbol e.*

5. So far, we have supposed the whole process between states 1 and 2 to be specified. We now show by reference to the First Law that only the end-states 1 and 2 need be specified, and so that E is a property. The *form* of the proof should be given especial attention.

Proof that energy is a property

Fig. 6.2 is a diagram having arbitrary properties of a system, x and y, as abscissa and ordinate. (For definiteness, x and y might be thought of as

* Where relevant and possible, lower-case type will be used for the value per unit mass of any quantity signified by capital type. Shortage of letters precludes complete consistency with this convention.

volume and pressure respectively.) The points marked 1 and 2 represent two states of the system.

Consider a cyclic process executed by the system: it starts from state 1, proceeds to state 2 along the path marked A on Fig. 6.2, and returns to state 1 along the path C. From the First Law we can write

$$\oint (dQ - dW) = \int_{1 \atop A}^{2} (dQ - dW) + \int_{2 \atop C}^{1} (dQ - dW) = 0 \qquad (6.6)$$

wherein the summation around the cyclic process has been split into its component parts. The two integrals on the right-hand side of eq. (6.6) represent, respectively, the summation of the heat and work during the change of state from 1 to 2 along path A, and the summation from 2 to 1 along path C.

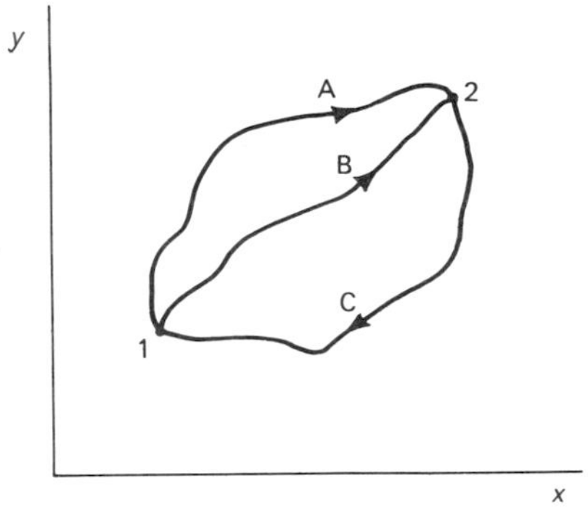

Fig. 6.2 Paths of cyclic processes passing through two common state-points.

Now consider a second cyclic process differing from the first only in that the outward path is the different one marked B on Fig. 6.2. Applying the First Law to this cyclic process, we obtain

$$\int_{1 \atop B}^{2} (dQ - dW) + \int_{2 \atop C}^{1} (dQ - dW) = 0 \qquad (6.7)$$

Combination of eq. (6.6) and eq. (6.7), yields

$$\int_{1 \atop A}^{2} (dQ - dW) = \int_{1 \atop B}^{2} (dQ - dW) \qquad (6.8)$$

signifying that the integral of $(dQ - dW)$ from state 1 to state 2 is the same for path A as for path B.

But all that has been specified about paths A and B is that they are different. It follows that

$$\int_{1}^{2} (dQ - dW) \text{ has the same value for } any \text{ path between 1 and 2.} \qquad (6.9)$$

If we now write

$$\int_1^2 dQ \equiv Q,$$ the net heat transfer as the system changes from state 1 to state 2,

(6.10)

and

$$\int_1^2 dW \equiv W,$$ the net work done during the same change,   (6.11)

statement (6.9) becomes:

$(Q - W)$ has the same value for *any* path between 1 and 2.

But from the definition of energy, eq. (6.4),

$$Q - W \equiv E_2 - E_1$$

Therefore

$E_2 - E_1$ has the same value for *any* path between 1 and 2.   (6.12)

This means that the value of $E_2 - E_1$ depends only on the end-states. From what has been said earlier about the "properties of properties" (consequence (ii) p. 102), it follows that *energy E is a property.*

Remarks about energy, E

Relation to the "energy" concept of mechanics. In mechanics, energy is sometimes defined as "the capacity-for-doing-work" of a system. This "mechanical energy", as we may call it, is not the same thing as the energy of thermodynamics, although the two concepts have features in common. An example will illustrate similarities and differences.

The ball in the bowl: Fig. 6.3 shows a system comprising a hemispherical bowl and a steel ball. Suppose that initially the ball is at rest at the lowest point of the bowl, and that then it is raised by an external force (for example a hand) until it is at rest near the rim. The work done by the system is negative in this process since the surroundings have done work on the system; heat transfer being absent, it follows from eq. (6.4) that $E_2 - E_1$ is positive, i.e. the energy of the system increases. The "mechanical energy" has also increased, and indeed by the same amount, for the ball is higher than it was.

Fig. 6.3 Bowl containing
a steel ball.

Now suppose that the ball is released so that it rolls backwards and forwards within the bowl. Since there are neither heat nor work interactions at the boundary of the system during the oscillations, it is seen from eq. (6.4) that there can be no change of energy. If friction may be ignored, this is true also of the "mechanical energy"; for, regarding only the ball (the bowl does not change), the decrease of gravitational potential energy which the ball suffers in reaching the bottom of the bowl is exactly balanced by its increase in kinetic energy: the sum of the two energy terms remains constant.

If friction is present on the other hand, the height reached by the ball in each successive oscillation is reduced; the maximum velocity attained is also lower at each oscillation. Finally the ball comes to rest again in the bottom of the bowl. Since still no heat or work interactions are present at the system boundary, the energy of the system has the same value as immediately after the ball was first raised. The "mechanical energy" however has *decreased* to the value prevailing *before* the ball was first raised. In mechanics it is said that "the energy has been dissipated".

An alternative phrase used in mechanics to describe the events of the last paragraph is: "the energy has been converted into heat". This phrase is not in accordance with thermodynamic usage of the word "heat"; for, although the ball and the bowl no doubt have a higher temperature than in the beginning, this has not been accomplished by heat transfer between the system and its surroundings.

A legitimate form of the above phrase is: "the mechanical energy has been transformed into that mode of energy which can be directly altered by heat transfer". This latter mode of energy is termed internal energy* (see p. 120).

In summary, energy is seen to include the energy which is defined in mechanics, but to be more general than it. For example, it covers internal energy also. Processes may occur in which the energy of the system remains constant, but transformations occur from one form of mechanical energy to another and from mechanical to internal energy (see for example pp. 107 108, 109).

It may be mentioned that the "capacity-for-doing-work" idea plays an important part in more advanced thermodynamic texts than the present. It forms the core of the concept of "availability".

The Law of Conservation of Energy. Eq. (6.4) and eq. (6.5) may be regarded as alternative statements of the First Law of Thermodynamics, coupled with the statement that E is a property. A particular case of some interest is that of a system in which both heat and work interactions are absent:

when $Q = 0$ and $W = 0$: $\Delta E = 0$ (6.12)

---

* Many students find it helpful to retain a "mechanical" picture of this mode of energy by associating it with the energies of translation, rotation, and intermolecular attraction of the microscopic structure of the system. This interpretation is substantially correct but it is not necessary. Eq. (6.4) suffices for the thermodynamicist.

This equation represents a corollary of the First Law which is known as the *Law of Conservation of Energy*. In words, this is:

The energy of a system remains constant if the system is isolated from its surroundings as regards heat transfer and work.

This statement is of course less general than the First Law, since it does not state how the energy changes when Q and W are *not* zero.

Other modes of energy. Gravitational potential energy, kinetic energy, and internal energy have all been mentioned above as modes of energy. These three forms of energy will be dealt with quantitatively in Chapter 7; the third, internal energy, which can be altered by heat transfer alone, will be given a special symbol. In this section we mention qualitatively some of the other modes of energy.

A system may change its chemical state, as for example when gasoline and air react so as to form carbon dioxide, water vapour and other combustion products. Often, if the system is isolated from its surroundings as regards heat and work, the temperature of the system changes greatly as a result of such a chemical-reaction process; in the gasoline-air case the temperature increases. It is sometimes helpful to regard such processes, which are at constant energy (see eq. (6.13)), as involving a transformation of "chemical energy" into internal energy.

A compressed spring or elastic structure may be regarded as having part of its energy in the form of "strain energy". A related form of energy is that associated with the phenomenon of capillarity, or surface tension; systems exhibiting this phenomenon may be said to have part of their energy in the form of "surface energy".

A system comprising electrically charged elements may be considered, by reason of the attractions and repulsions existing between these elements, to have part of its energy in the form of "electrical energy". "Magnetic energy" correspondingly is present if the system comprises magnetic poles. If both magnetic and electric effects are simultaneously present, their interactions also produce a contribution to the energy of the system.

The above remarks are of interpretative value only. When any one of the above modes of energy is to be used in a calculation, strict definition and derivation in terms of the First Law are necessary (see page 379 for example). The important point is that the First Law and the energy concept are entirely general, and cover any sort of process that may be encountered.

Some Examples Involving Energy

Block-and-plane problems

Although somewhat academic, the following two problems bring out important features of heat, work and energy.

EXAMPLE 6.1: *Non-conducting block; conducting plane.*

Fig. 6.4 shows a rough inclined plane on the sloping face of which stands a rectangular block. The block is a poor conductor of heat while the plane is a good one; for example the block might be of wood and the plane of copper.

108   ENGINEERING THERMODYNAMICS

Problem. The block slides slowly down the plane, which rises in temperature as a result. State whether the heat, work and increase in energy are positive, negative, or zero, for

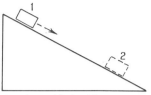

Fig. 6.4   Block sliding down a rough inclined plane.

each of the following systems: (*a*) the block, (*b*) the plane, (*c*) the block and the plane. *Answer.* The solution is contained in the following table:

System	Q	W	ΔE
(*a*) Block	0	+ve	−ve
(*b*) Plane	0	−ve	+ve
(*c*) Block + plane	0	0	0

Comments. System (a). Q equals 0 because the system is a non-conductor of heat; the block acts as a thermal insulator to the surface of the plane with which it is in contact. W is +ve because, although the actual external effect is the rise in temperature of the plane, the block could have carried out the *same process* and had as its sole effect the raising of a weight (for example if the plane had been lubricated and the block connected to a suitable pulley mechanism).

Since $Q - W$ is thus negative, it follows by definition that ΔE is −ve, eq. (6.5). The mode of energy which has suffered a decrease is clearly the gravitational potential energy, for only the position of the block has changed.

System (b). Q equals 0 for the same reason as before. W is −ve for the plane because it is +ve for the block. So $Q - W$ is +ve, and therefore ΔE is +ve by definition. It is the internal energy of the plane which has increased.

System (c). The combined system, block + plane, does not interact with its surroundings in any way. Therefore Q equals 0 and W equals 0 for this system. By definition therefore ΔE equals 0 also. The decrease in the gravitational potential energy of one part of the system (the block) is exactly counterbalanced by the increased internal energy of another part of the system (the plane).

EXAMPLE 6.2: *Conducting block; non-conducting plane*

Suppose, this time, that the block is made of some good conductor such as copper, while the inclined plane is made of such a poor conductor of heat that its conductivity may be taken as zero.

Problem. As before, except that this time it is the block which becomes warmer.

Answer.

System	Q	W	ΔE
(*a*) Block	0	0	0
(*b*) Plane	0	0	0
(*c*) Block + plane	0	0	0

Comments. System (a). Q equals 0, because a heat interaction requires two systems in communication, and in the present case one of them (the plane) is non-conducting. W equals 0, because the system has no effect external to itself whatsoever. More specifically W equals 0, because the block, in executing the same process, (i.e. its fall in level and its rise of temperature) could not have the rise of a weight as its *sole* external effect. This may be seen by supposing the real process to be replaced by the following processes. First the block slides frictionlessly down the plane to its final position, so raising a weight; the temperature of the block is unchanged. Second, the temperature of the block is raised to the value attained in the real process by heat transfer from a hot body. Thus, in completely establishing the final state of the real process, two effects have occurred in the imaginary surroundings, namely, the rise of a weight and heat transfer from a hot body. The rise of a weight has not been the *sole* effect and therefore we cannot conclude that work has been done. Consequently ΔE equals 0 also, the decrease in gravitational potential energy of the block being exactly compensated by the increase in its own internal energy.

System (b). Q equals 0 and W equals 0 for the plane, because they are zero for the block. ΔE is consequently zero, which is easily understood since the plane suffers no change whatsoever.

System (c). Q, W, and ΔE are all 0, because the combined system has no interaction with its surroundings.

Final remark: It may be thought puzzling that in the two cases the work done, by the block for example, differs, even though the mechanical aspects of the process are apparently the same. There is not much to be said about this other than to confirm that it is a logical consequence of the definitions of heat and work, which are themselves not arbitrary but merely express the distinction which will be needed to formulate the Second Law of Thermodynamics below. It will be remembered that the "force-moving-through-a-distance" definition of work leads to difficulties in the case of solid friction, since the "distance" is different for the two sides of the boundary* (see p. 71).

Problems involving chemical reaction

We conclude with an application of the First Law to systems which undergo chemical reaction. As well as illustrating the generality of the energy concept, the examples have engineering importance and are also relevant to Chapter 16 on Combustion.

EXAMPLE 6.3: *The constant-volume calorimeter (bomb calorimeter)*

Fig. 6.5 illustrates a rigid sealed "bomb" immersed in a bath of water which is kept at a uniform temperature by stirring. The bomb contains a crucible holding 0·000 65 kg of fuel oil, and an atmosphere of oxygen at a pressure of 20 atm. The oil can be ignited electrically from outside.

Problem. The oil is ignited. For 30 s after ignition no detectable temperature rise occurs in the water bath. Thereafter the water temperature begins to rise and after some time becomes steady at 2 K above its original value. If the mass of water is 3·3 kg, by how much has the energy of the system comprising bomb and contents increased, (a) 30 seconds after ignition, (b) when the temperatures have become steady?

Neglect the work associated with the igniting current and with the stirring, and the heat transfer from the water to its surrounding container and to the atmosphere.

---

* Someone is sure to ask, "What happens when both block and plane are non-conducting?" The simplest answer is that the question has no meaning, since "non-conducting" means "a very poor conductor relative to the other system". The question partakes of the nature of that involving an irresistible force and an immovable object.

Solution (*a*). 30 s after ignition, the system comprising bomb and contents has had no external effects whatever. Specifically therefore Q and W equal 0, so that from eq. (6.4),

$$\Delta E = 0. \qquad\qquad Answer\ (a)$$

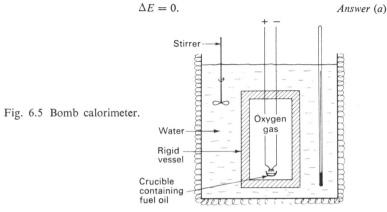

Fig. 6.5 Bomb calorimeter.

Solution (*b*). In the whole process, the heat transfer from the bomb to the water bath is equal to the mass of water times the number of degrees temperature rise. Thus, for the system specified, Q is given by:

$$
\begin{aligned}
Q &= -3{\cdot}3 \times 2 \\
&= -6{\cdot}6\ \text{kcal} \\
&= -6{\cdot}6 \times 4{\cdot}1868 \times 10^3\ \text{N m} \\
&= -27{\cdot}6 \times 10^3\ \text{N m} \\
&= -27{\cdot}6\ \text{kJ}.
\end{aligned}
$$

Since the bomb is rigid, W equals 0, so that

$$\Delta E = Q - W$$

$$= -27{\cdot}6 - 0$$

$$= -27{\cdot}6\ \text{kJ} \qquad\qquad Answer\ (b)$$

or
$$\Delta E = -\frac{27{\cdot}6}{0{\cdot}000\ 65} = -42\ 500\ \text{kJ/kg fuel}$$

Comments: (*a*) It is noteworthy that during the initial period the energy change is zero, even though the bomb contents will have a very high temperature as a result of the combustion. This may be interpreted as an increase in internal energy being exactly compensated by a decrease in "chemical energy".

(*b*) Note that finally the energy of the system is lower than initially, even though the final system temperature is slightly above the initial temperature.

An estimate can be made of the value that ΔE would have if the final system temperature had been exactly equal to the initial temperature, as would occur if the water bath were very large. The *negative* of this new ΔE, when expressed on a basis of unit mass of fuel, is known* as the *constant-volume heating value* of the fuel. For hydrocarbon fuels its value is around 44 000 kJ/kg fuel. This topic is discussed in greater detail in Chapter 16, p. 392.

* Among other names.

The internal-combustion engine. Gasoline and diesel engines burn hydro-carbon-oil fuels with air under approximately constant-volume conditions. If we suppose that the ΔE for the combustion process is a constant of the fuel, when the final combustion-product temperature equals the initial fuel and air temperature, a calculation can be made of the specific fuel consumption of a hypothetical engine which effects such energy changes with work interactions only.

EXAMPLE 6.4

Problem. An oil of constant-volume heating value 44 000 kJ/kg is burned with air in a hypothetical engine in such a way that: (i) the oil-air mixture reacts adiabatically at constant volume; (ii) the final temperature of the combustion products equals the initial temperature of the oil-air mixture. What is the specific fuel consumption of the engine in kg fuel/kWh?

Solution. Since Q equals 0 by hypothesis, we have for a system comprising 1 kg of oil with associated air, from eq. (6.5)

$$0 - W = -44\ 000 \text{ kJ/kg fuel}$$

i.e.
$$W = 44\ 000 \text{ kJ/kg fuel}$$

$$= \frac{44\ 000}{3600} \text{ kWh/kg fuel}$$

Therefore the specific fuel consumption

$$= \frac{3600}{44\ 000} \text{ kg fuel/kWh}$$

$$= 0\cdot0818 \text{ kg fuel/kWh.} \qquad\qquad \textit{Answer}$$

Comment. Actual gasoline and oil engines have specific fuel consumptions three or four times as large as this figure. Two reasons for this are: (i) in real engines the heat transfer between the reacting mixture and its surroundings is not zero; and (ii) the temperature of the outflowing products of combustion greatly exceeds that of the inflowing fuel and air. A full discussion of the differences is, however, too advanced to be given here.

Alternative statements of the First Law and of the definition of heat transfer

In the present book we have introduced work, heat and the First Law of Thermodynamics in a way which accords closely with the historical development of the subject. There exist, however, several other ways, chiefly distinguished by the order in which the concepts are defined. One such alternative, adopted for example by Shapiro, Baehr, and Zemansky (see Bibliography), runs as follows:

First, work is defined, much as we have done above.

Secondly, the *adiabatic* system is defined as one which is subject only to work interactions with its surroundings.

Then, the First Law is stated as: For any prescribed change of state of an adiabatic system, the net work is found experimentally to be independent of the path of the process. Hence, the energy of the system, defined as increasing by an amount equal to the net work in the adiabatic process, is shown to be a property. Energy so defined is of course entirely equivalent to the energy as introduced above.

Finally, in recognition of the fact that work is not the only sort of interaction possible between a system and its surroundings, heat is defined as: the sum of W and ΔE in any process. Heat defined in this way is in all respects equivalent to heat as defined in the present book.

Other entirely legitimate procedures can be found in various textbooks on thermodynamics; which is to be preferred is very much a matter of taste. In favour of the procedure just outlined is the fact that it accords with present metrological practice, involving only one unit quantity, that of work; there is thus no need to think in addition about the temperature rise of unit mass of water through one degree. On the other hand, the definition of energy by reference to adiabatic processes, followed immediately by its use in connexion with non-adiabatic processes, is sometimes found difficult to understand. However, even readers of the present book will have to surmount a similar obstacle when they reach Chapter 12, so this objection has only limited force.

BIBLIOGRAPHY

BAEHR, H. D., *Thermodynamik*. Springer, Berlin, 1962.
SHAPIRO, A. H., *Dynamics and Thermodynamics of Fluid Flow*. Ronald Press Co., New York. Vol. I 1953, Vol. II 1954.
ZEMANSKY, M. W., *Heat and Thermodynamics*. McGraw-Hill, New York and Maidenhead 4th Ed., 1957.

CHAPTER 6—PROBLEMS

6.1 Two identical mixtures of ice and water contained in two similar vessels, A and B undergo identical changes of state at atmospheric pressure as follows:
(i) A heat transfer of 30 Btu to the contents of vessel A causes some of the ice to melt
(ii) Stirring work done on the contents of vessel B causes an equal quantity of ice to melt. The stirrer rotates at a speed of 1485 rev/min and operates for 40 minutes. The mean torque imposed by the mixture on the stirrer is 0·75 lbf in. Vessel B is well insulated
Evaluate: (a) the stirring work done on the contents of vessel B in ft lbf.
(b) the mechanical equivalent of heat in ft lbf/Btu.
(c) the change in the energy of the mixture in each vessel.
Neglect the change of volume due to the melting of the ice.

6.2 A vertical cylinder fitted with a frictionless leakproof piston contains a quantity of gas. The piston is free to move and its mass is such that the gas pressure is 200 × 10^3 N/m²; the upper surface of the piston is exposed to the atmosphere. The gas executes a cycle by undergoing the following processes in sequence;
(i) With the cylinder well-insulated, 1200 N m of stirring work are done on the gas by a paddle wheel projecting through the cylinder wall. As a result the gas temperature rises and the piston moves slowly upwards; the increase in the gas volume is 2·80 dm³
(ii) With the insulation removed and the paddle wheel stationary, heat transfer from the gas of 0·64 kJ restores the gas on its initial state.

Evaluate:—(a) the displacement work and hence the net work done by the gas during process (i),

(b) the net work done by the gas in process (ii),

(c) the increase in energy of the gas in process (i) and in process (ii),

(d) the increase in energy of the gas for the combined process (i) plus (ii).

6.3 In an experiment to determine the mechanical equivalent of heat, a paddle wheel was fixed to the shaft of an engine and rotated in water contained in a well-insulated closed vessel. The vessel was mounted freely on the shaft and prevented from rotating by weights attached to its side.

At an engine speed of 270 rev/min it was found that 173 kg of water had increased in temperature by 88 K in 51 min. The weights attached to the vessel exerted a force of 492 N at a distance of 1·5 m from the axis of rotation of the engine shaft. Sketch the arrangement and evaluate the mechanical equivalent of heat. The heat transfer required to raise the temperature of 1 kg of water by 1 K is $4·19 \times 10^3$ J.

6.4 The working fluid in an engine continuously executes a cyclic process. During one cycle the fluid engages in two work interactions: 15 000 N m to the working fluid and 44 000 N m from the working fluid. Also during the cycle there are three heat transfers, two of which are known: 75 kJ to the working fluid, and 40 kJ from the working fluid. Determine the magnitude and direction of the third heat transfer.

6.5 Reconsider problem 5.6 and state whether the increase in energy, ΔE, is positive, negative or zero.

6.6 Reconsider problem 5.7 and state whether the increase in energy, ΔE, is positive, negative or zero.

6.7 Reconsider problem 5.8 and state whether the increase in energy, ΔE, is positive, negative or zero.

6.8 A dashpot is a device for controlling the motion of mechanisms. It consists of a cylinder, closed at both ends, fitted with a piston; a rod, which protrudes through one of the cylinder covers, connects the piston to the mechanism. Both ends of the cylinder contain oil and are in communication, via a small hole drilled through the piston. As the piston is pushed inward, the oil in one end of the cylinder is forced through the hole to the other end.

In a particular dashpot, the mechanism connected to the piston has a kinetic energy of 4000 N m initially; finally the mechanism is at rest. Assuming heat transfers to be negligible evaluate the increase in the energy of each of the following systems:

(i) The mechanism-piston combination.

(ii) The oil.

(iii) The mechanism-dashpot combination.

6.9 A gyroscope is set spinning and is placed inside a well-insulated rigid box; initially the gyroscope has a kinetic energy of 1000 N m. Evaluate the increase in energy of the contents of the box when the speed of the gyroscope has fallen to zero.

6.10 For tests on a turbine rotor to determine the effect of fluid friction on its rotational motion, the rotor is mounted in the turbine casing with the inlet and exhaust flanges blanked-off. After the casing has been charged with steam, the rotor is accelerated to a specified speed, the drive is disconnected, and the rate at which the rotor decelerates is observed.

In a particular test the rotor has a kinetic energy of 200 000 N m when the drive is disconnected. Evaluate the increase in energy of the contents of the turbine casing as the rotor decelerates to rest. Neglect bearing friction and assume zero heat transfer.

6.11 (a) A system undergoes a process in which the heat transfer to the system is 40 kJ and the work done by the system is 45 000 N m. Evaluate the increase in the energy of the system.

(b) In a second process between the same initial and final states, the same system does 35 000 N m of work; there is also heat transfer during this second process. Determine the magnitude and sign of the heat transfer.

6.12 (a) A system comprising 3 kg of a mixture of air and nitrogen is initially at a pressure of 100×10^3 N/m² and a temperature of 30 °C. The mixture undergoes a process to a pressure of 400×10^3 N/m² and a temperature of 90 °C. During the process the heat transfer from the mixture is 10 kJ and the work done on the mixture is 64 000 N m. Evaluate the increase in the energy of the mixture.

(b) Following the first process, the mixture undergoes a second process to a pressure of 100×10^3 N/m² and a temperature of 30 °C. The work done by the mixture during the second process is 50 000 N m. Determine the magnitude and direction of the heat transfer during the second process, assuming that the specification of the pressure and temperature fixes the state of the mixture completely.

(c) What is the increase in the energy of the mixture after it has undergone both processes in sequence?

6.13 In a closed-circuit wind tunnel, air delivered by a fan circulates in succession through a cooler and the test section and then returns to the suction side of the fan. An electric motor outside the air duct drives the fan by means of a shaft protruding through the air-duct walls. Heat transfer through the air-duct walls is negligible. Assuming that the supply of coolant to the coolers is shut off, and considering the contents of the air duct as the system, state whether the heat, the work and the increase in energy are positive, negative or zero for the following cases:

(i) when the fan is running continuously;

(ii) when the current is switched-off and the fan speed gradually falls to zero.

6.14 Reconsider problem **6.13** with the motor mounted within the air duct; in this case the electric supply cables pass through the air-duct walls.

6.15 State whether the heat, the work, and the increase in energy are positive, negative or zero in each of the following cases. The system in question is the cell in each case.

(a) A lead-acid secondary cell, discharging adiabatically, supplies current to a resistor.

(b) A lead-acid secondary cell, discharging adiabatically, supplies current to an electric motor.

(c) The lead-acid cell of (a) undergoes the same change from the same initial state as in (a) while standing on open circuit for a long time.

6.16 A cyclic process, known as the air-standard Otto cycle, is often used as a standard of comparison for reciprocating internal-combustion engines. The cycle is executed by unit mass of air contained in a cylinder closed by a frictionless leakproof piston; it consists of four processes in sequence, as follows:

(i) Air initially at a pressure of 100×10^3 N/m² and a specific volume of 1·3 m³/kg (state 1) undergoes a fully-resisted, adiabatic compression to a specific volume of 0·18 m³/kg (state 2). During the process the pressure p is related to the corresponding specific volume v by $pv^{1.4} = a$ constant.

(ii) Heat transfer to air of 840 kJ, at constant volume, causes the air pressure to rise to 3500×10^3 N/m² (state 3).

(iii) The air undergoes a fully-resisted adiabatic expansion according to $pv^{1.4} =$ constant, to the initial specific volume (state 4).

(iv) Heat transfer from the air, at constant volume, causes the air pressure to fall to its initial value, 100×10^3 N/m², to complete the cycle.

Sketch the cycle on a pressure ~ specific-volume state diagram and evaluate:

(a) the work done by the air during each process;

(b) the heat transfer from the air during process (iv);

(c) the increase in the energy of the air in each process.

6.17 (a) An insulated rigid vessel contains some powdered coal and air at a pressure of 10 atm and a temperature of 20 °C. The coal is ignited; there results a rise in the pressure and temperature of the contents of the vessel; the final temperature is 550 °C. Taking the vessel and its contents to be the system under consideration, evaluate the increase in the energy of the system.

(b) The insulation is now removed. A heat transfer of 45 kJ from the system causes the temperature to fall to the initial value, 20 °C. Evaluate the increase in the energy of the system during this process.

(c) Taking the initial energy of the system as 30 kJ, write down the energy values after process (a) and after process (b).

6.18 A 100 mm-diameter vertical cylinder, closed by a piston, contains a combustible mixture at a temperature of 15 °C. The piston is free to move and its mass is such that the mixture pressure is 240×10^3 N/m^2; the upper surface of the piston is exposed to the atmosphere. The mixture is ignited. As the reaction proceeds, the piston moves slowly upwards and heat transfer to the surroundings takes place. When the reaction is complete and the temperature of the contents has been reduced to the initial value, 15 °C, it is found that the piston has moved upwards a net distance of 85 mm and that the magnitude of the heat transfer to the surroundings is 4 kJ.

Evaluate the increase in the energy of the contents of the cylinder.

6.19 (a) A system consisting of a mixture of air and gasoline vapour at an initial temperature of 15 °C is contained in a rigid vessel. The mixture undergoes the following processes in sequence.

(i) The mixture temperature is raised to 200 °C by a heat transfer of +3 kJ.

(ii) The mixture is ignited and burns completely; this process is adiabatic and the temperature rises to 1500 °C.

(iii) The temperature of the products of combustion is reduced to 120 °C by a heat transfer of −32 kJ.

Evaluate the energy of the system after each process given that the initial energy of the system is 10 kJ.

(b) An equal mass of the same mixture is contained in a cylinder closed by a piston. The mixture undergoes the following processes in sequence.

(i) Adiabatic compression to a temperature of 200 °C.

(ii) Adiabatic combustion at constant volume until burning is complete.

(iii) Expansion to a temperature of 120 °C, the work done by the system being 31 000 N m.

(iv) Cooling at constant volume until the temperature is again 15 °C.

On the assumption that the energy of the system depends only on its temperature and chemical aggregation (i.e. the way in which the elements are chemically combined), evaluate the work done during process (i) and heat transfer during process (iii). Also state, with reasons, whether or not the system has executed a cyclic process.

7   The Pure Substance

Introduction

In the previous chapter, the First Law of Thermodynamics was applied to systems of considerable complexity, involving chemical, mechanical, electrical and other effects. Fortunately the systems encountered in thermodynamics are frequently less complex, and comprise fluids which do not change chemically or exhibit significant electrical, magnetic or capillary effects. Because of this, and because important statements can be made about such relatively simple systems, the present chapter will be devoted to explaining the nature of these systems. They will be given the generic name of *the pure substance*.

One of the important features of the pure substance is that its state can be represented as a point on a plane diagram and so, in some circumstances, changes of state can be represented by curves on the diagram. The latter part of the present chapter will be devoted to this matter. The opportunity will be taken to define three new properties: the specific heat at constant volume, the specific heat at constant pressure, and the enthalpy.

Symbols

a, b Constants.

c_p Specific heat at constant pressure.

c_v Specific heat at constant volume.

E Energy of a system.

e Specific energy of a system.

F Force.

g Acceleration due to gravity.

g_c Constant in Newton's Second Law

H Enthalpy of a system.

h Specific enthalpy of a system.

m Mass.

p Pressure.

Q Net heat transfer to a system.

t Temperature.

U Internal energy of a pure substance.

u Specific internal energy of a pure substance.

V System volume. Velocity

v Specific volume.

W Net work done by a system.

x, y, z Arbitrary properties of a system.

z Elevation of a system above arbitrary datum.

Δ Increase in ($\equiv$ final value – initial value).

The Pure Substance

Definition of the pure substance

A pure substance is a particular sort of system, characterized by the following definition:

A pure substance is a system which is: (a) homogeneous in composition, (b) homogeneous in chemical aggregation, (c) invariable in chemical aggregation.

Explanation of the definition. (*a*) "Homogeneous in composition" means that the composition of each part of the system is the same as the composition of every other part. "Composition" means the relative proportions of the chemical elements into which the sample can be analysed. It does not matter how these elements are combined.

In Fig. 7.1 for example, system (i), comprising steam and water, is homogeneous in composition, since chemical analysis would reveal that hydrogen and oxygen atoms are present in the ratio 2 : 1 whether the

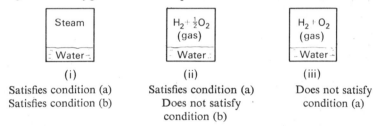

(i)	(ii)	(iii)
Satisfies condition (a)	Satisfies condition (a)	Does not satisfy
Satisfies condition (b)	Does not satisfy	condition (a)
	condition (b)	

Fig. 7.1 Illustrating the definition of a pure substance.

sample be taken from the steam or from the water. The same is true of system (ii), containing uncombined hydrogen and oxygen gas in the atomic ratio 2 : 1 in the upper part, and water in the lower part. System (iii) however is not homogeneous in composition; for the hydrogen and oxygen are present in the ratio 1 : 1 in the upper part, but in the ratio 2 : 1 (as water) in the lower part.

(*b*) "Homogeneous in chemical aggregation" means that the chemical elements must be combined chemically in the same way in all parts of the system. Consideration of Fig. 7.1 shows that system (i) satisfies this condition also; for steam and water consist of identical molecules. System (ii) on the other hand is not homogeneous in chemical aggregation since, in the upper part of the system, the hydrogen and oxygen are not combined chemically (individual atoms of H and O are not uniquely associated), whereas in the lower part of the system the hydrogen and oxygen are combined in the form of water.

Note, however, that a uniform mixture of steam, hydrogen gas, and oxygen gas would be regarded as homogeneous in both composition and chemical aggregation, whatever the relative proportions of the components.

(*c*) "Invariable in chemical aggregation" means that the state of chemical combination of the system does not change with *time* (condition (*b*) referred to variations with *position*). Thus a mixture of hydrogen and oxygen, which changed into steam during the time that the system was under consideration, would not be a pure substance.

EXAMPLES

Fig. 7.2 contains some examples of systems, some of which are pure substances. The reader should check his understanding of the three qualifications of a pure substance by reference to this diagram.

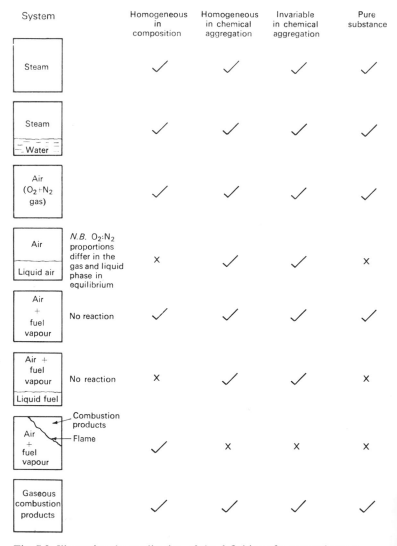

System	Homogeneous in composition	Homogeneous in chemical aggregation	Invariable in chemical aggregation	Pure substance
Steam	✓	✓	✓	✓
Steam / Water	✓	✓	✓	✓
Air (O_2+N_2 gas)	✓	✓	✓	✓
Air / Liquid air — *N.B.* O_2:N_2 proportions differ in the gas and liquid phase in equilibrium	×	✓	✓	×
Air + fuel vapour — No reaction	✓	✓	✓	✓
Air + fuel vapour / Liquid fuel — No reaction	×	✓	✓	×
Air + fuel vapour — Combustion products / Flame	✓	×	×	×
Gaseous combustion products	✓	✓	✓	✓

Fig. 7.2 Illustrating the application of the definition of a pure substance.

The Two-Property Rule

The definition of the pure substance has been given this precise form because of an important feature of systems which obey the above three conditions. This feature may be expressed by what we shall call the *Two-Property Rule*, namely:

The state of a pure substance of given mass can be fixed by specifying two properties, provided that (i) the system is in equilibrium, and (ii) gravity, motion, electricity, magnetism and capillarity are without significant effects.

Remarks on the Two-Property Rule. 1. To give a concrete example, if measurements are made of the pressure and temperature of a system comprising 1 kg of gaseous air, all the properties of that system can be deduced; there is only one possible value of the volume of the air, for instance. This does not mean that knowledge of the pressure and temperature alone tells us what the volume is; but it does mean that every time the pressure and temperature are caused to have these particular values, the volume will be the same.

2. The Rule states only: "*can* be fixed", because not every pair of properties suffices to specify the state. For example, consider a system comprising water and steam. If the pressure is found to be 1 atm and the temperature is 100 °C, reflection will show that a number of states could exist: the system might consist of pure steam, or of a large proportion of steam with a small proportion of water, or of a little steam with a lot of water, or of pure water. This is because, when water is changed into steam at 1 atm pressure, the temperature remains at 100 °C throughout the complete process. In this case the two properties, pressure and temperature, are not independent and therefore count as one property

The state of a steam-water system at 1 atm pressure *would* be fixed however, if the *volume* of the system were measured instead of the temperature; the volume changes greatly according to whether steam or water predominates in the mixture. Pressure and volume (or specific volume) are independent properties in this case.

3. It is necessary for the system to be in equilibrium, as will be seen if we consider a system comprising steam and water, of which the volume and the internal energy are the specified properties. If the steam and water are not initially at the same temperature for example, then, at constant volume and constant energy (physically realized by enclosing the system with a rigid, thermally-insulated container), the pressure will gradually change as temperature equilibrium is brought about by heat transfer between the steam and the water. Clearly, if several pressures and temperatures can be observed, the state of the system is not fixed. The equilibrium requirement also excludes such conditions as "supercooling"* from the range of validity of the Two-Property Rule.†

* See page 180 for explanation of this term.

† Actually a mixture of H_2 and O_2 gas at room temperature is not in thermodynamic equilibrium, for it tends spontaneously, though very slowly, to react chemically. Nevertheless, for many purposes, such mixtures can be treated as though they were in equilibrium; the Two-Property Rule can be applied to them.

4. Although gravity is always present in terrestrial experiments, its effects can be neglected provided, firstly, that the system is not so large that differences of hydrostatic pressure are significant, and, secondly that, when the system moves, the observer moves with it. This second proviso simultaneously excludes other significant effects of motion, like kinetic energy.

5. Capillarity (surface tension), electrical and magnetic effects are all of negligible importance for most of the systems encountered in engineering: for example, steam, air and combustion products. The restriction involved in condition (ii) of the Two-Property Rule is therefore not a practically hampering one.

6. The Two-Property Rule applies only to liquid and gaseous systems; for solids can have pressures (positive or negative) which differ according to the direction of measurement. More than two properties are therefore required to specify completely the state of a system comprising a solid.

7. The Two-Property Rule is also obeyed by some systems which are not pure substances. Obedience to the rule is therefore not an infallible sign that a system is homogeneous in composition, etc.

8. A symbolic representation of the Two-Property Rule is that, for these systems, three properties (x, y, z) of a system can be related by expressions of the form

$$z = f(x,y) \tag{7.1}$$

where f( ) signifies some function of z. For example, the equation

$$p = \frac{(a + bt)}{v} \tag{7.2}$$

where p, v and t stand for pressure, specific volume and temperature and a and b are constants, might represent the relation between p, v and t for a particular substance. In general however, the function f( ) will not have any recognizable algebraic form. Eq. (7.1) is then simply another way of saying: "if x and y are fixed, so is z".

The energy of a pure substance: internal energy

It is convenient to have a new symbol for the energy of a pure substance under the conditions required by the Two-Property Rule. This symbol is U; when the system is of unit mass, the lower-case symbol, u, will be used.

For a pure substance we can relate E and U by:

$$E = U + \text{terms accounting for gravity, motion, capillarity,}$$
$$\text{electricity and magnetism.} \tag{7.3}$$

U happens to be that mode of energy which can be directly influenced by heat interactions, though not of course by these alone; it is therefore the internal energy referred to in Chapter 6. We continue to refer to it as internal energy below; the proviso that we are dealing with a pure substance in the absence of gravity, etc. will be re-iterated from time to time. The symbol u stands for the *specific internal energy* of the system; the word "specific" is often omitted, however.

The First Law. For a pure substance under these conditions, the First Law of Thermodynamics, eq. (6.5), may now be written as:

$$Q - W = \Delta U \tag{7.4}$$

or, if the system is of mass m,

$$\frac{Q - W}{m} = \Delta u. \tag{7.5}$$

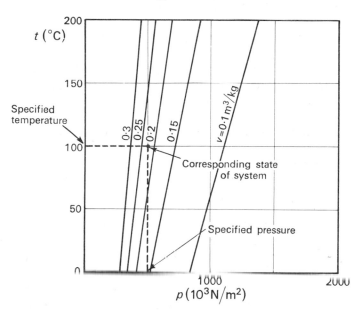

Fig. 7.3 Illustrating interpolation on a property chart.

Tables and diagrams

One reason for the importance of pure substances is that paper is two-dimensional; for, if two properties specify the state of a system, a table or diagram on a single sheet of paper can represent all possible states of the system. As examples of what is meant, consider Table 7.1 and Fig. 7.3, both of which convey information about the specific volume of air at various pressures and temperatures. In each case, specification of a pressure and a temperature, i.e. of two properties, defines a location on the paper and so a particular value of the specific volume; e.g. at $t = 100\,°C$ and $p = 500 \times 10^3\ N/m^2$, the corresponding v is $0.214\ m^3/kg$. In the case of the table, the specific volume is indicated by the number situated in the appropriate pressure column and temperature row. In the case of the graph, the specific volume is found by interpolation between the lines of constant specific volume nearest to the point with the appropriate t-ordinate and p-abscissa.

TABLE 7.1 *Specific Volume of Air* (m³/kg)

$t(^{\circ}C)$ \ p (kN/m²)	10	50	100	500	1000	2000
0	7·84	1·57	0·784	0·154	0·078	0·039
50	9·28	1·86	0·928	0·186	0·093	0·046
100	10·71	2·14	1·071	0·214	0·107	0·054
150	12·15	2·43	1·215	0·243	0·122	0·061
200	13·60	2·72	1·360	0·272	0·136	0·068

Examples of tables of properties of a pure substance will be found in Appendix B. They will be discussed in more detail in Chapter 9. The remainder of the present chapter will be devoted to discussing some features of processes represented on property diagrams.

Some Processes Executed by Pure Substances

Conditions for validity of diagrammatic representation of processes

Before discussing particular processes, it is important to clarify the conditions under which states and processes can be represented on diagrams.

The first point is that the conditions of the Two-Property Rule must be satisfied; that is to say that we must be dealing with a pure substance, it must be in equilibrium, and effects due to gravity, motion, capillarity, electricity and magnetism must be absent. The most easily overlooked requirement is that for equilibrium.

When these conditions are satisfied, a given point on the diagram represents a fixed state of the system; that is to say that all the system properties are determinate there.

Now consider the question of representing a change of state, i.e. a process on the diagram. The path of this change of state is the succession of states passed through by the system. *If each of these states is an equilibrium state* each may be represented by a point on the diagram. Since the states merge continuously, one into the next, the corresponding points form a continuous line joining the point representing the initial state of the process to that of the final state.

Below, examples will be given of processes of which the paths may be represented by continuous lines. They represent what in Chapter 3 were termed "fully-resisted processes" (p. 51). An example will also be given of a process of which only the initial and final states can be represented on a property diagram; this is an "un-resisted expansion"; the intermediate states of the system are not equilibrium states, so no line representing the path of the process can be drawn.

The constant-volume (isochoric) process

Fig. 7.4 serves to illustrate processes characterized by constancy of volume: the container enclosing the system is rigid. A practical example

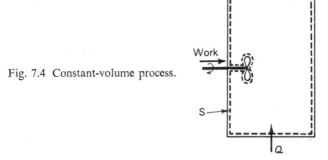

Fig. 7.4 Constant-volume process.

of such a process is that occurring in the cylinder of the Newcomen engine when heat transfer between the injected water and the steam causes the latter to condense; in this case the steam initially in the cylinder is to be taken as the system.

Fig. 7.4 has an arrow indicating heat interaction with the surroundings and another arrow (and symbolic paddle-wheel) indicating that stirring

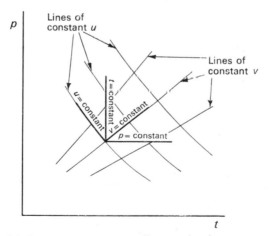

Fig. 7.5 Pressure~temperature diagram showing processes in which p, v, u and t remain constant.

work may in general be done. Displacement work is excluded however by the rigidity of the system boundary. The fluid within the system boundary can be any pure substance.

Figs. 7.5, 7.6 and 7.7 represent respectively $p{\sim}t$, $v{\sim}u$, and $p{\sim}v$ diagrams for the substance in question. Each of them has two families of property lines: the $p{\sim}t$ diagram has lines of constant specific volume and constant internal energy; and so on.

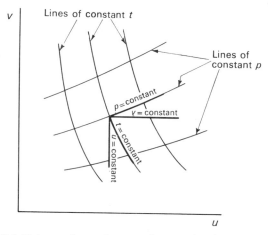

Fig. 7.6 Volume~internal-energy diagram showing processes
in which p, v, u and t remain constant.

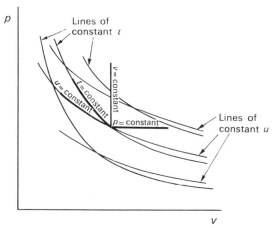

Fig. 7.7 Pressure~volume diagram showing processes
in which p, v, u and t remain constant.

Provided that the constant-volume process involves a continuous succession of equilibrium states, which is possible even with stirring work provided that the stirring rate is low, the process may be represented on each of the property diagrams by lines such as those marked "v = constant" on Figs. 7.5, 7.6, and 7.7. On Fig. 7.6 the line is horizontal because v is the ordinate of that diagram; on Fig. 7.7 the line is vertical since v is the abscissa.

A constant-volume line on a property diagram is sometimes called an *isochore*; correspondingly a process at constant volume is called *isochoric*.

The specific heat at constant volume. The opportunity will now be taken to define a new property. Although the name will be familiar to those instructed in elementary physics, the definition used in thermodynamics should be carefully noted.

Definition. The specific heat at constant volume, c_v is the rate of change with temperature of the specific internal energy of the system when the specific volume is held constant.

Symbolically, this definition runs

$$c_v \equiv \left(\frac{\partial u}{\partial t}\right)_v \qquad (7.6)$$

where $\partial u/\partial t$ means the rate of change of u with respect to t, as is usual in differential calculus, while the bracket with the suffix v signifies that the latter property is constant during the change.

Remarks on c_v

1. The significance of c_v is most easily recognized by reference to a property diagram such as Fig. 7.8, with internal energy as ordinate, temperature as abscissa, and lines of constant volume drawn on the body of the

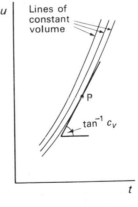

Fig. 7.8 Illustrating the definition of specific heat at constant volume.

graph. Eq. (7.6) shows that for any state of the system, e.g. that represented by point P in Fig. 7.8, c_v is equal to the slope of the constant-volume line passing through the point.

2. Evidently there is a unique value of c_v at each point of the diagram. This means that c_v is a *property*, i.e. a function of the state. No matter how one approaches the point P, the constant-volume line always has the same slope there. The slope differs from point to point however; to specify c_v we must specify a point on the diagram, which requires of course two coordinates.

3. The great usefulness of the property c_v is that, not withstanding what has just been said about its variability in general, for many substances over a restricted range of conditions c_v can be taken as constant. For example, over an important range of temperatures and pressures, the value of c_v for water may be taken as constant and equal to 4·186 kJ/kg K; this simplifies calculations.

4. The name "specific heat" and its commonly-used alternative, "heat capacity", must now be regarded as misnomers hallowed by tradition. They derive from a "caloric" view of heat, which was thought to be "absorbed" by the material as a sponge soaks up water.

5. The definition given above, in terms of internal energy, is preferable to that given in elementary physics texts, in terms of "the heat to raise the temperature one degree at constant volume"; for the temperature rise might be brought about by (stirring) work in the absence of heat transfer. Beside this, the fact that c_v is a property is brought out more clearly if it is defined in terms of the two other properties, u and t.

6. It should not be forgotten that u is the internal energy of a pure substance, and that it is therefore only for such systems that c_v is defined. In systems which react chemically, it is possible to have appreciable temperature changes at constant volume and energy (which might suggest that c_v was zero), and also large energy changes at constant volume and temperature (which might suggest that c_v was infinite); but such systems are not of course pure substances, and therefore have no c_v.

The constant-pressure (isobaric, or isopiestic) process

Often processes take place at substantially constant pressure. For example the water in a boiler is at almost the same pressure on entry as when it leaves as steam. Constant-pressure processes in general are illustrated by Fig. 7.9, wherein the fluid system is supposed to be contained in a

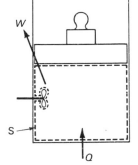

Fig. 7.9   Constant-pressure process.

cylinder beneath a frictionless leak-proof piston surmounted by a constant weight. For generality, a paddle-wheel is indicated, but this time of course displacement work is done as well as stirring work; for the piston must move in order to keep the pressure constant.

The lines marked "$p = $ constant" in Figs. 7.5, 7.6, and 7.7 indicate the path of a constant-pressure process which passes through a continuous succession of equilibrium states. They are commonly called *isobars*, and sometimes *isopiests*.

The specific heat at constant pressure. We now define another specific heat, namely the specific heat at constant pressure, c_p. Its definition may at first appear even more arbitrary than that for c_v; and it involves the introduction of yet another new property; the *specific enthalpy, h*. We give only symbolic definitions at this stage, namely:

Definitions: $$h \equiv u + pv \tag{7.7}$$

$$c_p \equiv \left(\frac{\partial h}{\partial t}\right)_p \tag{7.8}$$

Here $\partial h/\partial t$ signifies the rate of change of h with respect to t, while the bracket with the suffix p indicates that the pressure is held constant during the differentiation.

Remarks on the enthalpy* h

1. The convenience of having a name and symbol for the quantity $(u + pv)$ will not become apparent until steady-flow processes are dealt with in Chapter 8. Nevertheless, it is evident that we are free to define h in accordance with eq. (7.7), and moreover that this quantity represents a *property*; for u, p and v are all properties, and when their values are fixed the value of h is uniquely determined also.

2. The word "enthalpy" is pronounced with the stress on the second syllable: enthálpy. This helps to prevent confusion with "éntropy", introduced in Chapter 13.

3. To be meaningful, the internal energy u and the product pv must be expressed in like units, for example in N m/kg or Btu/lb$_m$. The enthalpy h is then in the same units.

4. Since enthalpy is a property, it may be used like other properties as ordinate or abscissa of a property diagram. Fig. 7.10 illustrates an enthalpy$\sim$ temperature diagram for a particular pure substance; the lines drawn on it represent states of constant pressure.

5. The symbol H is used to denote the enthalpy of a system of mass m; thus $H \equiv mh$. If the system volume is V, it follows from eq. (7.7) that

$$mh = mu + mpv$$

i.e. $$H = U + pV$$

Remarks about c_p

1. The definition of c_p (eq. (7.8)) indicates that this quantity is a slope, namely that of a line of constant pressure on an $h\sim t$ diagram, e.g. Fig. 7.10.

* As in the case of *specific* internal energy, the qualifying adjective is usually omitted except where emphasis is required.

As in the case of c_v, this observation shows that c_p is a property, the value of which varies, in general, according to the state of the substance.

2. In words, the definition of c_p is:

The specific heat at constant pressure, c_p, is the rate of change with temperature of the specific enthalpy of the system when the pressure is held constant.

3. Elementary physics texts ordinarily define c_p as "the heat transfer necessary to raise the temperature of unit mass of the substance through one degree at constant pressure"; the absence of stirring work is understood. While preferring our present definition, which is in terms of properties

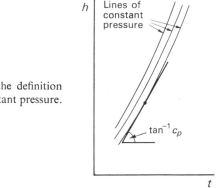

Fig. 7.10 Illustrating the definition of specific heat at constant pressure.

rather than of a process, we have to demonstrate that it includes the "physics" definition.

Consider the system of Fig. 7.9 which will be supposed to comprise unit mass of a pure substance; let it undergo a small constant-pressure expansion in the absence of stirring work. Let the heat transfer be dQ, and the work be dW. Application of the First Law, eq. (7.5), to the system leads to:

$$dQ - dW = du \qquad (7.$$

Since the only work is displacement work, the net work done by the system dW, is simply equal to the displacement work, $p\,dv$, where dv is the small increase of the (unit mass) system volume; eq. (7.9) then becomes

$$dQ = du + p\,dv \qquad (7.10$$

We wish to prove that dQ in eq. (7.10) is equal to c_p when the temperature change is unity.

Now, by differentiating the definition of enthalpy, eq. (7.7) and combining that result with eq. (7.10) we obtain

$$dQ = dh - v\,dp \qquad (7.1$$

Further, the two-property rule tells us that h can be expressed as a function of t and p. The application of the rules of partial differentiation to this statement, to obtain dh, then permits us to write eq (7.11) as

$$dQ = \frac{\partial h}{\partial t}\, dt + \frac{\partial h}{\partial p}\, dp - v\, dp \qquad (7.12)$$

When the pressure is held constant, as in this case, the last two terms on the right-hand side of eq. (7.12) are each zero. Eq. (7.12) therefore reduces to

$$dQ = \left(\frac{\partial h}{\partial t} \cdot dt\right)_p$$

$$= c_p,\ \text{when the temperature increment is unity.}$$

This completes the required proof.

4. Further remarks on c_p parallel those on c_v. Its usefulness derives from the approximate constancy of c_p for many substances over restricted ranges. It has meaning only for pure substances.

The constant-temperature process

Fig. 7.11 illustrates a system which undergoes changes at constant temperature. Such processes occur in practice whenever the system changes its

Fig. 7.11 Constant-temperature process.

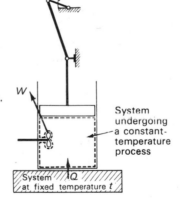

state slowly while in thermal contact with a larger system of uniform temperature. Boiling and condensation processes also occur at constant temperature if the pressure is constant, as has already been mentioned.

If the process is fully-resisted, and so involves solely a succession of equilibrium states, constant-temperature processes can be represented by curves drawn on property diagrams, as illustrated by the curves marked "$t = $ constant" on Figs. 7.5, 7.6 and 7.7.

Constant-temperature processes are often called *isothermal processes;* the constant-temperature lines on the property diagrams are called *isotherms*.

The constant-internal-energy process

Processes involving only equilibrium states and in which the heat and the work interactions exactly balance at all times are rare in practice. However, if they do occur, they are processes of constant internal energy (since $du = dQ - dW = 0$); they may be represented by lines on property diagrams. Figs. 7.5, 7.6 and 7.7 show examples. Further, in such a process, if all the work is done against a slowly-moving piston, the work is given by $W = \int p \, dv$; it is equal therefore to the area under the constant-u line on the $p{\sim}v$ diagram.

The equal-internal energy process

A process which bears some resemblance to the constant-u process just discussed is the equal-internal-energy process. In the equal-u process, as

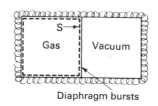

Fig. 7.12 Illustrating a process in which the initial and final internal energies are equal.

in the constant-u process, the initial and final values of the internal energy are equal; but in contrast to the constant-u process, the intermediate values of u, during the process, are not the same. A typical example of an equal-u process is the diaphragm-bursting process of Chapter 3, p. 65. Fig. 7.12 shows the system in its initial position to the left of the diaphragm; the space to the right of the diaphragm is evacuated. The

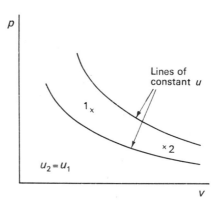

Fig. 7.13 Pressure${\sim}$volume diagram showing the initial and final state-points of the process illustrated in Fig. 7.12.

process involves the bursting (or removal) of the diaphragm which allows the fluid to fill the whole space. The container is insulated and rigid so both heat and work are absent. It follows from eq. (7.4) that the initial and final internal energies are equal. The *energy* of the system remains constant throughout, because the heat and the work are both zero; but the *internal* energy differs from the energy *during* the process because effects of motion are not negligible then.

Fig. 7.13 shows, by means of the points marked 1 and 2, all that can be represented of the process on a $p{\sim}v$ diagram: the initial and final state points. They have been drawn between the constant-internal-energy curves of the diagram for clarity. States between 1 and 2 cannot be represented on the diagram by a continuous constant-u line through 1 and 2, because they are not in equilibrium: the pressure, for example at any intermediate instant in the process, is not uniform throughout the system; so what ordinate could be allotted to the state-point?

Something approaching a continuous line can be obtained by the modified process, with the same initial and final states, shown in Fig. 7.14, where a

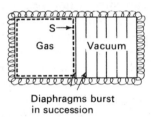

Fig. 7.14 Modified process with the same initial and final states as that shown in Fig. 7.12.

Diaphragms burst
in succession

large number of diaphragms are caused to burst successively, enabling the whole volume to be filled by the fluid in stages. Then shortly after each diaphragm bursts, but while the next diaphragm is still intact, an equilibrium state is arrived at. Successive states are shown as crosses on Fig. 7.15;

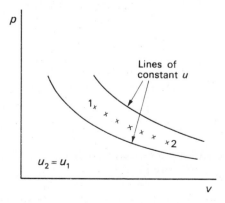

Fig. 7.15 Showing successive equilibrium states of the process illustrated in Fig. 7.14.

they all lie on the same curve of constant internal energy. Nevertheless since each equilibrium state is always separated from the rest by states in which equilibrium does not exist, no matter how many diaphragms are used, the points on Fig. 7.15 can never join up to form a line. One consequence of this is that $\int_1^2 p \, dv$ for the process has no meaning; for p is not prescribed for each v. However, if anyone with sufficiently blurred vision does evaluate the area under the "line" joining 1 and 2, this will certainly not equal the work done in the process; for the work, it will be remembered, is zero (p. 66).

A new expression for E

It will be noted that the enumeration of the intermediate states during the processes discussed in the previous section required that the fluid motion be taken into account; we remarked that the energy and the internal energy were different, but we were not then in a position to make quantitative statements about their relation. Processes in which fluids are in motion occur frequently in practice; we now turn to the relation of E to U. For the moment, only *pure substances* will be considered; the extension to substances in which chemical reaction occurs is made on p. 378.

Equation (7.3) expresses E as the sum of U and terms accounting for motion, gravity, capillarity, electricity and magnetism. For most fluids of interest to the mechanical engineer, capillarity, electricity and magnetism make negligible contributions to the sum; their influences will therefore be neglected. In flow problems, however, motion must obviously be considered; and, since motion implies change of position (elevation), the influence of gravity must be considered also. We therefore now consider, in turn, the contributions of motion and gravity to the energy. It is possible to treat them separately, and then to add the resulting expressions; for both are determined by the relative positions of the system and the observer, which can obviously be altered by choosing a different observer, without altering properties of the system like pressure, temperature and specific volume. Both the results to be derived will be familiar from elementary mechanics; they are given here for formal completeness.

Kinetic energy. Consider the rigid system illustrated in Fig. 7.16. It is acted upon by a single force F; in the process to be considered it moves a

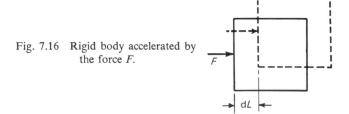

Fig. 7.16 Rigid body accelerated by the force F.

small distance dL in the direction of the force. In this process the velocity of the system increases from V to $V + dV$, but there are no changes in other system properties. Heat interactions are absent ($Q = 0$).

The work done by the system is $-F\,dL$, because the surroundings could have carried out the same process and done work $F\,dL$ by raising weights instead of accelerating the system. The First Law, eq. (6.5), for the process can therefore be written

$$0 - (-F\,dL) = m\,de \qquad (7.13)$$

where m is the mass of the system and de is the increase in its specific energy.

Now F may be related to the velocity change dV by Newton's Second Law of Motion, eq. (2.6)*, which runs

$$F = \frac{m}{g_c}\frac{dV}{d\tau} \qquad (7.14)$$

where τ stands for time, so that d$V/$dτ is the acceleration.

From the theory of kinematics, we have

$$\frac{dV}{d\tau} = V\frac{dV}{dL} = \frac{1}{2}\frac{d(V^2)}{dL} \qquad (7.15)$$

whence

$$F = \frac{m}{g_c} \cdot \frac{1}{2}\frac{d(V^2)}{dL} \qquad (7.16)$$

Substituting eq. (7.16) in to eq. (7.13) and re-arranging, we obtain the required relation between the changes in specific energy and in the velocity; it is

$$de = \frac{1}{2g_c}\,d(V^2) \qquad (7.17)$$

For a large change of the above type (adiabatic, system boundary rigid), eq. (7.17) integrates to

$$e_2 - e_1 = \frac{V_2^2}{2g_c} - \frac{V_1^2}{2g_c} \qquad (7.18)$$

where suffixes 1 and 2 denote the initial and final states.

Gravitational potential energy. To account for the contribution of gravity, we consider the rigid system of mass m shown in Fig. 7.17, executing an

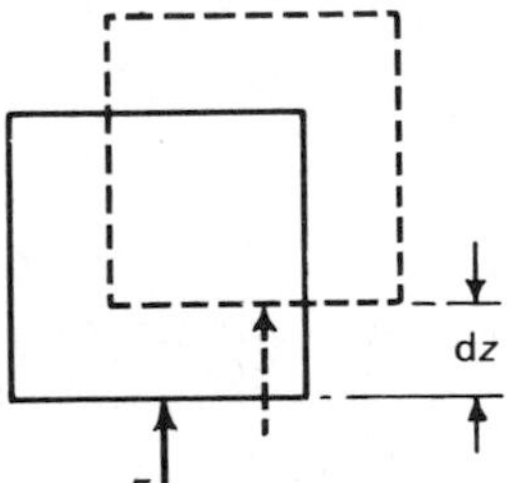

Fig. 7.17 Rigid body raised in a gravitational field.

---

* If SI units are to be used, and Newton's Second Law is written as $F = ma$, then g_c can be dropped from all equations, i.e. in these circumstances g_c equals unity. (See page 31.)

adiabatic process; the mass is raised slowly and vertically by a force F through a small distance dz; there exists a gravitational field which is such that freely-falling bodies have an acceleration g

The First Law for the process is

$$0 - (-F\,dz) = m\,de \qquad (7.19)$$

by the same reasoning as before.

From Newton's Second Law and the statement about the gravitational field (p. 33), we know that the force executed by the latter, i.e., the weight-force of the body, is numerically equal to F and is given by

$$F = \frac{mg}{g_c} \qquad (7.20)$$

Combination of eq. (7.19) and eq. (7.20) yields the required relation between the increase in specific energy and the increase in elevation, namely

$$de = \frac{g\,dz}{g_c} \qquad (7.21)$$

For a large adiabatic change of elevation of a rigid body in a *uniform* gravitational field, eq. (7.21) integrates to

$$e_2 - e_1 = \frac{gz_2}{g_c} - \frac{gz_1}{g_c} \qquad (7.22)$$

The new expression for e. By reason of the two results just derived, we can now re-write eq. (7.3) for unit mass of a pure substance as

$$e = u + \frac{V^2}{2g_c} + \frac{gz}{g_c} + \begin{array}{l}\text{terms accounting for capillarity,}\\ \text{electricity and magnetism.}\end{array} \qquad (7.23)$$

Here z is to be measured, like e itself, from some arbitrary base level. The velocity, V, is measured relative to the observer. The terms $V^2/2g_c$ and gz/g_c are known respectively as the *kinetic energy* and the *gravitational potential energy* of the system per unit mass.

The First Law applied to systems in motion. The foregoing analysis now enables us to write the First Law of Thermodynamics, eq. (6.5), for a *moving system* comprising a pure substance: substituting the new expression for e, eq. (7.23), in eq. (6.5) and remembering that E equals me, we obtain

$$\frac{Q - W}{m} = \Delta \left(u + \frac{V^2}{2g_c} + \frac{gz}{g_c} + \begin{array}{l}\text{terms accounting for capillarity,}\\ \text{electricity and magnetism.}\end{array}\right) \qquad (7.24)$$

In many engineering problems, the terms accounting for capillarity, electricity and magnetism are negligible; they can therefore be dropped from eq. (7.24), to give:

$$\frac{Q - W}{m} = \Delta \left(u + \frac{V^2}{2g_c} + \frac{gz}{g_c}\right) \qquad (7.25)$$

The results expressed in equations (7.23) and (7.25) assume considerable importance in the next chapter where the First Law is applied to fluids in motion.

CHAPTER 7—PROBLEMS

7.1 A fluid system comprising 2·5 kg of a pure substance is initially at a pressure of 700×10^3 N/m² and a temperature of 200 °C; the corresponding specific volume of the substance is 0·2 m³/kg. The system undergoes a process to a final pressure of 700×10^3 N/m² and a final specific volume of 0·2 m³/kg. For the pure-substance fluid in question, the pressure and specific volume are independent properties.

(a) What are: the final temperature of the substance, and the increase in its specific internal energy?

(b) During the process, the work done by the system is 12 000 N m. Determine the magnitude and direction of the heat transfer.

7.2 A system comprising a pure substance undergoes a process in which its internal energy decreases by 10×10^3 J. The heat transfer to the system during the process is 6×10^3 J.

(a) Determine the magnitude and direction of the work done.

(b) If, in undergoing a different process between the same end-states, the work done had been zero, what would have been the magnitude and sign of the heat transfer?

7.3 A fluid system consisting of 4·2 kg of a pure substance has an energy E of $85·7 \times 10^3$ J. The kinetic energy of the system is 13·6 N m and its gravitational potential energy is 4·9 N m. The system undergoes an adiabatic process in which the final specific internal energy is 159 kJ/kg, the final kinetic energy is $1·15 \times 10^3$ N m and the final gravitational potential energy is 850 N m. The effects due to electricity, capillarity and magnetism are assumed to be absent.

(a) Evaluate the initial value of the specific internal energy of the fluid.

(b) Determine the magnitude and sign of the work done during the process.

(c) The fluid system undergoes a second adiabatic process between the same initial and final pressures and temperatures as above. In this second process the effects of gravity and motion are negligible. Assuming pressure and temperature to be independent properties, determine the magnitude and sign of the work done during the process.

7.4 A mass of 0·3 kg of a pure substance at a pressure of 90×10^3 N/m² and a temperature of 40 °C occupies a volume of 120 dm³. Given that the internal energy of the substance is $30·4 \times 10^3$ J, evaluate the specific enthalpy of the substance.

7.5 A pure substance is contained in a cylinder closed by a piston. The substance undergoes a fully-resisted, constant-pressure process in which the only work done is the displacement work at the slowly-moving piston face. Show that the heat transfer during the process is equal to the increase in the enthalpy of the substance.

7.6 A pure substance is contained in a cylinder closed by a piston. A paddle wheel, rotated by means of a shaft protruding through the cylinder wall, causes the substance to undergo a fully-resisted constant-pressure process as the piston moves outwards. There is no heat transfer during the process. Show that the stirring work done on the substance is equal to the increase in the enthalpy of the substance.

7.7 The internal energy and the enthalpy of certain pure substances may be considered to be functions of (i.e. dependent upon) temperature only. Further, over restricted ranges of conditions, the specific heats at constant volume and constant pressure may be assumed to be constant. Show that for *any* process executed by a pure substance satisfying these conditions:

$$\Delta u = c_v \Delta t$$

and
$$\Delta h = c_p \Delta t$$

7.8 Examine the following situations, all of which satisfy the conditions set down in problem **7.7**.

(a) The specific heat of water at constant *volume* may be taken as 4186 J/kg K. Evaluate the increase in internal energy of 1 kg of water at atmospheric pressure as its temperature is increased from 0 °C to 100 °C at constant *pressure*.

(b) The specific heat of ice at constant *pressure* may be taken as 2093 J/kg K. Evaluate the increase in enthalpy of 2 kg of ice at atmospheric pressure as its temperature is increased from −20 °C to 0 °C at constant *volume*.

(c) 3 kg of air at a pressure of 400×10^3 N/m² and a temperature of 60 °C are contained in a cylinder closed by a piston. As a result of expansion and heat transfer the temperature falls to 5 °C and the pressure to 300×10^3 N/m². For air the specific heat at constant pressure may be taken as 1005 J/kg K and the specific heat at constant volume may be taken as 718 J/kg K. Evaluate (i) the increase in the internal energy of the air, (ii) the increase in the enthalpy of the air. Can the work or the heat transfer be evaluated for this process?

7.9 The relation between the properties of oxygen gas may be expressed over a restricted range by

$$pv = 260t + 71 \times 10^3$$

and
$$t = 1.52u - 273$$

where p is in N/m², v in m³/kg, t in °C and u in kJ/kg.

(a) Evaluate the specific heat at constant volume and the specific heat at constant pressure in J/kg K.

(b) Show that for *any* process executed by unit mass of oxygen $\Delta u = c_v \Delta t$ and $\Delta h = c_p \Delta t$.

(c) 1 kg of oxygen at a pressure of 600×10^3 N/m² and a temperature of 280 °C, is contained in a rigid vessel. Heat transfer to the oxygen increases its enthalpy by 28 kJ/kg. Evaluate the final temperature, the heat and the final pressure.

7.10 The following data have been extracted from Table IV of the Steam Tables (Appendix B). They show the variation in the enthalpy of steam with temperature at a fixed pressure, 200×10^3 N/m².

t °C	150	200	300	400	500	600	700	800
h kJ/kg	2768·5	2870·5	3072·1	3276·7	3487·0	3704·0	3927·6	4157·8

Plot h versus t and hence obtain values of the specific heat at constant pressure at $t = 200$ °C and at $t = 650$ °C for $p = 200 \times 10^3$ N/m².

7.11 The properties of a certain gas are related by

$$pv = 310 (t + 273)$$

and
$$u = u_0 + 0.84t$$

where p is in N/m², v in m³/kg, t in °C and u in kJ/kg.

A cylinder fitted with a piston contains 0·02 m³ of this gas at a pressure of 350×10^3 N/m² and a temperature of 80 °C. As the gas expands to a lower pressure the work done by the gas is 2900 N m and the heat transfer from the gas is 1.9×10^3 J.

(a) Determine the temperature of the gas after expansion.

(b) If the gas undergoes an adiabatic process between the same end states, evaluate the work done by the gas in this case.

7.12 (a) A system comprising 0·95 kg of a pure substance has a specific internal energy of 16 kJ/kg. The system is moving with a velocity of 120 m/s at an elevation of 1500 m above sea level. Evaluate the energy of the system relative to an observer at rest at sea level.

(b) The system undergoes a process to a final specific internal energy of 20 kJ/kg, a final velocity of 200 m/s and a final elevation of 270 m. The work done on the system during the process is 2200 N m. Evaluate the magnitude and direction of the heat transfer during the process.

8 The First Law Applied to Flow Processes

Introduction

In the discussion of the First Law of Thermodynamics, we focused attention on the processes undergone by a fixed body of material, the *thermodynamic system*; we thereby related the heat and the work at the system boundary to the changes of energy of the material within the boundary.

In many engineering problems, however, the natural focus of attention is a piece of equipment through which material flows continuously. Examples are:

(a) *A hydro-electric plant*, comprising a water turbine coupled to a dynamo. Water from a high-level reservoir flows into the turbine continuously, emerging at reduced pressure at the outlet. The flow within the turbine rotor and associated pipes is turbulent and confused, so that it would be difficult to follow the history of an individual particle (system) of water in order to compute its energy changes and the magnitudes of the heat and work interactions at its boundaries. What we are more directly interested in is how much electrical work is delivered at the terminals.

(b) *A steam turbine* in which hot, high-pressure steam enters the casing, passes between alternate rows of fixed and moving blades, and emerges, partially condensed, with a lower internal energy but higher velocity than at entry. Once again the processes undergone by a given particle of steam are too complex to be observed and analysed; indeed they will not be the same for every particle. Yet we need to determine how the average change of state of the steam in its passage through the turbine is related to the shaft work which the turbine does on its surroundings, and to the heat transfer from the turbine casing to the atmosphere.

(c) *A gasoline engine* running at constant speed. Air and gasoline enter the engine in a more or less pulsating fashion, while combustion products (nitrogen, carbon dioxide, carbon monoxide, steam, etc.) pass out of the exhaust pipe, usually pulsating even more violently. The cylinders will not all receive combustible mixture of quite the same mixture ratio; nor will the ignition and cylinder-cooling arrangements operate uniformly between cylinders. Again, however, we need to find some way of applying the First Law of Thermodynamics so as to relate the changes which have taken place in the fuel and air to the work performed by the engine and to the heat transfers to the cooling water and elsewhere.

Despite the complexity of these examples, it will be shown below that, if the system boundaries are chosen judiciously, the analysis of Chapter 6 can be applied directly and in a simple fashion. Moreover, problems of these types occur so frequently in engineering practice that the application warrants a slight re-formulation of the First Law.

An important characteristic of each of the examples cited is that, as far as the inlet and outlet pipes are concerned, the flow of material may be taken as *steady*, i.e., invariant with time. This is true even of the gasoline engine, and of other reciprocating machines, provided that conditions are considered at points sufficiently far upstream and downstream for the pulsations to have died away. Our re-formulation of the First Law is therefore called the *Steady-Flow Energy Equation*. In deriving it, a new concept will be introduced, the *control volume*; this is a counterpart to, and as important as, the concept: *system*. A new sub-division of work will also be introduced, namely: *flow work* and *external work*.

The chapter comprises the derivation of the new equation, followed by important examples of its use. In the course of one of these, the *Continuity Equation* is introduced (p. 153). There follows a demonstration of how the control-volume concept can be used also in processes where the flow is not steady. Finally, we explain the similarities and differences between the Steady-Flow Energy Equation and two other equations which arise in fluid mechanics and which are based on a quite different physical principle: *Newton's Second Law of Motion*.

Symbols

A	Cross-sectional area.	$\dot{m}$	Mass flow rate of fluid.
c_v	Specific heat at constant volume.	p	Pressure.
E	Energy of a system.	Q	Heat transfer.
E_C	Energy of the material within a control surface.	$\dot{Q}$	Heat-transfer rate.
e	Specific energy.	t	Temperature.
F	Force.	u	Specific internal energy of a pure substance.
g	Gravitational acceleration.	u'	Specific internal energy of a chemical substance.
g_c	Constant in Newton's Second Law.	V	Velocity.
h	Specific enthalpy of a pure substance.	v	Specific volume.
h_t	Stagnation enthalpy of a pure substance.	W	Net work done by a system.
h'	Specific enthalpy of a chemical substance.	W_x	External work.
L	Displacement.	$\dot{W}_x$	External-work rate.
m	Mass.	z	Elevation of a system above an arbitrary datum.
		α, β	Gas angles (Figs. 8.12b and c).
		σ	Volume enclosed by a control surface.

τ Time, time interval.
Δ Increase in ($\equiv$ final value − initial value).

Subscripts
1, 2 Inflow, outflow sections of a control surface.

i, f Initial, final state of the material within a control surface.
A Air.
F Fuel.
G Flue gas.
P Products.
S Steam.

The Steady-Flow Energy Equation

The control volume

Consider, as an example, the triple-expansion reciprocating steam engine illustrated in Fig. 8.1. We suppose it to be operating at constant rotational

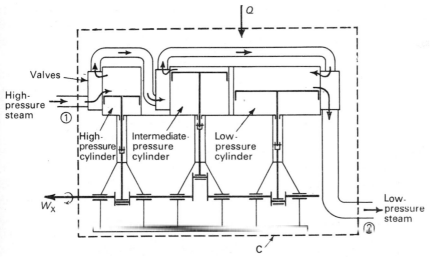

Fig. 8.1 Triple-expansion reciprocating steam engine.

speed and to have been running long enough to be warmed right through.

The first and most important step in the analysis is the drawing of an imaginary envelope around the engine, shown dotted in Fig. 8.1, and marked C. This cuts the steam entry and exhaust pipes at the sections marked 1 and 2 respectively, which have been chosen sufficiently far from the valve-gear for the pulsations in steam velocity to be damped out. The steam mass-flow rates therefore can be regarded as steady, and of course equal. The space bounded by the imaginary envelope, C, is called a *control volume*; C is often called a *control surface*.

Definition. A control volume is any volume of fixed shape, and of fixed position and orientation relative to the observer.

Comparison between control volume and system. The control-volume and system concepts have in common that they are both defined by *boundaries*. Further, as will be seen, both concepts are useful because they permit the interactions at the boundaries to be related to the changes occurring within them.

However, the concepts are distinguished by two features:

(i) The system boundary may, and usually does, change shape, position and orientation relative to the observer. The control-volume boundary does not, by definition.

(ii) Material may, and usually does, flow across the control-volume boundary. No such flow takes place across a system boundary, by definition.

These similarities and distinctions have caused the control volume to be termed an *open system*. This term will not however be used in this book.

Interactions at a control surface. In addition to the flow of steam across the boundary of the control volume in Fig. 8.1, we must consider two other interactions with the surroundings. The first is the *external work* indicated by an arrow and marked W_x; the second is the *heat transfer* likewise indicated by an arrow and marked Q.

Heat and work have so far been defined only as interactions between systems. However, parts of the control-volume boundary across which no material flows can be locally regarded as parts of a system boundary; the previous definitions of heat and work apply there without change.

At parts of the boundary where material is flowing, greater care is necessary. Usually we compute the heat transfer and shear work there in the same way as for a system. The displacement work, here often called *flow work*, is dealt with separately, as shown below (p. 143). The algebraic sum of all the work interactions, less the flow work, is called the *external work*.

Definition. External work is all the work done at the control surface other than that associated with normal forces at places where material crosses the surface.

In engineering thermodynamics the only sorts of external work of importance are *shear (shaft or stirring) work and electrical work*. In Fig. 8.1, the only external work occurs where the control surface cuts the engine shaft; in this case, therefore, W_x is the shaft work delivered by the engine.

The heat transfer Q, is indicated by a single arrow, although the interactions may actually spread over the whole control surface. The sign convention for Q and W_x is the same as for systems (pp. 89 and 47). Thus the Q arrow points inwards, even though in this example the heat transfer is certainly *from* the engine casing *to* the engine-room.

Derivation of the Steady-Flow Energy Equation

In order to make use of the First Law as stated in Chapter 6, a system must be chosen. Referring to Fig. 8.2, which shows the control volume C of Fig. 8.1 once more, with its contents omitted for clarity, we choose the system which initially has the boundary marked S_i. S_i is identical with C except at section 1, where it also encloses some material in a short length of the inlet pipe. Let the extra material, in this case steam, have mass m_1.

We now consider the later instant of time, at which the mass m_1 has just completely entered the control volume C; the system boundary now coincides with C at the section 1. Simultaneously, however, material has

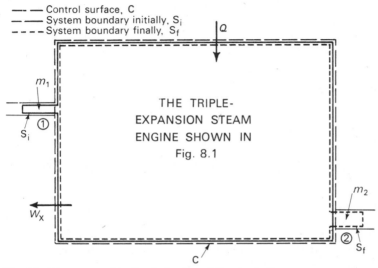

Fig. 8.2 Diagram used in the derivation of the steady-flow energy equation.

flowed along the exhaust pipe: the system, which by definition always comprises the same collection of matter, has therefore moved its boundary to the new position marked S_f in Fig. 8.2. S_f coincides with C except at the section 2, where it encloses an additional mass of material m_2. We shall shortly postulate that the flow is steady, so that m_1 equals m_2; for the time being however, we consider the more general non-steady case.

Application of the First Law. The system has now undergone a process; Q and W_x will be taken as the heat and external-work interactions during this process. The First Law may now be applied. We have, from eq. (6.5),

$$Q - W = \Delta E \qquad (8.1)$$

Here the heat transfer Q requires no comment. Evaluation of the net work, W, done by the system requires more thought; for, in addition to the external work W_x, the net work W must include the displacement work done by moving parts of the system boundary at sections 1 and 2. We apply eq. (3.6) to obtain expressions for these displacement-work quantities. If the specific volumes of the fluid at sections 1 and 2 are respectively v_1 and v_2, the volumes swept out by the corresponding parts of the system boundary are $v_1 m_1$ and $v_2 m_2$. If the (constant) local pressures are p_1 and p_2, the corresponding displacement-work terms are $-p_1 v_1 m_1$ and $+p_2 v_2 m_2$, the signs being ascribed in accordance with the previous convention. We can therefore write, by reference to the definition of external work,

$$W = W_x - p_1 v_1 m_1 + p_2 v_2 m_2 \qquad (8.2)$$

We now seek an expression for ΔE in eq. (8.1); it is convenient to treat separately the parts of the system within the control volume and the parts outside it. We use the symbols $E_{C,i}$ and $E_{C,f}$ respectively for the initial and final energies of the contents of the control volume; the *specific* energies of the fluid in the pipes at sections 1 and 2 respectively are denoted by e_1 and e_2. It follows that:

the initial energy of the system is

$$e_1 m_1 + E_{C,i};$$

the final energy of the system is

$$E_{C,f} + e_2 m_2.$$

Eq. (8.1), the First Law for the system in the specified process, can thus be written as

$$Q - (W_x - p_1 v_1 m_1 + p_2 v_2 m_2) = E_{C,f} + e_2 m_2 - E_{C,i} - e_1 m_1 \qquad (8.3)$$

Re-arrangement of the equation. Eq. (8.3) will now be simplified by introducing: (i) the condition that the flow is steady; and (ii) the Conservation-of-Mass principle.

If the flow is steady, so that changes in all measurable quantities within the control volume are either non-existent or cyclic, we can write

$$E_{C,f} = E_{C,i} \qquad (8.4)$$

wherein it is assumed that, if cyclic changes are in question, an integral number of cycles has taken place during the process considered. Thus, in the reciprocating-steam-engine example, the crank-shaft must be in the same position at the end of the process as at the beginning. Eq. (8.4) holds, even though each non-solid part of the control volume will be occupied by different particles of material at the two instants by reason of the flow; for the states of the particles are identical.

The Conservation-of-Mass principle is now invoked to show that the two mass quantities, m_1 and m_2, must be equal; for otherwise there would be an accumulation of material within the control volume, which is contrary to the requirement of steady flow. We can therefore write

$$m_1 = m_2 = m, \qquad \text{say.} \qquad (8.5)$$

Insertion of eq. (8.4) and eq. (8.5) in eq. (8.3) simplifies the latter equation, which now becomes

$$Q - W_x - m(p_2 v_2 - p_1 v_1) = m(e_2 - e_1) \qquad (8.6)$$

It is convenient to collect all the fluid properties on one side of the equation and the two interactions at the control surface on the other side; the equation then becomes

$$Q - W_\mathrm{x} = m(e_2 + p_2 v_2 - e_1 - p_1 v_1) \qquad (8.7)$$

The pv product. The pv terms in the above equations have arisen as expressions for the displacement work done at the moving system boundaries; for this reason they are sometimes called "flow work", though we shall not use this name here. A reason for separating them from W_x in the equation is that it is the latter, the *external work*, that the engineer is interested in practically. It should be noted that pv and $p\,dv$ are *not* related to each other as integral and differential.

A separate misleading line of thought is sometimes provoked by calling pv "pressure energy", which carries with it the suggestions either that pv is already included within e, or that pv ought always to be added to e in order completely to describe the energy, whenever p and v are finite. *Both these suggestions are false.* The only legitimate physical interpretation of pv in this context is that, though expressible entirely in terms of fluid properties, it represents the work done by the adjacent fluid in forcing unit mass of the fluid into or out of the control volume as the flow proceeds. The name "pressure energy" should never be used.

The Steady-Flow Energy Equation for a pure substance. To proceed with the derivation of the Steady-Flow Energy Equation, it is necessary to introduce the expression for e, eq. (7.23), namely:

$$e = u + \frac{V^2}{2g_c} + \frac{gz}{g_c} + \begin{array}{l}\text{terms accounting for capillarity,*}\\ \text{electricity and magnetism.}\end{array} \qquad (8.8)$$

We shall, however, omit the usually negligible capillarity, electricity and magnetism terms. Eq. (8.7) then becomes

$$Q - W_\mathrm{x} = m\left(u_2 + p_2 v_2 + \frac{V_2^2}{2g_c} + \frac{gz_2}{g_c} - u_1 - p_1 v_1 - \frac{V_1^2}{2g_c} - \frac{gz_1}{g_c}\right) \qquad (8.9)$$

Here the terms u and pv have been placed together in order to make obvious the next step, their substitution by the enthalpy h, defined by eq. (7.7) as

$$h \equiv u + pv \qquad (8.10)$$

Eq. (8.9) thus becomes

$$Q - W_\mathrm{x} = m\left[\left(h_2 + \frac{V_2^2}{2g_c} + \frac{gz_2}{g_c}\right) - \left(h_1 + \frac{V_1^2}{2g_c} + \frac{gz_1}{g_c}\right)\right]$$

or

$$\frac{Q - W_\mathrm{x}}{m} = \Delta\left(h + \frac{V^2}{2g_c} + \frac{gz}{g_c}\right) \qquad (8.11)$$

where $\Delta(\ldots)$ stands for "increase in $\ldots$" as before,

* Note that g_c equals unity when SI units are used, and may be dropped from the equations.

Eq. (8.11) is the *Steady-Flow Energy Equation* for a pure substance. It is one of the most useful results of engineering thermodynamics and is important and simple enough to be learned by heart. Its name is abbreviated below to S.F.E.E.

An example using the S.F.E.E.

EXAMPLE 8.1

Problem. Suppose that measurements made at the inlet and outlet to the triple-expansion steam engine of Fig. 8.1, together with reference to tabulated properties of steam, establish the pressures, temperatures, enthalpies, velocities, elevations and mass-flow rate as those shown on Fig. 8.3. The shaft power is also known. We wish to calculate the heat transfer from the engine casing and associated ducting as part of an analysis of the economic benefit which may result from the use of better insulating materials.

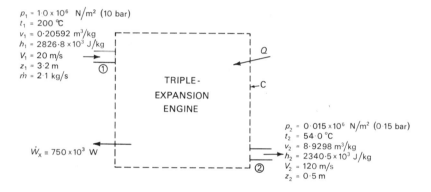

Fig. 8.3

Solution. We use the Steady-Flow Energy Equation, eq. (8.11), to solve this problem. The first step is to decide on the units which it is proposed to use. Here we choose J/kg. Then W_x/m is given by:

$$\frac{W_x}{m} = \frac{750 \times 10^3}{2 \cdot 1} = 356 \cdot 3 \times 10^3 \text{ J/kg}$$

With the insertion of the appropriate values shown on Fig. 8.3, and g_c being given the appropriate value of unity (see page 31), the S.F.E.E., eq. (8.11), becomes

$$\frac{Q}{m} - 356 \cdot 3 \times 10^3 = \left(2340 \cdot 5 \times 10^3 + \frac{120^2}{2 \times 1 \cdot 0} + \frac{g \times 0 \cdot 5}{1 \cdot 0} \right)$$

$$- \left(2826 \cdot 8 \times 10^3 + \frac{20^2}{2 \times 1 \cdot 0} + \frac{g \times 3 \cdot 2}{1 \cdot 0} \right)$$

Taking $g = 9 \cdot 81$ m/s², we obtain

$$\frac{Q}{m} - 356 \cdot 3 \times 10^3 = (2340 \cdot 5 \times 10^3 + 7 \cdot 2 \times 10^3 + 4 \cdot 9) - (2826 \cdot 8 \times 10^3 + 200 + 31 \cdot 4)$$

i.e.
$$\frac{Q}{m} = - 123 \times 10^3 \text{ J/kg}$$

Since the required answer should be the heat transfer per unit time, we multiply Q/m by the steam mass flow rate giving

$$\text{Heat transfer rate} = -123 \times 2\cdot1$$

$$= -258\cdot3 \times 10^3 \text{ J/s} \qquad \textit{Answer}$$

Remarks. 1. The negative sign denotes that the heat transfer is from the steam to the surroundings.

2. Examination of the individual terms in the equation shows that the contribution of the gravitational potential energy terms is quite negligible. This is almost always the case when gases and vapours are in question.

The kinetic-energy terms are also small; one is negligible. However, in many cases this is not so. For example, the kinetic energy of the steam leaving a turbine is usually appreciable.

3. The pressures and temperatures do not feature in this calculation. They are listed on the diagram because the enthalpies would normally be evaluated by their aid; the specification of pressure and temperature would not suffice if the steam were wet, however (see p. 186).

Basis and units: various forms of the S.F.E.E.

Time basis. In Example 8.1 the computation would have been slightly shorter had the basis of the equation been one second. For such analyses it is convenient to write the Steady-Flow Energy Equation on a unit-time basis. To do this we merely note that, for conditions to be steady, the quantities m, W_x and Q occur in the same time interval τ, and that division of these quantities by τ gives respectively the mass flow rate $\dot{m}$, the external-work rate (or external power) $\dot{W}_x$ and the heat-transfer rate $\dot{Q}$. Eq. (8.11) may therefore be expressed as:

$$\frac{\dot{Q} - \dot{W}_x}{\dot{m}} = \Delta\left(h + \frac{V^2}{2g_c} + \frac{gz}{g_c}\right) \qquad (8.12)$$

In Example 8.1, $\dot{Q}$ and $\dot{W}_x$ of eq. (8.12) would have been expressed in J/s and $\dot{m}$ in kg/s; each term within the brackets, including h, would still have the units of J/kg of steam. It should be noted that it is necessary to examine the units in each case considered. An example of another unit system in common use now follows.

Practice in mechanics and fluid mechanics. In subjects involving a large amount of algebra, anything that reduces the number of symbols is advantageous. Rigid-body and fluid mechanics are such subjects; and moreover they are largely concerned with Newton's Second Law of Motion. As a result it is common to use unit systems in which g_c is numerically unity. If this practice is followed, g_c may be omitted from the equation. The S.F.E.E. eq. (8.11) then becomes

$$\frac{Q - W_x}{m} = \Delta\left(h + \frac{V^2}{2} + gz\right) \qquad (8.13)$$

Each term then has the units of, for example, N m/kg (see Chapter 2).

Final general remarks on the S.F.E.E.   Before we consider particular engineering applications, a few more points need to be made about the validity of the S.F.E.E.

Identification of steady flow.   For purposes of applying the S.F.E.E., the foregoing discussion shows that steady flow can be deemed to exist if:
   (i) The states and rates of the material streams crossing the control-volume boundaries do not change with time; *and*
   (ii) *either,* the state at each point within the control volume does not change with time,
   or, only cyclic variations of these states occur; *and*
   (iii) *either,* the heat and work rates do not change with time,
   or the heat and work rates, averaged over a single cycle, do not change with time.

Control volume with many streams of material.   In the derivation of the S.F.F.E., a single entering stream and a single leaving one were considered. The form for the more general case, where there are many such streams, is readily deduced; in this case it is convenient to adopt a time basis. By reference to eq. (8.12), it is seen that the enthalpy fluxes in the "single-stream" case can be expressed by a term of the form $\dot{m}(h_1 + V_1^2/2g_c + gz_1/g_c)$ for the inflow, and a similar one for the outflow stream. It follows that in the "multi-stream" case there are terms corresponding to each inflow and each outflow stream; and it is the sum of the inflow terms minus the sum of the outflow terms which equals the difference $\dot{Q} - \dot{W}_x$. The S.F.E.E. thus becomes:

$$\dot{Q} - \dot{W}_x = \sum_{\text{out}} \dot{m} \left(h + \frac{V^2}{2g_c} + \frac{gz}{g_c} \right) - \sum_{\text{in}} \dot{m} \left(h + \frac{V^2}{2g_c} + \frac{gz}{g_c} \right) \quad (8.14)$$

where $\sum_{\text{out}}$ and $\sum_{\text{in}}$ stand respectively for the summations of all the out-flowing and in-flowing fluxes of enthalpy, kinetic energy and gravitational potential energy.

The significance of non-uniform entering streams.   When deriving the S.F.E.E., we have supposed that, for example, the velocity of the entering stream at the control-volume boundary has a unique value. On reflection, however, we remember that the velocity varies across the pipe cross-section, as seen in Fig. 3.23. The usual practice is to insert in the kinetic-energy term the mean value of velocity V obtained from the equation

$$V = \frac{\dot{m}v}{A} \quad (8.15)$$

where $\dot{m}$ is the mass flow rate, A is the cross-sectional area, and v is the specific volume (which is also *assumed* uniform). This involves a negligible error in most cases.

In fluid-mechanics problems, however, for example in the study of the flow of air from a cascade of compressor or turbine blades, the non-uniformity is often all-important. We then imagine an infinite number of streams to be crossing the control-volume boundary. The summations of eq. (8.14) now become integrals, and the S.F.E.E. is written as

$$\dot{Q} - \dot{W}_x = \int_C \left(h + \frac{V^2}{2g_c} + \frac{gz}{g_c} \right) d\dot{m} \qquad (8.16)$$

where $d\dot{m}$ is the mass flow rate in an infinitesimal stream, positive if outwards, negative if inwards, and $\int_C$ means that the integral is carried out over the whole control-volume boundary.

Extension to flow processes involving chemical reaction. So far we have supposed the flowing material to be a pure substance. Provided however that an extended definition of enthalpy is used, a form of the S.F.E.E. can be derived which also covers chemically-reacting substances. This is done in connexion with one of the particular examples below (p. 158).

Particular Examples of Steady Flow

We now apply the Steady-Flow Energy Equation to a number of common engineering examples. Incidentally more general matters will be touched on, namely the *Continuity Equation* and the aforesaid extension to reacting substances.

Throttling

When a fluid flows through a restriction, such as an almost-closed valve or a plug of porous material placed in a pipe-line, there is an appreciably lower pressure on the downstream than on the upstream side of the restriction. The flow is then said to have been *throttled*. Fig. 8.4 illustrates the situation in an insulated pipe. The S.F.E.E. can be applied to relate the downstream conditions to those upstream, provided that the control volume chosen is sufficiently large for the outgoing stream, at section 2 in Fig. 8.4, to be reasonably uniform.

Since in this example the process may be taken as adiabatic and there is no external work, the S.F.E.E., eq. (8.11) becomes

$$0 = h_2 + \frac{V_2^2}{2g_c} - h_1 - \frac{V_1^2}{2g_c} \qquad (8.17)$$

wherein the gravitational-potential-energy terms have been ignored.

Often the pipe velocities in throttling are so low that the kinetic-energy terms are also negligible. We then have

$$h_2 = h_1 \tag{8.18}$$

signifying that the enthalpy is unchanged.

Of course, the friction forces on the wall of a long insulated pipe cause a fall of pressure even in the absence of restrictions like valves; this

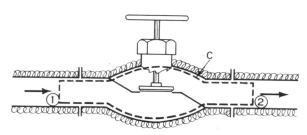

Fig. 8.4 Throttling: a partially-open valve in a pipe line.

phenomenon can also be classified as throttling. In such a case however it is often not permissible to neglect the change in kinetic energy, for the reduced pressure causes the specific volume of the fluid to increase; this increase may be so large, in the case of a vapour or gas, that the fluid must flow much faster at the downstream end than does the relatively dense in-flowing material. In an extreme case, the velocity at the downstream end can reach the velocity of sound. Taking this at a typical value of 300 m/s, we have

$$\frac{V_2^2}{2g_c} = \frac{(300)^2}{2 \times 1 \cdot 0} = 45 \times 10^3 \text{ N m/kg}$$

$$= 45 \times 10^3 \text{ J/kg}$$

In general, if the downstream state has been calculated from eq. (8.18), it is always necessary to check the downstream kinetic energy to make sure that the change of kinetic energy is indeed negligible.

Adiabatic machines

Under this title we consider such apparatus as turbines and reciprocating engines which give a positive power output, and rotary or reciprocating compressors which require power *in*put. Often these are well lagged, so that the heat transfer can be neglected.

With *reciprocating machines*, the flow velocities are often small and the kinetic-energy terms can be neglected. The S.F.E.E., eq. (8.11), then becomes

$$-\frac{W_x}{m} = h_2 - h_1 \tag{8.19}$$

wherein the gravitational term is omitted as usual. It is evident that, if the power output is positive, the enthalpy of the leaving fluid is less than that of the entering fluid by the amount of the work done per unit mass of fluid. With compressors, on the other hand, the leaving fluid has the higher enthalpy.

EXAMPLE 8.2

Problem. At entry to the reciprocating compressor of a refrigerator, the refrigerant, Freon-12, has a pressure and temperature of 200×10^3 N/m² and -10 °C respectively. At exit from the compressor, the Freon has a pressure of 900×10^3 N/m² and a temperature of 55 °C. The flow is steady.

Evaluate the external work, in J/kg of Freon, assuming the compressor to be adiabatic and changes in the kinetic and potential energies to be negligible.

Solution. Since the machine is adiabatic and changes of velocity and height are negligible, the form of S.F.E.E. given in eq. (8.19) is appropriate. To obtain W_x/m we therefore require the values of h_1 and h_2, the enthalpy of the Freon at entry to and exit from the compressor respectively. These are obtained from tables of properties of Freon and are as follows:

At: $p_1 = 200 \times 10^3$ N/m² and $t_1 = -10$ °C: $h_1 = 183 \cdot 2 \times 10^3$ J/kg of Freon-12;

$p_2 = 900 \times 10^3$ N/m² and $t_2 = 55$ °C: $h_2 = 215 \cdot 8 \times 10^3$ J/kg of Freon-12.

Inserting these values in eq. (8.19), we obtain

$$ -\frac{W_x}{m} = (215 \cdot 8 - 183 \cdot 2) \times 10^3 = 32 \cdot 6 \times 10^3 $$

or   External work $= -32 \cdot 6 \times 10^3$ J/kg   *Answer*

Rotary machines usually cause the fluid to flow at high velocity, so the kinetic energies cannot be neglected. We then have, from eq. (8.11) with $Q = 0$ and $\Delta z = 0$, that

$$ -\frac{W_x}{m} = h_2 + \frac{V_2^2}{2g_c} - h_1 - \frac{V_1^2}{2g_c} \tag{8.20} $$

In the case of steam turbines, the velocity of the dense in-flowing steam is often small; then $V_1^2/2g_c$ is negligible.

EXAMPLE 8.3

Problem. The turbine of a jet engine receives a steady flow of gases at a pressure of 720×10^3 N/m², a temperature of 870 °C and a velocity of 160 m/s. It discharges the gases at a pressure of 215×10^3 N/m², a temperature of 625 °C and a velocity of 300 m/s.

Evaluate the external work output of the turbine in J/kg of gas. The process may be assumed to be adiabatic.

Solution. In this case the effect of kinetic energy is important. We must therefore use the form of S.F.E.E. given in eq. (8.20) to analyse the conditions in this adiabatic machine. To evaluate W_x, we need the enthalpies of the gases entering and leaving the turbine; these are obtained from property tables:

At $p_1 = 720 \times 10^3$ N/m² and $t_1 = 870$ °C: $h_1 = 1000 \times 10^3$ J/kg of gas.

At $p_2 = 215 \times 10^3$ N/m² and $t_2 = 625$ °C: $h_2 = 719 \times 10^3$ J/kg of gas.

Insertion of these values in eq. (8.20), together with the values of velocity and the appropriate value of g_c, unity, yields

$$-\frac{W_x}{m} = \left(719 \times 10^3 + \frac{(300)^2}{2 \times 1 \cdot 0}\right) - \left(1000 \times 10^3 + \frac{(160)^2}{2 \times 1 \cdot 0}\right)$$

$$= (719 \times 10^3 + 45 \times 10^3) - (1000 \times 10^3 + 12 \cdot 8 \times 10^3)$$

$$= -248 \cdot 8 \times 10^3 \text{ J/kg of gas.}$$

i.e.   External work $= 248 \cdot 8 \times 10^3$ J/kg of gas.   *Answer*

Heat-transfer equipment

Boilers and condensers have, as their main purpose, the effecting of heat transfer to or from a steadily-flowing material. An example is the evaporator of a refrigerating plant shown in Fig. 8.5, in which liquid Freon enters a coil in contact with the air in a refrigerator cabinet and leaves as vapour.

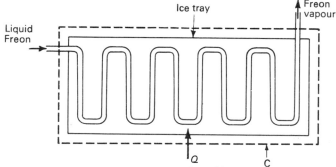

Fig. 8.5 Evaporator of a refrigerator.

The velocities are small, we suppose; there is no external work; and the gravitational terms are negligible. The S.F.E.E. therefore may be written as

$$\frac{Q}{m} = h_2 - h_1 \qquad (8.21)$$

It should be noted that there is no need to specify that the pressure should be uniform; indeed, friction at the pipe walls causes the pressure at outlet to be somewhat below that at inlet. The frictional shear stresses on the wall of course do no external work; for their points of application do not move.

The superheater. To illustrate the use of the S.F.E.E. in a situation where there are more than two entering streams, Fig. 8.6 shows the superheater of a boiler (N.B. a superheater raises the temperature of the steam above that of the boiling water). If a control volume were drawn to fit tightly over the superheater tubes, eq. (8.21) would be applicable. We choose however to consider a larger control volume, enclosing the insulated duct through which pass the flue gases from the furnace. Examination of the control-volume boundary shows that there is no heat or external work, but that there are four material streams: steam enters at S_1 and leaves at S_2; gases enter at G_1 and leave at G_2.

On the assumption that the stream properties are uniform at the four cross-sections, that the steam and gas mass-flow rates are $\dot{m}_S$ and $\dot{m}_G$

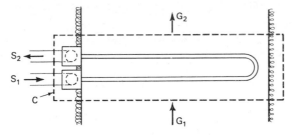

Fig. 8.6 Superheater of a boiler.

respectively, and that kinetic and gravitational potential energies are small, the S.F.E.E. (multi-stream form, eq. (8.14)) gives

$$0 = \dot{m}_S(h_{S_2} - h_{S_1}) + \dot{m}_G(h_{G_2} - h_{G_1}) \qquad (8.22)$$

where the subscripts have obvious meanings.

Re-arranging eq. (8.22), we have

$$\dot{m}_S(h_{S_2} - h_{S_1}) = \dot{m}_G(h_{G_1} - h_{G_2}) \qquad (8.23)$$

which shows that the increase in enthalpy of the steam flow is exactly equal to the enthalpy decrease of the gas stream. This is to be expected because the right- and left-hand sides of eq. (8.23) are each equal, numerically, to the heat transfer between the steam and gas streams; this may be demonstrated by the application of the S.F.E.E. to each of these streams in turn.

The steam de-superheater. As a final example of equipment of this character, chosen to show that there is no necessity for the two streams to remain separate, Fig. 8.7 shows a steam de-superheater, in which the steam

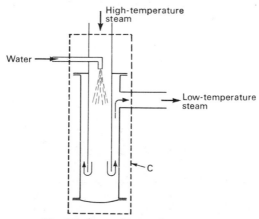

Fig. 8.7 Steam de-superheater.

has its temperature reduced by being mixed with water which is sprayed in from a hot-water main. This practice is adopted in chemical processing plant where accurate control of temperature is required, and also in boiler plant.

In this case it is impossible to draw a control volume around either of the two streams which enter; for the steam and injected water become completely mixed. However, the control volume C shown in Fig. 8.7 provides all the information which is ordinarily required. If the mass flow rates of the injected water, of the steam entering, and of the steam leaving are $\dot{m}_w$, $\dot{m}_{S_1}$ and $\dot{m}_{S_2}$, respectively, and if kinetic and gravitational energies are neglected as before, the S.F.E.E. gives

$$0 = \dot{m}_{S_2} h_{S_2} - (\dot{m}_{S_1} h_{S_1} + \dot{m}_{w_1} h_{w_1}) \tag{8.24}$$

The mass flow rates are obviously connected, by the mass-conservation principle, in the form

$$\dot{m}_{S_2} = \dot{m}_{w_1} + \dot{m}_{S_1} \tag{8.25}$$

Eq. (8.24) and eq. (8.25) are used, in conjunction with tabulated properties of water and steam, to determine how much water must be supplied to bring about a given temperature change in the steam.

The adiabatic nozzle

An example in which the kinetic-energy terms are definitely not negligible is provided by the *nozzle*, of which the whole purpose is to cause the fluid to leave with a higher velocity than that with which it enters. A nozzle is a specially-shaped duct. If only small velocity changes are to be caused, such that the associated pressure changes cause only small density changes, the nozzle is *convergent*; this means that its cross-sectional area only decreases in the direction of flow. For larger pressure differences, such as are encountered in steam turbines or in rocket motors, the cross-section at first decreases and later increases; this is the case of the *convergent-divergent nozzle*, also called the *Laval nozzle* after its inventor. Fig. 8.8 shows an example.

If the nozzle is insulated ($Q = 0$) and if the gravitational terms are neglected, the S.F.E.E. relates the entering and leaving conditions through the equation

$$0 = h_2 + \frac{V_2^2}{2g_c} - h_1 - \frac{V_1^2}{2g_c} \tag{8.26}$$

wherein W_x has of course been put equal to zero because there is no external work.

The meaning of eq. (8.26) can be seen more clearly by re-arranging it as

$$\frac{V_2^2}{2g_c} - \frac{V_1^2}{2g_c} = h_1 - h_2 \tag{8.27}$$

which shows that the increase of kinetic energy is equal to the decrease of enthalpy.

It should be noted that the equation holds whether the friction at the wall is appreciable or not. It is indeed identical with that which holds for an adiabatic throttling process in which the kinetic energy is not neglected.

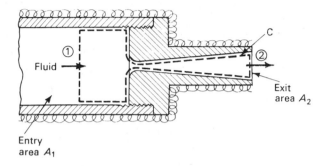

Fig. 8.8 Convergent-divergent nozzle.

Stagnation enthalpy. It is sometimes convenient to unite h and $V^2/2g_c$ in a single term. We therefore define the *stagnation* or *total enthalpy*, h_t, by

$$h_t \equiv h + \frac{V^2}{2g_c} \tag{8.28}$$

The epithet "stagnation" derives from the fact that h_t and h become identical if the fluid is made stagnant, i.e., if it is brought to rest in an adiabatic steady-flow process without external work.

The S.F.E.E., can be re-written in terms of h_t as

$$\frac{Q - W_x}{m} = \Delta \left(h_t + \frac{gz}{g_c} \right) \tag{8.29}$$

Further, the results derived for adiabatic throttling and for the adiabatic nozzle can be summarized in the statement:

In an adiabatic steady-flow process, with zero external work and negligible gravitational-potential-energy change, the stagnation enthalpy of the fluid remains constant. In symbols:

$$Q = 0, \quad W_x = 0, \quad \Delta \left(\frac{gz}{g_c} \right) = 0: \qquad h_{t_2} = h_{t_1} \tag{8.30}$$

The continuity equation. We have mentioned above the need to check the magnitude of the kinetic-energy term even when neglecting it, but we have not yet indicated how to do so. The influence of the cross-sectional area has also been mentioned. This is an appropriate point to introduce the *continuity equation*, which is an expression of the *conservation-of-mass principle.*

Consider the duct shown in Fig. 8.9; at the section 1, this has the cross-sectional area A_1. Let the velocity of flow have the uniform value V_1; let the specific volume of the fluid be v_1 and let its mass flow rate be $\dot{m}$. Then the above quantities are related by

$$\dot{m} = \frac{V_1 A_1}{v_1} \tag{8.31}$$

The truth of this is most readily perceived by imagining the fluid to be at rest while the section 1 travels to the left at velocity V_1. Then in time

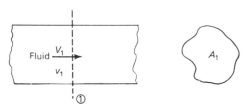

Fig. 8.9 Steady flow in a duct.

interval τ it travels a distance τV_1. Since the area is A_1, the volume swept out is $\tau V_1 A_1$. The mass of fluid contained in this volume is $\tau V_1 A_1/v_1$. The mass of fluid crossing the imaginary surface 1 in unit time is therefore $(\tau V_1 A_1)/\tau v_1$), i.e., it is $V_1 A_1/v_1$. This is the required result, eq. (8.31).

Considering now steady flow through a control volume such as that of Fig. 8.8 for example, and noting that the conservation of mass requires that the mass flow rates at sections 1 and 2 must be identical, we have

$$\frac{V_1 A_1}{v_1} = \frac{V_2 A_2}{v_2} \tag{8.32}$$

This equation can also be applied at any section intermediate between and 2. We note that the following equation holds at any section

$$A = \frac{\dot{m} v}{V} \tag{8.33}$$

In a nozzle, the velocity V increases continuously; this tends to require decreasing area in the flow direction, a requirement that predominates at first. But the pressure decreases continuously. When the pressure reduction becomes large enough, the consequent increase in the specific volume becomes more important, so that the ratio v/V begins to increase. This is the reason for the convergent-divergent form of nozzles for gases when the ratio of inlet pressure to outlet pressure is large.

The impulse-turbine wheel

In this section an example is chosen which, as well as being immediately interesting to engineers, also illustrates that a control volume can be in motion relative to the earth and that the magnitude of the external-work term depends on this motion.

Description. Fig. 8.10*a* illustrates a section through a turbine wheel. This consists of a rotating disc (rotor), to the periphery of which are fitted blades R which are curved so that the gas, which flows from left to right through the annular gap, has its angle of swirl (about the rotor axis) altered. The initial swirl is given to the gas by fixed blades or nozzles N, situated to the left, i.e. upstream, of the rotor.

Fig. 8.10*b* shows more clearly the geometry of the nozzles N and rotor blades R. It represents a developed section along the cylindrical surface marked A in Fig. 8.10*a*. The nozzles N are to be regarded as at rest, while the rotor blades R are moving vertically upwards.

The relationships between the velocities are shown by the velocity diagram of Fig. 8.10*c*, which employs the convention commonly adopted in studies of the kinematics of machines. Points r and n represent respectively the velocity of the rotor and the velocity of the nozzles; the line joining them therefore represents the magnitude and direction of the velocity of R relative to N. It is assumed that the flowing gas leaves both the nozzles and the rotor-blade passages in directions parallel to the adjacent surfaces. Then if nf_i and nf_o represent respectively the velocity vectors of the gas entering and leaving the rotor, these points must be situated such that $\angle f_i nr$ equals β and $\angle f_o rn$ equals α in Fig. 8.10*c*, corresponding to the angles β and α shown on Fig. 8.10*b*.

The turbine chosen for illustration is of the *impulse* type and has negligible friction; for present purposes, this means that the gas experiences no change of pressure or temperature in passing through the rotor passages, and that the speed *relative to the rotor* is unchanged, i.e. that distance $f_i r$ equals distance $f_o r$. We choose the inlet and outlet gas angles, α, to be equal, which means that the axial velocities of the entering and leaving gas are identical; therefore on Fig. 8.10*c* the point f_o is vertically below f_i.

The reader is asked to accept the above brief description as sufficient for the time being. Fuller discussion can be found in more advanced texts.

Application of the S.F.E.E. We first consider the control volume marked C_n in Figs. 8.10*a* and *b*. This is supposed to be at rest relative to the nozzles and relative to the earth. Examination of its boundaries shows that there is an external-work interaction $\dot{W}_x$ and a steady flow of gas. Since heat transfer is absent (we assume), and since the enthalpies of the entering and leaving gases are equal (because it is an impulse turbine), the S.F.E.E. eq. (8.12) becomes

$$-\frac{\dot{W}_x}{\dot{m}} = (V_{nf_o}^2 - V_{nf_i}^2)/2g_c \qquad (8.34)$$

Here the suffix nf_o means "leaving gas relative to nozzles (rest)" and nf_i means "entering gas relative to nozzles (rest)". The velocities are represented respectively by the lengths nf_o and nf_i in Fig. 8.10c, where it is evident that the outlet velocity is considerably smaller than the inlet velocity. Eq. (8.34) shows that the external work per unit mass of gas is equal to the decrease in its kinetic energy measured relative to the control volume at rest (C_n).

We now consider a second control volume: that marked C_r in Fig. 8.10b. This encloses one passage between the rotor blades, *and moves with the rotor*. Examination of its boundary shows that not only is there no heat transfer, but there is *no external work either*; for no shaft protrudes through the boundary of C_r as it did through that of C_n. How has this result, apparently contradictory to eq. (8.34), come about?

The explanation becomes clear when we apply the S.F.E.E. to the moving control volume C_r. As before, $\dot{Q}$ and the enthalpy change are zero. With $\dot{W}_x$ also zero we have

$$0 = (V_{rf_o}^2 - V_{rf_i}^2)/2g_c \tag{8.35}$$

i.e.
$$V_{rf_o} = V_{rf_i} \tag{8.36}$$

Here the *relative* velocities between rotor and gas have been used because the control volume moves with the rotor.

Eq. (8.36) shows that, relative to the rotor, the gas leaves with the same velocity* as that with which it enters. But this we already knew: it is a consequence of having an impulse or constant-pressure turbine wheel. There is therefore no contradiction between the two applications of the S.F.E.E.: the work terms are different because of the differing frames of reference (see page 48); but so are the kinetic-energy changes for the same reason.

Further analysis in terms of the momentum change. The reader who is still worried by the difference between the above two results may find the following analysis helpful:

Application of Newton's Second Law of Motion to control volume C_r for the direction of blade movement (see fluid-mechanics texts for justification) yields

$$F = \frac{\dot{m}}{g_c} [V_{rf_i} \cos \alpha - V_{f_{or}} \cos (-\alpha)] \tag{8.37}$$

where F is the force exerted on the blades in the direction of motion, $\dot{m}$ is the mass flow rate of the gas, and the cosines are introduced to give the velocity components of the gas in the direction of rotor movement. The right-hand side is the difference between the entering and leaving momentum fluxes.

---

* In magnitude, not direction.

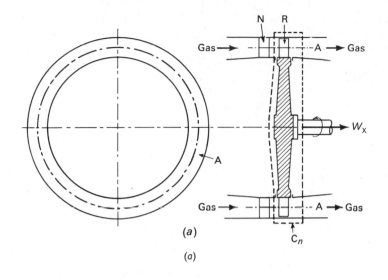

(a)

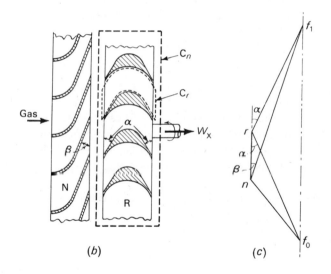

(b)   (c)

Fig. 8.10

(a) Single-stage impulse turbine.
(b) Impulse turbine blading.
(c) Velocity-vector diagram for impulse blading.

157

We now shift to a co-ordinate system at rest relative to the earth, i.e., relative to the nozzles, and introduce the blade speed V_{nr}; the rate at which the gases do work on the rotor becomes

$$\text{rotor power} = FV_{nr} \qquad (8.38)$$

Expressing F in terms of gas velocities from eq. (8.37) and dividing by $\dot{m}$ to get the work per unit mass of gas, we obtain

$$\text{work on rotor per unit mass of gas} = \frac{V_{nr}}{g_c}(V_{rf_i} + V_{rf_o})\cos\alpha \qquad (8.39)$$

But this must equal $\dot{W}_x/\dot{m}$, which has already been expressed by eq. (8.34). Equating the right-hand sides of the two equations, we obtain

$$-\frac{V_{nf_o}^2 - V_{nf_i}^2}{2g_c} = \frac{V_{nr}}{g_c}(V_{rf_i} + V_{rf_o})\cos\alpha \qquad (8.40)$$

Each of the velocities in eq. (8.40) is represented by a length in Fig. 8.10c. Application of trigonometry to the velocity diagram should enable us to check the correctness of our analysis. This is left as an exercise for the student.

Steady-flow combustion processes

The First Law, as introduced in Chapter 6, is valid for all types of systems and processes. When the Steady-Flow Energy Equation was introduced, however, only pure substances were considered. We now remove this restriction to the extent of allowing chemical reaction to occur. This may be done simply by replacing u and h by the new symbols u' and h', where the prime signifies that we are concerned with substances which may react chemically, i.e., with materials which are not pure substances (see definition on p. 378).

We now have,* in the absence of capillarity, electricity and magnetism,

$$e = u' + \frac{V^2}{2g_c} + \frac{gz}{g_c} \qquad (8.41)$$

and

$$h' = u' + pv' \qquad (8.42)$$

For such substances, more than two independent properties will be needed to specify the state; but no such specification is required in the derivation of the S.F.E.E., the form of which for unit mass of reacting mixture is similar to that for a pure substance, eq. (8.11), and runs

$$\frac{Q - W_x}{m} = \Delta\left(h' + \frac{V^2}{2g_c} + \frac{gz}{g_c}\right) \qquad (8.43)$$

* These equations should be compared with the corresponding ones for pure substances viz., eq. (7.23) and eq. (7.7).

Correspondingly, a "multi-stream" form may be developed similar to that for pure substances, eq. (8.14). In this case the outflow streams are the

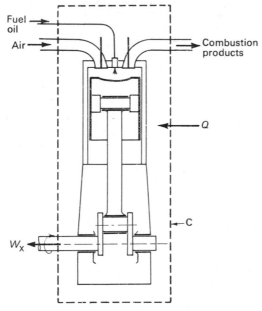

Fig. 8.11   Diesel engine

products of combustion and the inflow streams the reactants; the equation thus becomes

$$Q - \dot{W}_x = \sum_{\text{products}} \dot{m}\left(h' + \frac{V^2}{2g_c} + \frac{gz}{g_c}\right) - \sum_{\text{reactants}} \dot{m}\left(h' + \frac{V^2}{2g_c} + \frac{gz}{g_c}\right)$$

$$(8.44)$$

In this form the S.F.E.E. may be applied to an internal-combustion engine for example. Consider the diesel engine shown in Fig. 8.11, and note that the materials crossing the boundaries are air and fuel oil flowing inwards, and combustion products outwards; then eq. (8.44) becomes

$$\dot{Q} - \dot{W}_x = \dot{m}_P h'_P - (\dot{m}_A h'_A + \dot{m}_F h'_F) \qquad (8.45)$$

where the suffixes P, A and F refer to products, air and fuel respectively, $\dot{m}$ signifies mass flow rate, $\dot{Q}$ and $\dot{W}_x$ are heat and external work per unit time, and the kinetic and gravitational terms have been neglected. For all internal-combustion engines, $\dot{Q}$ will be negative; it comprises the heat transfers from the reacting mixture to the cooling water and atmosphere.

The subject of combustion will be returned to in Chapter 16. In the meantime we emphasize that $\dot{Q}$ is *not* the "calorific value" of the fuel.

EXAMPLE 8.4

Problem. A gasoline engine develops a brake power of 50 kW. The mass flow rate of fuel into the engine is 15 kg/h and that of air is 215 kg/h; the temperature of the fuel-air stream entering the engine is 15 °C and the temperature of the exhaust gases leaving the engine is 900 °C. The heat-transfer rate to the jacket cooling water is 42 kJ/s and that to the surrounding atmosphere is 15 kJ/s.

15×10^3 W.

Evaluate the increase in the specific enthalpy of the mixture stream as it flows through the engine, assuming the kinetic and potential energies to be negligible.

Solution. The problem requires the evaluation of $\Delta h'$ in eq. (8.43). Basis: 1 lb$_m$ of mixture entering the engine as fuel and air and leaving it as combustion products.

Mass flow rate of mixture through engine, $\dot{m} = \dot{m}_{\text{fuel}} + \dot{m}_{\text{air}}$

$$= = (15 + 215) \frac{1}{3600} \text{ kg/s}$$

$$\text{or} \left(\frac{15 + 215}{3600}\right) = 63{\cdot}9 \times 10^{-3} \text{ kg/s}$$

$$\dot{W}_x = 50 \text{ kW} = 50 \times 10^3 \text{ J/s}$$

and so

$$\frac{W_x}{m} \equiv \frac{\dot{W}_x}{\dot{m}} = \frac{50 \times 10^3}{63{\cdot}9 \times 10^{-3}} = 782 \times 10^3 \text{ J/kg of mixture}$$

Now

$$\dot{Q} = \dot{Q}_{\text{coolant}} + \dot{Q}_{\text{atmosphere}}$$

$$= -42 \times 10^3 - 15 \times 10^3 = -57 \times 10^3 \text{ J/s}$$

$$\therefore \quad \frac{Q}{m} = \frac{\dot{Q}}{\dot{m}} = \frac{-57 \times 10^3}{63{\cdot}9 \times 10^{-3}} = -892 \times 10^3 \text{ J/kg of mixture}$$

where the negative sign shows that the heat transfer is *from* the mixture stream as it passes through the engine. On insertion of these values into eq. (8.43), we get

$$-892 \times 10^3 - 782 \times 10^3 = \Delta h' + 0 + 0$$

whence

$$\Delta h' = -1674 \times 10^3 \text{ J/kg of mixture} \qquad \qquad Answe$$

Remarks. 1. The negative sign indicates that the enthalpy of the mixture stream *decreases*, during the change from fuel and air to products, even though the temperatur *increases.*

2. The "calorific value" of the fuel does not appear in the analysis.

Comparison of the Steady-Flow Energy Equation with the Euler and Bernoulli Equations of Fluid Mechanics

Several mentions of the science of *fluid mechanics* have already been made and it is clear that the thermodynamics and the mechanics of flow processe are intimately related. In this section a short digression will be made, i order to clarify the distinction between the Steady-Flow Energy Equatio and two equations which belong to fluid mechanics: the *Euler Equation* an the *Bernoulli Equation*. Emphasis is given to this matter because th S.F.E.E. is often confused with, i.e., thought to be identical with, one o other of these equations.

The Euler Equation. Fig. 8.12 shows a fluid flowing steadily along a "stream-tube", which for our purposes can be regarded as an *imaginary* pipe enclosing the flow, of sufficiently small cross-section for the velocity to be taken as uniform. We consider the thin control volume C, shown by the dotted line. This has two faces at right angles to the flow direction, while

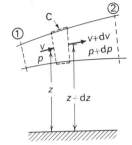

Fig. 8.12 Steady flow in a stream-tube.

the remainder of the control-volume boundary coincides with the surface of the part of the stream-tube which it encloses.

At the upstream face, the stream velocity is V and the pressure is p; the mean elevation of the fluid above some arbitrary datum is z. At the downstream face, the values of the corresponding quantities are $V + \mathrm{d}V, p + \mathrm{d}p$, and $z + \mathrm{d}z$. Shear forces on the control-volume boundaries are supposed not to be present.

Application of Newton's Second Law of Motion. It is shown in fluid-mechanics texts that application of Newton's Second Law, i.e., the momentum principle, to this control volume yields the relationship

$$v\,\mathrm{d}p + \frac{V\,\mathrm{d}V}{g_c} + \frac{g\,\mathrm{d}z}{g_c} = 0 \qquad (8.46)$$

The content of this equation, expressed in words, is that the sum of the forces acting on the fluid in the control volume, resolved in the direction of motion, is equal to the rate of flow of momentum out of the control volume minus the rate of flow of momentum into the control volume.

Eq. (8.46) is the *Euler Equation.*

Restrictions on the validity of the Euler Equation. The conditions under which eq. (8.46) is valid will now be summarized. They are:

(i) Steady flow.

(ii) No shear forces, i.e., no friction, and no shaft or stirring work.

(iii) No electrical, magnetic or capillary forces.

It is not necessary that heat transfer should be absent however. For that matter, an electric current may also be present, if it acts in a purely dissipative (ohmic) manner.

The Bernoulli Equation. The thickness of the control volume of Fig. 8.12 in the stream direction was infinitesimal; however, a control volume of finite thickness may be considered. Newton's Second Law for such a control volume is given by *integrating* the Euler Equation in the stream direction. We note first that the term $V\,dV/g_c$ can be regarded, according to the rules of differentiation, as $d(\frac{1}{2}V^2)/g_c$. Denoting the upstream face of the finite control volume by suffix 1 and the downstream face by suffix 2 (Fig. 8.12), we may integrate eq. (8.46) and obtain

$$\int_1^2 v\,dp + \frac{V_2^2}{2g_c} - \frac{V_1^2}{2g_c} + \frac{gz_2}{g_c} - \frac{gz_1}{g_c} = 0 \qquad (8.47)$$

This is known as the *Bernoulli Equation*. Its conditions of validity are the same as those of the Euler Equation.

The Bernoulli Equation for an incompressible fluid. The Bernoulli Equation is frequently required for a fluid such as water, which can be regarded as *incompressible*; that is to say that the fluid density may be regarded as constant. Gases may also be treated as incompressible if the pressure changes are small. Since the specific volume v is the reciprocal of the density, it too is constant; v can therefore be taken outside the integral sign in eq. (8.47), which can consequently be written as

$$p_2 v - p_1 v + \frac{V_2^2}{2g_c} - \frac{V_1^2}{2g_c} + \frac{gz_2}{g_c} - \frac{gz_1}{g_c} = 0 \qquad (8.48)$$

For comparison with the S.F.E.E., we write eq. (8.48) as

$$0 = \Delta\left(pv + \frac{V^2}{2g_c} + \frac{gz}{g_c}\right) \qquad (8.49)$$

wherein v is constant and $\Delta(\ldots)$ means "increase in ..." as before.

Comparison of the Bernoulli Equation and the S.F.E.E. In order to show how the confusion of eq. (8.49) with the S.F.E.E. has come about, the latter will be written with $(u + pv)$ substituted for h. Eq. (8.11) then becomes

$$\frac{Q - W_x}{m} = \Delta\left(u + pv + \frac{V^2}{2g_c} + \frac{gz}{g_c}\right) \qquad (8.50)$$

Comparison of eq. (8.49) with eq. (8.50) shows that they have several terms in common. This has led to the suggestion that the S.F.E.E. is merely a form of the Bernoulli Equation, extended by the addition of energy and boundary-interaction terms. That this is not so may be seen by noting the restricted conditions under which the Bernoulli Equation is valid. The most striking are:

(i) shear forces must be absent; and

(ii) The flow must be incompressible (i.e., v = constant).

The S.F.E.E. on the other hand is valid whether friction is present or not, and regardless of how the density of the fluid changes.

The S.F.E.E. is therefore not an extended form of the Bernoulli Equation.

Combination of the Euler Equation with the S.F.E.E.

To emphasize the distinction between the Euler Equation, which is based on Newton's Second Law of Motion only, and the S.F.E.E., which is based on Newton's Second Law of Motion *and* the First Law of Thermodynamics, the equations will be combined. Since the Euler Equation, eq. (8.46), is more general than the Bernoulli Equation for an incompressible fluid, the former will be used. It is therefore necessary to express the S.F.E.E. in differential form, i.e., to consider the form valid for a control volume of infinitesimal thickness in the flow direction. Differentiating eq. (8.50), we have

$$\frac{dQ - dW_x}{m} = du + p \, dv + v \, dp + \frac{V \, dV}{g_c} + \frac{g \, dz}{g_c} \qquad (8.51)$$

For comparison with eq. (8.46), dW_x must be put equal to zero. We then note that the Euler Equation states that the last three terms of eq. (8.51) are zero. *If the Euler Equation is valid*, therefore, the S.F.E.E. reduces to

$$\frac{dQ}{m} = du + p \, dv \qquad (8.52)$$

At first sight, eq. (8.52) appears merely to be the First Law of Thermodynamics. Further thought reveals that eq. (8.52) is the particular form of the First Law valid for a pure substance, in the absence of gravity, motion, electricity, magnetism and capillarity, in a process where the only work is displacement work.

Comments on equation (8.52)

1. The restriction to a pure substance was of course introduced in the way the S.F.E.E. was written. A more general form could have been used.

2. The required "absence" of gravity and motion appears more surprising. This condition merely means however that, in deriving eq. (8.52) from the First Law, the evaluation of the energy change and the work is done *with reference to a frame which moves with the fluid*.

3. The significant implication of eq. (8.52) is that the conditions for validity of the Euler Equation imply that only displacement work is done by the fluid particles in their travel. This is another way of saying that friction is absent and that all expansions of the fluid particles are fully-resisted.

4. In Chapter 11, p. 224, where we return to this subject, it will be shown that the condition that friction should be absent signifies that the flow is "reversible".

5. The most important lesson to learn from the present section is: *The Steady-Flow Energy Equation is general; the Euler and Bernoulli Equations are not.*

Further Uses of the Control-Volume Concept

In the final section of this chapter we discuss the extension of the control-volume concept to unsteady-flow processes. So far, the First Law has been applied to general processes by focussing attention on a fixed body of material, the system; only for steady-flow processes has the control volume been used. This distinction of procedure is still recommended for most problems encountered in engineering thermodynamics. However, in more advanced work, for example the theory of unsteady compressible fluid flow, the control-volume analysis is simpler than the system analysis. We lay the foundations for such advanced work in the subsequent paragraphs.

The First Law of Thermodynamics for unsteady processes: control-volume analysis

We return to eq. (8.3), which expresses the First Law written for the control volume shown in Fig. 8.2. After re-arrangement and the writing of $(u + V^2/2g_c + gz/g_c)$ for e, this equation becomes

$$Q - W_x = E_{C,f} - E_{C,i} + m_2 \left(h_2 + \frac{V_2^2}{2g_c} + \frac{gz_2}{g_c} \right)$$

$$- m_1 \left(h_1 + \frac{V_1^2}{2g_c} + \frac{gz_1}{g_c} \right) \qquad (8.53)$$

Since the flow is not steady, we cannot this time write $E_{C,f} = E_{C,i}$ and $m_2 = m_1$. Instead we generalize eq. (8.53) to the case where there are arbitrary numbers of entering and leaving streams instead of the two streams of Fig. 8.2. Then eq. (8.53) becomes

$$Q - W_x = E_{C,f} - E_{C,i} + \sum_{\text{out}} m \left(h + \frac{V^2}{2g_c} + \frac{gz}{g_c} \right)$$

$$- \sum_{\text{in}} m \left(h + \frac{V^2}{2g_c} + \frac{gz}{g_c} \right) \qquad (8.54)$$

Here $\sum_{\text{out}} m(\ldots)$ represents the sum of all the out-going fluxes of enthalpy, kinetic energy and gravitational potential energy; each h, V^2 and z is appropriated to the particular m which it multiplies, while $\sum_{\text{in}} m(\ldots)$ represents the sum of all the in-going fluxes in the same way. $E_{C,f}$ and E_C represent, as before, respectively the final and initial energy contents of the control volume, it being understood that the material in the control volume at the end of the process differs from that there initially, both in its identity and its total mass. E_C includes the kinetic and gravitational potential energies of the material in the control volume, which in general comprises solids (pistons, springs, etc.) as well as fluid ; but *there is no pv term* for either the solid or the fluid material. Thus, if only a pure substance is present within the control volume, we should evaluate the energy E_C of the material in the control volume at any instant from

$$E_C = \int_\sigma \left(u + \frac{V^2}{2g_c} + \frac{gz}{g_c} \right) \frac{d\sigma}{v} \qquad (8.55)$$

Here $d\sigma$ is a volume element of the control volume; v is the local specific volume so that $d\sigma/v$ represents the mass of substance in the element $d\sigma$; u, V and z are the local values of these quantities; and $\int_\sigma$ means that the integral is carried out over the whole volume.

A warning example: the thermostatically-controlled tank

Use of eq. (8.54) is not recommended for elementary problems of unsteady flow: the system analysis is safer and just as simple. We conclude discussion of the First Law for the present by an illustration of the dangers of confusing the two approaches.

Fig. 8.13 shows a hot-water tank containing an electrical-heating element which is controlled by a thermostat so as to maintain the water temperature

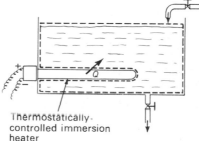

Fig. 8.13 Thermostatically-controlled hot-water tank.

Thermostatically-controlled immersion heater

t the constant value t_1. Cold water can enter from the supply pipe where he temperature is t_0. We wish to calculate the heat transfers from the eating element which ensue: (*a*) when water is supplied to the tank; and *b*) when it is drawn off.

How not to do it. A careless reading of the foregoing chapter might lead he student to write

$$\text{``First Law: } \dot{Q} = \frac{d}{d\tau} (mu) \text{''} \qquad (8.56)$$

here $\dot{Q}$ is the rate of heat supply, m is the mass of water in the tank, u is its nternal energy per unit mass and τ is time. Loosely, eq. (8.56) might be put nto words as "heat-transfer rate equals rate of change of internal energy", /hich has a convincing ring.

Since for water we can write

$$u = c_v(t - t_0) \tag{8.57}$$

where c_v is the (nearly constant) specific heat at constant volume, where t is the water temperature at any time, and where t_0 is the temperature adopted for the base of enthalpy, eq. (8.56) can be written as

$$\dot{Q} = c_v \frac{d}{d\tau} m(t - t_0)$$

$$= c_v(t - t_0) \frac{dm}{d\tau} + c_v m \frac{dt}{d\tau}$$

Now $dt/d\tau$, the time variation of tank-water temperature, is made zero by the action of the thermostat. Our false form of the First Law therefore reduces to

$$\dot{Q} = c_v(t - t_0) \frac{dm}{d\tau}$$

Since in the present case the tank temperature is t_1, we have

$$\dot{Q} = c_v(t_1 - t_0) \frac{dm}{d\tau} \tag{8.58}$$

Case (a): In a filling process, the rate of increase of mass in the tank $dm/d\tau$, is positive. We therefore find that the input heat-transfer rate is also positive, as is to be expected; its magnitude can be evaluated from eq. (8.58). The heat-transfer rate evaluated in this way is indeed *correct*, if the small pv term is neglected. Perceiving this, the student may gain confidence in this formulation, eq. (8.56), of the First Law. A surprise awaits him however.

Case (b): Consider now the drawing off of water from the tank, replacement by fresh water being provided. In this case $dm/d\tau$ is *negative*. Since $(t_1 - t_0)$ is still positive, the evaluation of $\dot{Q}$ from eq. (8.58) shows that this quantity is also negative, i.e., drawing off water from the tank apparently causes heat transfer *from* the tank *to* the heater element. This definitely does not correspond to reality!

The absurd result derives from the incorrect formulation of the First Law given in eq. (8.56).

Correct method. The problem may be correctly solved by either the system or the control-volume analysis. In the former case, it is necessary to draw a system boundary, which of course moves during filling and emptying and to apply the First Law directly to the material within this boundary. In the latter case, a suitable fixed control surface is chosen, for example surrounding the tank, and eq. (8.54) is applied; watch has to be kept on the boundaries of this control volume in order to note what material crosses it, a precaution omitted in writing eq. (8.56).

Analysis of the problem by these two correct methods is left as an exercise for the reader.

Final remarks on control volume and system

In conclusion it must be emphasized that the *explicit* use of the system and control volume is among the most important techniques, not only of thermodynamics, but of engineering analysis generally. The extent to which they assist in clarifying thought and separating the relevant from the irrelevant can be fully appreciated only by those who were taught thermodynamics without ever being enjoined to relate the fluxes across and the interactions at a boundary to the changes within.

Two remarks will indicate the wide relevance of the concepts:

(i) In *structural analysis*, the "method of sections" is nothing but the drawing of a system boundary around a part of the structure and requiring that the forces at its boundaries should be in equilibrium with the gravitational forces on the material within the boundary.

(ii) In *economics*, a control-volume boundary drawn around a country permits the rate of internal consumption of goods to be related to their rate of production in the country's factories and to their transport across the frontier. Subsidiary boundaries enclosing various classes of the community help in the analysis of the internal transfer of goods. Explicit use of the boundary concept can go far to demolish unsound economic (and thermodynamic) arguments.

CHAPTER 8—PROBLEMS

Note. Unless otherwise advised, assume changes in elevation to be negligible and the local gravitational acceleration to be 9.81 m/s.

8.1 Oil flows into a cooler at a temperature of 80 °C at the rate of 0.08 kg/s and flows out at a temperature of 40 °C. Cooling water enters the cooler at 15 °C and leaves at 25 °C. The cooler is adiabatic and changes in kinetic energy of the oil and of the water are negligible. For both the oil and the water the enthalpy, h, is related to the temperature, t, by the equation: $h = c(t - t_0)$; for water, $c = 4.18 \times 10^3$ J/kg K, while for oil, $c = 1.88 \times 10^3$ J/kg K; t_0 is a constant.

Calculate the mass flow rate of the cooling water.

8.2 A steady flow of water at a temperature of 59 °C and a pressure of 300×10^3 N/m^2 ($h = 246.9 \times 10^3$ J/kg) enters a section of the heating plant of a building in which there are no pumps. The water leaves the section at a temperature of 47 °C and at a pressure of 280×10^3 N/m^2 ($h = 196.8 \times 10^3$ J/kg). The exit pipe is 30.5 m above the entry pipe. Evaluate the heat transfer from the water, per kilogram of water flowing, assuming changes in kinetic energy to be negligible.

8.3 The gas leaving the turbine of an aircraft jet engine flows steadily into the engine jet-pipe at a temperature of 900 °C, a pressure of 192×10^3 N/m^2 and a velocity of 300 m/s relative to the pipe. The gas leaves the jet-pipe at a temperature of 820 °C and a pressure of 105×10^3 N/m^2. Heat transfer from the gas is negligible. Using the following data, evaluate the velocity of the gas leaving the jet-pipe relative to the engine.

Data: The enthalpy of the gas depends on its temperature only. Extract from the tables of properties of the gas:

t, °C	820	900
h, J/kg	$862 \cdot 6 \times 10^3$	$951 \cdot 7 \times 10^3$

8.4 A gas flows steadily through a rotary compressor. The gas enters the compressor at a temperature of 16 °C, a pressure of 100×10^3 N/m² and an enthalpy of $391 \cdot 2 \times 10^3$ J/kg. The gas leaves the compressor at a temperature of 245 °C, a pressure of 600×10^3 N/m² and an enthalpy of $534 \cdot 5 \times 10^3$ J/kg. There is no net heat transfer to or from the gas as it flows through the compressor.

(*a*) Evaluate the external work done per unit mass of gas assuming the gas velocities at entry and exit to be negligible.

(*b*) Evaluate the external work done per unit mass of gas when the gas velocity at entry is 80 m/s and that at exit is 160 m/s.

8.5 Steam flows steadily into a condenser at the rate of 1·2 kg/s. The enthalpy of the steam at entry is $2 \cdot 3 \times 10^6$ J/kg and its specific volume is 18·5 m³/kg. The condensed steam ('condensate') has an enthalpy of 190×10^3 J/kg and leaves with negligible velocity. The heat transfer from the condensing steam to the atmosphere is 70×10^3 J/s.

(*a*) Given that the flow area at entry is 0·2 m² evaluate the steam velocity at entry.

(*b*) Evaluate the heat transfer to the cooling water per unit mass of steam condensed.

8.6 A valve is fitted in a 60-mm diameter horizontal pipe-line; the valve and the pipe-line are well insulated. The valve, which is partially open, throttles the steam flowing steadily along the pipe-line from a pressure of 2×10^6 N/m² to a pressure of 200×10^3 N/m². The enthalpy of the steam approaching the valve is $2 \cdot 77 \times 10^6$ J/kg and the steam mass flow rate is 0·03 kg/s.

(*a*) Assuming the change in the kinetic energy of the steam to be negligible, evaluate the enthalpy of the steam downstream of the valve.

(*b*) Verify that the kinetic energies are negligible by evaluating the steam velocity upstream and downstream of the valve. The specific volumes of the steam upstream and downstream of the valve are 0·0980 m³/kg and 0·9602 m³/kg respectively; these data are obtained from tables of properties of steam and correspond to the pipe-line states given above.

8.7 An air turbine forms part of an aircraft refrigerating plant. Air at a pressure of 295×10^3 N/m² and a temperature of 58 °C flows steadily into the turbine with a velocity of 45 m/s. The air leaves the turbine at a pressure of 115×10^3 N/m², a temperature of 2 °C and a velocity of 150 m/s. The shaft work delivered by the turbine is 54×10^3 N m/kg of air. Neglecting changes in elevation, determine the magnitude and sign of the heat transfer per unit mass of air flowing.

For air take $c_p = 1005$ J/kg K and assume that the enthalpy of air is a function of temperature only (see problem **7.7**).

8.8 A long, well-insulated pipe-line consists of two pipes connected in series, the internal diameters of which are 90 mm and 30 mm respectively. A steady flow of steam enters the 90-mm diameter pipe at a pressure of 350×10^3 N/m², a specific volume of 0·684 m³/kg, and an enthalpy of $2 \cdot 98 \times 10^6$ J/kg. At a point downstream in the 30-mm diameter pipe, the pressure is 300×10^3 N/m², the specific volume is 0·790 m³/kg and the enthalpy is $2 \cdot 968 \times 10^6$ J/kg.

Determine the velocity of the steam, at the two points in the pipe-line, and the mass flow rate of the steam.

8.9 A simple impulse turbine has a single nozzle and one row of rotating blades.

(*a*) Steam flows steadily into the nozzle with negligible velocity. Given that the decrease in the enthalpy of the steam in flowing through the nozzle is 120×10^3 J/kg and that the flow is adiabatic, evaluate the velocity of the steam leaving the nozzle.

(b) The nozzle is set at an angle of 20 degrees to the plane of rotation; the outlet angle of the rotor blades is 35 degrees measured from the plane of rotation; the blade speed is 221 m/s. Assuming ideal conditions across the rotor blades, i.e. no change in the steam speed relative to the rotor blades, construct the velocity triangles and hence evaluate the absolute velocity of the steam leaving the rotor blades.

(c) Apply the S.F.E.E. to a control volume rotating with the rotor blades, to find the increase in the enthalpy of the steam across the rotor blades. Assume the flow to be adiabatic.

(d) Apply the S.F.E.E. to a control surface enclosing the whole machine and fixed with respect to the nozzles, to evaluate the external work delivered by the turbine.

(e) Given that the mass flow rate of the steam is 0·04 kg/s, evaluate the power developed.

(f) Evaluate the stagnation enthalpies at each condition with respect to a stationary observer, given that the initial specific enthalpy of the steam is $2·6 \times 10^6$ J/kg.

8.10 A jet condenser consists of a vessel into which a steady stream of steam is discharged. A spray of water, injected at the top of the vessel, mixes with the steam and subsequently flows out with the condensed steam at the base of the vessel. The vessel is well lagged.

In a particular installation, steam enters the condenser with an enthalpy of $2·300 \times 10^6$ J/kg at the rate of 0·06 kg/s. The mixed stream of water leaves the condenser with an enthalpy of 340×10^3 J/kg.

Given that the enthalpy of the injected water is 70×10^3 J/kg and that changes in kinetic energy and in elevation are negligible, determine the mass flow rate of the injected water.

8.11 The steam supply to an engine comprises two streams which mix before entering the engine. One stream is supplied at the rate of 0·01 kg/s with an enthalpy of $2·952 \times 10^6$ J/kg and a velocity of 20 m/s. The other stream is supplied at the rate of 0·1 kg/s with an enthalpy of $2·569 \times 10^6$ J/kg and a velocity of 120 m/s. At exit from the engine the fluid leaves as two streams, one of water at the rate of 0·001 kg/s with an enthalpy of 420×10^3 J/kg and the other of steam; the fluid velocities at exit are negligible. The engine develops a shaft power of 25×10^3 W; the heat transfer is negligible. Evaluate the enthalpy of the second exit stream.

8.12 A steady stream of air is supplied to the combustion chamber of a gas-turbine engine at the rate of 15 kg/s. The temperature of the air is 190 °C, its velocity is 100 m/s and its enthalpy is $176·2 \times 10^3$ J/kg. Liquid fuel at a temperature of 15 °C flows into the combustion chamber with negligible velocity at the rate of 0·22 kg/s. The products of combustion leave the chamber at a temperature of 750 °C, a velocity of 200 m/s and an enthalpy of $787·5 \times 10^3$ J/kg. Heat transfer to the atmosphere is negligible. Evaluate the specific enthalpy of the entering fuel stream. The datum temperature for the enthalpy of the reactants and products is to be taken as 15 °C.

8.13 A gasoline engine has a specific fuel consumption of 0·3 kg/kWh. The stream of air and gasoline vapour, in the ratio 14 : 1 by mass, enters the engine at a temperature of 40 °C and leaves as combustion products at a temperature of 790 °C. The net heat-transfer rate from the fuel-air stream to the jacket cooling water and to the surroundings is 35×10^3 J/s. The shaft power delivered by the engine is 26 kW, Evaluate the increase in the specific enthalpy of the fuel-air stream assuming the changes in kinetic energy and in elevation to be negligible.

9 Properties of Pure Substances

Introduction

The First Law of Thermodynamics, either in its original or its steady-flow form, suffices for the solution of a great many practical problems in engineering. It involves, it will be recognized, the relation of the interactions heat and work, *at* the system or control-volume boundary, to the changes in energy or enthalpy occurring *within* the system or control volume.

Now energy and enthalpy are not amenable to direct measurement in the same way as are, for example, pressure and temperature. In order to make the First Law applicable to practical apparatus therefore, it is necessary to see how energy and enthalpy are related to more easily measurable properties. Provided that pure substances are in question, we are assured by the Two-Property Rule that relatively simple relations exist. Finding the relations however is an experimental matter which will now be discussed.

The present chapter starts with a description of experiments on a pure substance and of the results that are typically obtained. This leads to a more detailed discussion of property diagrams than was possible in Chapter 7, and to such concepts as *phase, saturation*, and *intensive* and *extensive property*. Thereafter Steam Tables are discussed.

Since experiments take time, knowledge of the thermodynamic properties of substances has always lagged behind engineering need. For example, it was not until 1760 that Black discovered the "latent heat" of vaporization of water; yet steam engines had been operating since the beginning of the century. Even now, knowledge of the properties of steam is insufficient to enable some advanced projects to be designed with certainty; experimental research on steam properties is being actively pursued. Knowledge of other substances is in a far less complete state.

Symbols

H	Enthalpy of a system comprising a pure substance.	U	Internal energy of a system comprising a pure substance.
h	Specific enthalpy of a pure substance.	u	Specific internal energy of a pure substance.
m	Mass.	V	System volume.
p	Pressure.	v	Specific volume.
Q	Heat transfer.	W_x	External work.
t	Temperature.	x	Dryness, dryness fraction.

τ Time.	fg Saturated liquid to saturated vapour.
Subscripts.	sf Saturated solid to saturated liquid.
f Saturated liquid (fluid).	
g Saturated vapour (gas).	sg Saturated solid to saturated vapour.
s Saturated solid.	

Facts About Pure Substances

Constant-pressure experiments: temperature∽pressure *(t∽p)* diagram

Consider the apparatus shown in Fig. 9.1, comprising a cylinder mounted

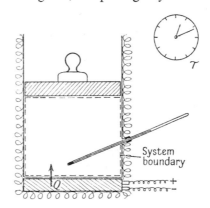

Fig. 9.1 Heat transfer at constant pressure.

System boundary

on an electrically heated hot-plate, a frictionless leak-proof piston surmounted by a weight, and a thermometer for measuring the temperature of the contents of the cylinder. The cylinder contains a pure substance; for definiteness this may be thought of as water,* though of course any substance can be chosen.

The experiment to be carried out is as follows: with a fixed weight on the piston, and so a fixed pressure beneath it, the heater is controlled in such a way that heat transfer to the cylinder contents occurs at a steady rate. Initially the system temperature is low, but it rises as a result of the heat transfer. The temperature is observed and plotted as a function of time.

The temperature∽time curves. Fig. 9.2 shows a series of temperature∽time curves obtained in this way. Their relative horizontal positions are without significance; the setting of the clock at the beginning of an experiment is, of course, arbitrary. The only differences between the conditions of the experiments result from the size of the weight on the piston, i.e. from the system pressure: the pressure is lowest for curve (a) on the left, and increases steadily for curves (b) through to (f).

* We use the word "water" to denote the chemical substance H_2O, regardless of whether it is in the form of ice, liquid water or steam.

The notable feature of all the curves is that they show horizontal steps, signifying that the temperature suddenly stops rising for a period and then picks up again. All the curves exhibit steps, but those at intermediate pressures (b, c and d) have two each, whereas the low-pressure (a) and high-pressure (e and f) curves have only one each.

If the temperatures at which the steps occur are noted and plotted against the appropriate pressures, the points lie on curves such as are shown in Fig. 9.3. There is a steeply sloping curve terminating suddenly at the point marked CP; another curve starts from the point marked TP on the first

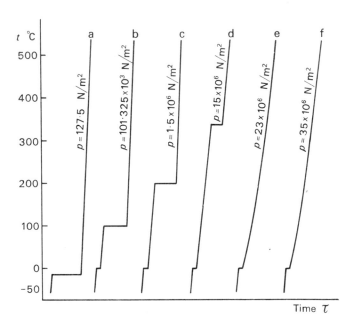

Fig. 9.2 Temperature~time curves for H_2O at constant pressure
(approximately to scale).

line and travels indefinitely to the right in a more or less horizontal direction (sloping downwards almost imperceptibly in the case of water). Comparison of Figs. 9.2 and 9.3 shows that experiment (a) must have been carried out at a pressure less than that of TP, while experiments (e) and (f) must have been at pressures higher than that of CP, in order that the number of temperature steps should be correct.

The three phases: solid, liquid, vapour. If a window in the cylinder wall permits observation of the cylinder contents, it is found that, during the time in which the system is traversing a temperature step, its appearance changes. The nature of the changes is best described by the statements that: in experiment (a), the substance changes gradually from a solid to a vapour state during the step; in experiments (b), (c) and (d), the lower step corresponds to a change from solid to liquid while the higher corresponds to a change from liquid to vapour; and in experiments (e) and (f) the step corresponds to a change from solid to liquid, but no subsequent vaporization occurs however high the temperature is raised.

The three conditions of the system, solid, liquid and vapour (or gas), are known as *phases*; we have the following formal, though scarcely informative, definition: **a phase is any physically homogeneous aspect of a system.**

The transitions from one phase to another, namely melting and freezing, vaporization and condensation, are known as *changes of phase*. They

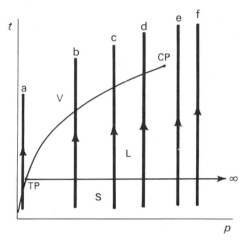

Fig. 9.3 Temperature~pressure diagram showing the phase boundaries.

correspond to changes in the types of force dominant between the individual molecules. The phase of the substance at any temperature and pressure is indicated on Fig. 9.3 by the letters S (for solid), L (for liquid) and V (for vapour) marking the various regions. The phase changes occur when the state-point crosses the lines on Fig. 9.3 mentioned above, which are therefore known as *phase boundaries*.

More careful experiments will indicate that, in general, small temperature steps occur along several lines within the general region marked S. This means that there are several solid phases, corresponding to the various crystalline states of matter. In engineering thermodynamics the solid phase is of relatively small importance, for the good reason that engines will not work if the fluid within them freezes. We shall therefore not distinguish between the various solid phases. The distinctions are however very important in the subject of *metallurgy*.

The solid-vapour phase change. Examination of Fig. 9.3 shows that, to the left of TP, the solid and vapour regions are contiguous: experiment (a) exhibits a change direct from the solid to the vapour phase. Such a change is known as *sublimation*.

Points on the line separating the solid and vapour regions correspond to the steps that would occur in the $t{\sim}\tau$ curves of Fig. 9.2 in a series of experiments such as (a). The line to the left of TP therefore shows how the sublimation temperature varies with pressure; it is known as the *solid-vapour saturation line*.

An example of sublimation is the gradual "disappearance" of fallen snow in prolonged dry, cold weather, without intermediate melting; this occurs because, even though the pressure of the atmosphere exceeds that of TP for water,* the *partial* pressure of the vapour, which is what matters in this case, is below that of TP. (See Chapter 15 for explanation of the term "partial pressure".) Another example is the sublimation of solid carbon dioxide ("dry ice") at atmospheric pressure.

An example of the reverse of sublimation is the formation of hoar frost, which is the transition direct from water vapour to ice without intermediate condensation and freezing.

The solid-liquid phase change. In Fig. 9.3 the line to the right of TP, which separates the solid and liquid regions, is the join of the points at which the lower steps occur in the $t{\sim}\tau$ curves (b), (c), (d), (e) and (f) of Fig. 9.2. It shows how the melting (or freezing) temperature varies with pressure; it is known as the *solid-liquid saturation line*.

The liquid-vapour phase change. The line joining TP and CP, in Fig. 9.3, separates the liquid and vapour regions; points on it correspond to the upper steps in the $t{\sim}\tau$ curves (b), (c) and (d) of Fig. 9.2 The line therefore shows the variation of boiling temperature with pressure and is known as the *liquid-vapour saturation line*. The reverse of boiling is known as *condensation*.

The triple point, TP. The point marked TP on Fig. 9.3 is known as the *triple point*. It corresponds to the only combination of pressure and temperature at which it is possible for the solid, liquid and vapour phases to exist side-by-side in equilibrium. For water, the triple point occurs at a temperature of 0·01 °C and a pressure of 611·2 N/m^2.

At the triple-point pressure, the freezing point is 0·01 K above the freezing point at atmospheric pressure. This is a consequence of the slight downward slope of the solid-liquid phase boundary of water as it travels to the right, a tendency which is also responsible for making ice-skating possible; for an increase of pressure (beneath the blades of the skates) at constant temperature causes a transition from the solid to the liquid phase, so that the water thus formed lubricates the motion.

---

* See footnote on page 171.

The critical point, CP. The upper termination of the liquid-vapour phase boundary, CP, is known as the *critical point*. For water, the relevant temperature and pressure are respectively 374·15 °C and 22·12 × 10⁶ N/m². The corresponding temperatures and pressures for substances which are vapours at atmospheric conditions are considerably lower, as is shown by Table 9.1.

TABLE 9.1 *Critical-Point Data*

Substance	Critical temperature	Critical pressure
Water	374·15 °C	$22·12 \times 10^6$ N/m²
Carbon dioxide	31·05 °C	$7·38 \times 10^6$ N/m²
Oxygen	−118·8 °C	$5·04 \times 10^6$ N/m²
Hydrogen	−239·8 °C	$1·30 \times 10^6$ N/m²

A consequence of the existence of the critical point which often puzzles students is that, at higher pressures and temperatures, *there is no demarcation between the liquid and the vapour states*. This means that it is possible, for example, to vaporize a substance repeatedly without ever condensing it. The

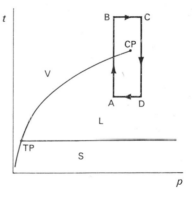

Fig. 9.4a Temperature∼pressure diagram showing a cyclic process which involves states above and below the critical point.

cyclic process ABDCA shown in Fig. 9.4a indicates how this may be done: Starting at state A, where the substance is a liquid, constant-pressure heating is carried out. The crossing of the liquid-vapour phase boundary is marked by all the features which we associate with boiling, e.g. bubbling and agitation. Let the heating be continued until the temperature of the substance, which is now entirely vapour, has risen above the critical temperature; the state-point is now that marked B on Fig. 9.4a. Let the system now be compressed at constant temperature to a pressure in excess of the critical pressure; the state-point shifts from B to C; no phase boundary is crossed and no change takes place in the appearance of the substance. If the system is now cooled to the original temperature at constant pressure, the state-point moves down the vertical CD; again no change takes place in the appearance of the substance, so the observer concludes that it is still vapour. To complete the cycle, the substance is expanded at constant temperature until the state A is reached once more at the original (lower) pressure; there has still been no change in the appearance. Is the substance still a vapour? The answer seems to be "yes"; but if we now repeat the constant-pressure heating process AB, the substance boils again! So in the process BCDA the substance has changed from vapour back to liquid without our noticing. Of course it is equally possible to carry out the reverse cyclic process ADCBA, and so to get repeated condensation without intermediate vaporization.

The apparent paradox results from the unwarranted expectation that hard-and-fast distinctions between phases must always exist. The following

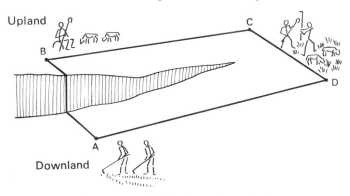

Fig. 9.4b Topographical analogy of conditions close
to the critical point.

analogy may prove helpful in understanding this: Fig. 9.4b illustrates an extensive escarpment which forms a boundary between a plateau called Upland and a low-lying country called Downland. Tribe A, living at the foot of the cliff, have no frontier disputes with tribe B, living on the plateau; for the cliff forms an obvious boundary and peace is ensured by the formula that, "Up is up, and down is down". In the course of time however tribesmen from B form a settlement at C in a region where the escarpment has tapered away and disappeared, while tribesmen from A form a nearby settlement at D. The consequences of continued attempts to adhere to the "Up is up" formula can readily be imagined.

It may be asked whether the solid-liquid phase boundary also disappears at sufficiently high pressure. The answer is: "not as far as is known". For water, this phase boundary is known to exist up to pressures of 4.4×10 N/m^2.

The enthalpy~pressure (h~p) diagram

A main purpose of the present chapter is to show how the "First-Law" properties of pure substances, viz. internal energy and enthalpy, are related to the easily measurable properties: pressure, temperature and specific volume. Enthalpy changes can be determined from the constant-pressure experiments already described, if measurement is made of the heat transfer to the system; this may be done by measuring the electrical work to the heater and allowing for the heat transfer to the cylinder walls and piston.

Since, in the processes under consideration, we have a pure substance in the absence of gravity, motion, etc., and since friction and stirring work are absent, the First Law, eq. (7.5), written for unit mass is

$$\mathrm{d}Q = \mathrm{d}u + p\,\mathrm{d}v \qquad (9.1)$$

where $\mathrm{d}Q$ is an elementary heat transfer, $\mathrm{d}u$ is the elemental change in the internal energy and $p\,\mathrm{d}v$ is the displacement work done in the fully-resisted expansion.

Eq. (9.1) may be written in terms of enthalpy, h as follows:

By definition, $\qquad\qquad\qquad h = u + pv \qquad\qquad\qquad (9.2)$

Differentiating eq. (9.2), we have

$$\mathrm{d}h = \mathrm{d}u + p\,\mathrm{d}v + v\,\mathrm{d}p$$

Eq. (9.1) then becomes

$$\mathrm{d}Q = \mathrm{d}h - v\,\mathrm{d}p \qquad (9.3)$$

Now pressure changes are absent, so $v\,\mathrm{d}p$ equals 0 in eq. (9.3), which then runs

$$\mathrm{d}Q = \mathrm{d}h$$

Thus the heat transfer equals the change in enthalpy of the substance in the constant-pressure process; so an enthalpy-pressure diagram can be constructed from the experimental observations.

Strictly, in order to construct the $h{\sim}p$ diagram, it is also necessary to calculate enthalpy changes during the shifts from one pressure to the next.

Phase boundaries. The results of such experiments for a typical pure substance are indicated in Fig. 9.5. It is evident that the $h{\sim}p$ diagram is more complicated than the $p{\sim}t$ diagram. This is a consequence of the fact that, whereas the system temperature does not vary during the change of phase, the enthalpy does (the heat transfer is continuous); the result is that now the phase boundaries have been broadened into bands of finite vertical height. These bands are indicated on Fig. 9.5 by the letters V + S, L + S, and L + V, signifying the two phases which are present in the corresponding regions. The regions of pure-solid, pure-liquid and pure-vapour phase are marked S, L and V, as before.

The triple and critical points. The triple point, like all other points (but one) on the $p{\sim}t$ phase boundaries, extends vertically when represented on the $h{\sim}p$ diagram, so that now it becomes a line. This is the boundary between the S + V, S + L and L + V regions of Fig. 9.5, these letters emphasizing that it is in the nature of the triple point to have all three phases in contact.

The critical point is the exception to the rule that phase-boundary points become extended vertically; or rather its extension is zero. The critical point is on the extreme right of the L + V envelope of Fig. 9.5; to the right

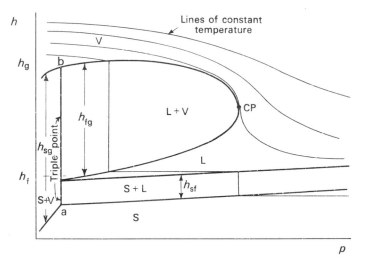

Fig. 9.5 Enthalpy~pressure diagram showing the
phase boundaries and isotherms.

of it there is no such envelope. It corresponds to the point in Fig. 9.4b at which the escarpment vanishes.

The "latent heats". The vertical height of a two-phase region (S + V, L + V or S + L) represents the enthalpy increase of the fluid when it changes from one phase to the other. This enthalpy increase is known as the *latent heat of sublimation, vaporization or melting* (fusion) as the case may be. The name "latent heat" is of course a legacy of the caloric theory which led to the idea that the "heat" was "hidden", and failed to cause the expected temperature rise. Somewhat better names for these "latent heats" are *enthalpy of sublimation, vaporization* and *melting* respectively. Still more precise, but more clumsy, would be to refer to the *increases* of enthalpy of sublimation, vaporization, and melting.

Taking the liquid-vapour transition as an example, we see from Fig. 9.5 that the latent heat of vaporization varies with pressure, decreasing rapidly to zero as the critical pressure is approached. This latter feature is an expression of the fact that the difference in enthalpy between the liquid and vapour states becomes smaller, the nearer is the critical state; it corresponds to the fact that the escarpment height must decrease gradually before it disappears altogether.

The latent heats, being enthalpy changes, are given the symbol h with the subscripts sg, sf or fg, where s stands for solid, f stands for liquid (i.e. fluid; the letter l might be mistaken for the numeral "one"), and g stands for vapour (or gas). Thus the enthalpy of vaporization has the symbol h_{fg}.

It may be seen from the geometry of Fig. 9.5 that, at the triple point conditions, $h_{sg} = h_{sf} + h_{fg}$.

Saturated liquid and vapour. Liquid at any state lying on the lower part of the L + V envelope is called *saturated liquid*. The name arises because liquid at a lower enthalpy than that of saturated liquid at the same pressure could "soak up" steam if it were added; the steam would condense. Saturated liquid, on the other hand, does not cause steam of the same temperature to condense; the two phases can co-exist in equilibrium. The enthalpy of saturated liquid is designated h_f.

Vapour at any state lying on the upper part of the L + V envelope is correspondingly called *saturated vapour*; for whereas vapour lying above this line would cause liquid added to it to vaporize, saturated vapour can "soak up" no more, but co-exists in equilibrium with the liquid. Sometimes saturated vapour is termed *dry* saturated vapour to emphasize that it carries no suspended liquid particles. The symbol h_g is used to denote the enthalpy of (dry) saturated vapour.

The upper boundary of the L + V envelope between the triple point and CP is called the *saturated-vapour line*; the lower boundary between these points is known as the *saturated-liquid line*.

Isotherms. A few lines corresponding to states of constant temperature are drawn on Fig. 9.5. Several features may be remarked:

(i) The isotherms are vertical (i.e. coincident with constant-pressure lines) in the two-phase regions S + V, S + L and L + V, signifying of course that no temperature change occurs during a constant-pressure phase change.

(ii) Outside the two-phase regions, the isotherms have a generally horizontal tendency except in the neighbourhood of the critical point. The isotherm through the critical point actually has a vertical tangent there.

(iii) The higher temperatures correspond to the higher enthalpies. Particularly at high temperatures it is notable that the isotherms become closely horizontal, signifying that the enthalpy at a given temperature is only weakly dependent on pressure. This will assume considerable significance in discussion of Ideal Gases (Chapter 14). The isotherms are also practically horizontal in the pure-liquid and pure-solid regions.

Superheat. In power-plant practice it is customary to call vapour at states lying above the saturated-vapour line: *superheated vapour.* This has arisen because for a long time boilers produced steam which, being in contact with water at the same pressure, was saturated. Later (1827) the *superheater* was devised; this caused the steam to be heated to a higher temperature, out of contact with the water. Sometimes the difference of temperature between the superheated vapour and saturated vapour at the same pressure is referred to as *the superheat* or *degrees superheat* of the vapour.

It may be as well to remark that "super-cooling" does not have the corresponding meaning. Instead, "super-cooled" vapour means vapour which has an enthalpy sufficiently low that, at the prevailing pressure, some of it ought to condense; yet it does not condense because of the absence of nuclei around which the droplets can form. Such states cannot be represented on property diagrams, for the latter describe only equilibrium states; super-cooled vapour is not in equilibrium, for it has a tendency to spontaneous change. This phenomenon is of some importance in the design of steam nozzles.

The pressure~volume ($p{\sim}v$) diagram

Like enthalpy, internal energy also has to be related to the other properties of the substance. One way of doing this is to make volume (piston-position) measurements during the course of the constant-pressure experiments described above. Then internal energy is deducible from the definitional relation, eq. (7.7), expressed as

$$u = h - pv \qquad (9.4)$$

We first consider the way in which the specific volume depends on other properties.

Diagram for a substance which contracts on freezing. If volume measurements are made during the constant-pressure experiments, a pressure~volume diagram can be plotted. Fig. 9.6 shows an example. In general shape it is similar to the $h{\sim}p$ diagram with the ordinate and abscissa interchanged it is conventional for p to be the ordinate in the present case. The single-phase and two-phase regions are indicated by letters as before. A constant pressure process is now represented by a horizontal line; so the isotherms are horizontal in the two-phase regions. Far from the critical point however, in contrast to their behaviour on the $h{\sim}p$ diagram, the isotherms tend to become rectangular hyperbolae with the p and v axes as asymptotes. This will also assume particular significance in discussion of the Ideal Gas (Chapter 14).

The triple point is again a line, and the critical point occurs at the apex of the L + V bulge, this time on the top instead of at the side.

In conformity with the symbols used earlier, v_f and v_g respectively denote the specific volumes of saturated liquid and of saturated vapour; and v_{fg} is defined to equal the difference ($v_g - v_f$).

Diagram for a substance which expands on freezing. Inspection of a horizontal line from left to right on Fig. 9.6 indicates that, when the system undergoes a constant-pressure heating process, the volume increases at each phase change. This is not always the case however. Water, for

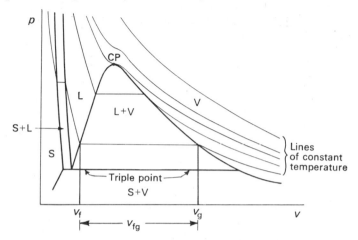

Fig. 9.6 Pressure~volume diagram for a substance which contracts on freezing.

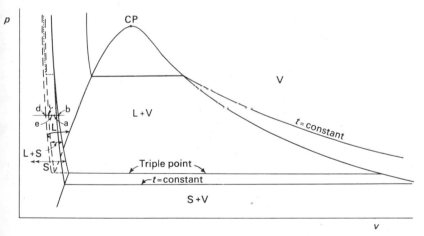

Fig. 9.7 Pressure~volume diagram for a substance which expands on freezing.

xample, *contracts* as it changes, at constant pressure, from the solid to he liquid state; indeed it continues to contract until (at atmospheric ressure) the temperature has risen to 4 °C.

Such behaviour makes the $p{\sim}v$ diagram considerably more complicated 1 the S + L region, for the solid- and liquid-phase regions overlap. Fig. .7 illustrates schematically the $p{\sim}v$ diagram for a substance like water; it is not to scale, for the reason that the specific volumes of the solid and the liquid are very much smaller than those of the vapour at a given pressure. The dotted line on the extreme left indicates that the liquid volume can be less than that of the liquid in equilibrium with the solid phase.

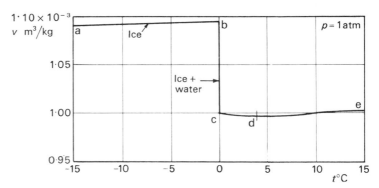

Fig. 9.8 Volume~temperature diagram for water showing
that water expands on freezing (to scale).

Fig. 9.8 shows, to scale, the specific volume of water plotted against temperature at a pressure of 1 atm. Corresponding points a,b,c,d,e are shown on Figs. 9.7 and 9.8.

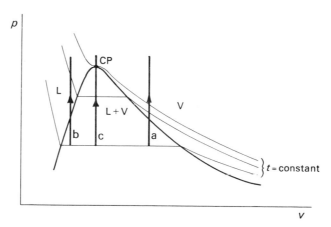

Fig. 9.9 Pressure~volume diagram showing constant-volume processes.

Constant-volume processes. A constant-volume process is represented by a vertical straight line on a $p{\sim}v$ diagram. Fig. 9.9 shows three such processes, a, b and c. In case a, we suppose that a mixture of liquid and vapour in a closed container (Fig. 9.10) is gradually heated. The initial ratio of liquid to vapour is relatively small. As the temperature rises (and the pressure with it), the state-point approaches and then crosses the saturated-vapour line on Fig. 9.9, signifying that the meniscus separating liquid and vapour descends to the bottom of the vessel. Finally only vapour is present; the liquid has boiled away.

In case b the initial ratio of liquid to vapour is large. This time, the heating process causes the state-point to approach and cross the *saturated-liquid* line, signifying that the meniscus rises to the top. It follows that finally only liquid is present; the heating has caused all the vapour to *condense*. This surprising result may be interpreted thus: the rise of pressure consequent on heating at constant volume outweighs, in its effect on the vapour condition, the rise of temperature.

In case c, the initial ratio of liquid to vapour has been so chosen that the system, when heated, passes through the critical point. In this case the

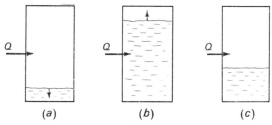

Fig. 9.10 Showing the direction of motion of the meniscus when different liquid and vapour mixtures are heated at constant volume.

meniscus neither falls to the bottom nor rises to the top but merely fades away, disappearing completely as the critical pressure is reached. The meniscus becomes hard to see even before this, because all the properties of the saturated liquid and saturated vapour approach each other; in particular the refractive-index difference, which alone enables the separation surface to be observed visually, becomes steadily smaller.

The internal-energy∼volume (u∼v) diagram

Combination of information about enthalpy and specific volume, by way of eq. (9.4), enables diagrams to be constructed with internal energy as one co-ordinate. The $u{\sim}p$ diagram is similar to that for enthalpy plotted against pressure. As our last example, we therefore show the $u{\sim}v$ diagram instead (Fig. 9.11). This exhibits some new features.

Fig. 9.11 holds for a substance which contracts on freezing (i.e. not water). The most notable feature is that this time the triple point has expanded to an *area* (marked S + L + V), the reason being that both u and v increase during a phase change. Constant-pressure and constant-temperature curves are indicated on the diagram; in the regions where two phases are present these curves become coincident straight-lines.

The whole diagram exhibits a "sloping" appearance, when compared for example with the $p{\sim}v$ diagram. In particular, the critical point is situated on the flank of the L + V envelope.

The isotherms tend to be horizontal in the vapour region far from the critical point; this tendency signifies that the internal energy tends to be a function of the temperature alone, a feature of relevance to the behaviour of Ideal Gases (Chapter 14).

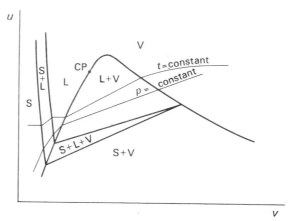

Fig. 9.11 Internal-energy~specific volume diagram for a substance which contracts on freezing.

Two sorts of properties: intensive and extensive

It will be recalled that the $p{\sim}t$ diagram was rather simple; when temperature was replaced by specific enthalpy or specific volume, more complex diagrams appeared; and substitution of u for p brought yet more complexity in the triple-point region. This arose because pressure and temperature are properties of a different sort from volume, enthalpy and internal energy.

The difference lies in the fact that the volume V, the enthalpy H and the internal energy U, of a system of fixed and uniform pressure, temperature, phase and composition, are *proportional to the mass* of the system; by contrast, the pressure and temperature are independent of the mass. The difference may be illustrated by considering a pure-substance system subdivided into two portions of equal mass. The volume, the enthalpy and the internal energy of each portion have values equal to *one half* of those of the corresponding quantities for the whole system; the pressure and temperature, however, have the *same* values in each portion of the system. Another way of putting it is that volume, enthalpy and internal energy are additive; pressure and temperature are not.

The difference is made explicit by calling p and t *intensive properties*, and V, H and U *extensive properties*. Other examples of intensive properties are viscosity, thermal conductivity and refractive index; other extensive properties are concentration (of some component of a mixture) and entropy (Chapter 13).

The additive nature of extensive properties makes it possible to write their values for a large system, through which the properties are not homogeneous, in the form of integrals. Thus

$$H = \int h \, dm \tag{9.5}$$

$$U = \int u \, dm \tag{9.6}$$

$$V = \int v \, dm \tag{9.7}$$

where H, U and V are respectively enthalpy, internal energy and volume of the whole system; h, u and v are the specific values of these properties for an element dm of the mass of the system; and the integral signs signify summation over all the mass elements.

Dryness

A particular example of this sort of addition is useful in relating the properties of a mixture of two phases to the properties of the pure saturated phases at the same pressure; such a mixture of liquid and vapour phases is often called a *wet* mixture. To describe this we introduce first the concept of *dryness*, or *dryness fraction*, as a measure of the relative proportions of vapour and liquid phases in a mixture.

Definition. The dryness of an equilibrium liquid-vapour mixture, x, is the mass fraction of vapour in the mixture.

It follows that the vapour-to-liquid mass ratio is given by

$$\frac{\text{mass of pure vapour phase}}{\text{mass of pure liquid phase}} = \frac{x}{1-x} \tag{9.8}$$

Saturated vapour correspondingly has a dryness of unity, or 100%; saturated liquid has a dryness of zero.

Extensive properties in the liquid-plus-vapour region. We can now write relations between h, u and v for a mixture of liquid and vapour, and h_g, u_g, v_g, h_f, u_f and v_f, the properties of the saturated phases. These are

$$h = xh_g + (1-x)h_f \tag{9.9}$$

$$u = xu_g + (1-x)u_f \tag{9.10}$$

$$v = xv_g + (1-x)v_f \tag{9.11}$$

It is sometimes more convenient to write these expressions in a way that brings the dryness in only once, namely

$$h = h_f + xh_{fg} = h_g - (1-x)h_{fg} \tag{9.12}$$

$$u = u_f + xu_{fg} = u_g - (1-x)u_{fg} \tag{9.13}$$

$$v = v_f + xv_{fg} = v_g - (1-x)v_{fg} \tag{9.14}$$

which is permissible since we have the definitions

$$h_{fg} \equiv h_g - h_f \tag{9.15}$$

$$u_{fg} \equiv u_g - u_f \tag{9.16}$$

$$v_{fg} \equiv v_g - v_f \tag{9.17}$$

For substances such as water, at pressures far below the critical, the specific-volume equations may often be simplified to

$$v \approx x v_g \tag{9.18}$$

because v_f is so much less than v_g as to be negligible. This is not of course permissible when x is very small.

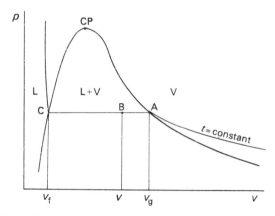

Fig. 9.12   Pressure—specific volume diagram showing the state point of a wet mixture.

Geometrical interpretation of mass ratio. Equations (9.8) to (9.11) are capable of geometrical interpretation. Fig. 9.12 will exemplify this. B is the state-point of a mixture of steam and water; the states of the saturated vapour and saturated liquid of which the mixture is composed are respectively represented by points A and C. Examination of the diagram shows that we can write

$$\frac{\overline{AB}}{\overline{BC}} = \frac{v_g - v}{v - v_f} \tag{9.19}$$

Substitution of v from eq. (9.11) leads to

$$\frac{\overline{AB}}{\overline{BC}} = \frac{1 - x}{x} = \frac{\text{mass of liquid in mixture}}{\text{mass of vapour in mixture}} \tag{9.20}$$

Thus the ratio of the distances of B from the saturation lines is equal to the *reciprocal* of the mass ratio of the respective pure phases; therefore nearness to the saturated-vapour line implies a high proportion of vapour, i.e. a large dryness.

This relation can be remembered if it is seen that B is in the right position for the fulcrum of a balance which carries the vapour mass at A and the liquid mass at C. Fig. 9.13 illustrates this.

Fig. 9.13 Mechanical analogy of the liquid-vapour mass ratio in a wet mixture.

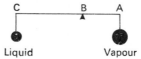

The throttling calorimeter

Provided that only one phase is present, a property diagram can be used for the determination of, say, enthalpy from measurements of pressure and temperature. For example, the $h{\sim}p$ diagram of Fig. 9.5 could be used: specification of p gives the horizontal position of the state-point; t locates the appropriate isotherm, and so fixes the point; then h can be read from the vertical scale.

However, if the state-point is within a two-phase region (the L + V region, for example), specification of pressure and temperature is insufficient: t gives the same information as p does, namely the horizontal position; but neither help in determining the required vertical position. The state could be determined by a volume measurement, but this is not always easy, particularly when the substance is flowing. This problem often arises in boiler practice, where it is necessary to measure the condition of the

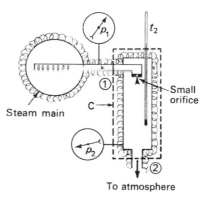

Fig. 9.14 Throttling calorimeter.

steam produced by a boiler lacking a superheater; this steam is always slightly wet, partly because of carry-over of water droplets from the boiler itself and partly because of condensation resulting from heat transfer through the walls of the steam main.

We now discuss a device which enables the condition of the wet steam to be determined, provided that the dryness is fairly close to unity. This is the *throttling calorimeter* illustrated in Fig. 9.14. A sampling tube, suitably positioned in the steam main for the steam entering it to be a representative sample, communicates through a throttle with a region of lower pressure. Upstream of the throttle, at point 1, either the pressure or the temperature is measured; downstream of the throttle, at point 2, both pressure and temperature are measured. Flow through the throttle is steady, but the velocities are sufficiently small for the kinetic-energy terms to be neglected. The whole calorimeter is thermally insulated.

Application of the Steady-Flow Energy Equation, eq. (8.11), to the throttling process leads to

$$\frac{Q - W_x}{m} = 0 = h_2 - h_1 \tag{9.21}$$

from which we deduce that the enthalpy at state 2 equals the enthalpy at state 1

$$h_2 = h_1 \tag{9.22}$$

Now examination of the shape of the saturated-steam line on an $h{\sim}p$ diagram shows that, in the range of common steam pressures, this curve slopes downwards to the left. Fig. 9.15 shows the relevant portion of the

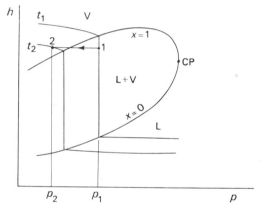

Fig. 9.15 Enthalpy~pressure diagram showing state points before and after throttling.

diagram; on it are drawn the two state-points, 1 and 2, of the throttling process. These are at the same height because of eq. (9.22). It is evident that, provided p_2 is sufficiently low, the state-point 2 lies in the superheat region.

Since state 2 is a single-phase state, the measurements of pressure and temperature, p_2 and t_2, suffice to specify the state as explained above. In particular the enthalpy h_2 is known. But this is the same as the enthalpy of the steam at state 1, h_1, which is supposed to be representative of the steam in the main. We now have two *independent* properties at state 1, h_1 and, say, p_1; state 1 is therefore completely specified. The geometry of Fig. 9.15

explains this more simply than words can do. In particular it is evident that the measurements determine the initial dryness of the steam.

It is not necessary to plot an $h{\sim}p$ diagram in order to evaluate measurements made on a throttling calorimeter: Steam Tables suffice. An example is given below (p. 195).

Throttling in general. In the case of the throttling calorimeter, when the steam enters wet and leaves superheated, there is an appreciable temperature difference between the two states. It should be realized that this is generally the case even if the throttle does not cause any phase change; the slope of the isotherms on an $h{\sim}p$ diagram ensures that a horizontal shift of the state-point brings it into a region of different temperature. Both rises and falls of temperature are observed in such experiments, which have been widely used in determinations of the properties of substances since their introduction by Joule and Thomson in 1852. This matter is referred to again in Chapter 14, p. 289.

The above analysis has neglected the effect of kinetic energy; if velocity changes are appreciable, allowance must be made as indicated in Chapter 8, pp. 147, 148.

Steam Tables

The most compact method of recording information about the thermodynamic properties of a substance is the use of tables. Examples are the Steam Tables given in Appendix B. Although insufficiently detailed for engineering use, these suffice for instructional purposes. The present section will be devoted to their explanation. It is important for an engineer to become thoroughly familiar with the use of these tables, in order that his attention should not be diverted from the meaning of his thermodynamic computations by the purely mechanical difficulty of looking up numbers in the tables.

The datum state

It has been explained above that only differences of internal energy are defined by the First Law of Thermodynamics; it is necessary to decide arbitrarily that the internal energy shall be zero at some definite condition, or *datum state*. Fixing the zero of internal energy fixes that of enthalpy. The internal energy of saturated water at the triple point ($t = 0{\cdot}01\ °C$) is arbitrarily chosen to be zero.

Since internal energy is related to enthalpy by:

$$h = u + pv \qquad (9.23)$$

it follows that the enthalpy of saturated water at $0{\cdot}01\ °C$ is slightly positive because of the (small) pv term.

The property data at $0\ °C$ given in Table II of the Steam Tables are for a fictitious state because the saturated-water line stops at the triple point which, as stated above, is at a temperature of 0·01 °C. So an extrapolation through 0·01 K is necessary.*

The *entropy* of saturated water (see Chapter 13) is also chosen as zero at 0·01 °C. However the entropy values in the Steam Tables will be ignored for the time being. After entropy has been discussed in Chapter 13, it will be seen that its treatment in the Tables is the same as that of h, u and v.

Table I: The triple point

Some of these data have already been discussed. Note that enthalpy is tabulated, but not internal energy; the reason is that the latter property can always be deduced from the former by way of eq. (9.23). Where a choice has to be made to economize in space, enthalpy is usually preferred because of the frequent occurrence of steady-flow processes.

EXAMPLE 9.1

Problem. (a) Check the value of the specific enthalpy of saturated water at the triple point given in Table I. (b) Check the specific internal energy of saturated steam at the triple point given in Table I.

Solution (a)

From equation (9.23)

$$h = u + pv$$

$$= 0 + 0·6112 \times 1·0002 \times 10^{-3}$$

$$= 0·000\ 611\ 3 \text{ kJ/kg} \qquad \qquad \textit{Answer (a)}$$

Solution (b)

$$u = h - pv$$

$$= 2501·6 - 0·6112 \times 206·1629$$

$$= 2375·6 \text{ kJ/kg} \qquad \qquad \textit{Answer (b}$$

Tables II and III: Saturated-water and saturated-steam data

These tables cover the same ground, but differ in that Table II temperature is the independent variable, whereas in Table III pressure is the independent variable. Of course, for saturated states, p and t are uniquely linked, so only one of the tables is strictly necessary. However interpolation is easier if both are available.

The data in these tables enable the liquid and vapour saturation lines to be plotted on a property diagram (see exercises at the end of this chapter).

* At one time this fictitious state was the adopted datum state. The effect of this change on the tabulated values is negligible for most purposes.

Remark. The tables contain redundant information, for example: v_{fg} as well as v_f and v_g; and u_g as well as h_g, p and v_g. The reader is advised to sample the data and check their consistency.

EXAMPLE 9.2

Problem. In Table III for $p = 60 \times 10^3$ N/m, u_g equals $2489 \cdot 7 \times 10^3$ J/kg and v_g equals $2 \cdot 731$ m³/kg. Evaluate h_g

Solution.
$$h_g = u_g + pv_g$$
$$= 2489 \cdot 7 \times 10^3 + 60 \times 2 \cdot 7317 \times 10^3$$
$$= 2489 \cdot 7 \times 10^3 + 163 \cdot 9 \times 10^3$$
$$= 2653 \cdot 6 \times 10^3 \text{ J/kg} \qquad\qquad Answer$$

This agrees with tabulated value of h_g.

2. For the saturated liquid, enthalpy and internal energy are nearly equal, so only the former is tabulated. In many cases it is permissible to neglect the difference between them, which is small because of the small value of the specific volume of water. The above example about the triple point exemplifies this.

3. u_{fg} and h_{fg} differ by the amount of the term pv_{fg}. This is definitely not negligible. For example at 400×10^3 N/m² it amounts to $400 \times 0 \cdot 462\,20 \times 10^3 = 184 \cdot 5 \times 10^3$ J/kg.

4. As the critical point is approached (bottom of the tables) v_{fg}, u_{fg} and h_{fg} all tend to zero.

5. Since the enthalpy of water is almost independent of pressure, and increases by approximately $4 \cdot 19 \times 10^3$ J/kg for each 1 K rise in temperature, we can write

$$h \approx u \approx 4 \cdot 19 \times 10^3 t$$

with h and u in J/kg and t in °C.

6. If data are required for intermediate temperatures or pressures, linear interpolation is normally sufficiently accurate.

EXAMPLE 9.3

Problem. Determine h_g at a pressure of 128×10^3 N/m².

Solution.

Table III is the more convenient in this case.

For 120×10^3 N/m² we have $h_g = 2683 \cdot 4 \times 10^3$ J/kg

For 140×10^3 N/m² we have $h_g = 2690 \cdot 3 \times 10^3$ J/kg

Interpolating linearly we have

at 128×10^3 N/m²: $h_g = \dfrac{128 - 120}{140 - 120} \times (2690 \cdot 3 - 2683 \cdot 4) \times 10^3 + 2683 \cdot 4 \times 10^3$

$$= \frac{2}{5} \times 6 \cdot 9 \times 10^3 + 2683 \cdot 4 \times 10^3$$

$$= 2686 \cdot 2 \times 10^3 \text{ J/kg} \qquad\qquad Answer$$

Note that it is never wise prematurely to drop the numbers after the decimal point, because in First-Law calculations *differences* of enthalpy or internal energy are involved. Though small compared with the actual values of h (or u), the numbers after the decimal point may not be negligible compared with the differences of h (or u).

7. There is no need to provide tables for the properties of steam in the two-phase (L + V) region: the saturation data suffice. Properties of mixtures of saturated steam and water, i.e. of wet steam, can be obtained by introducing the data of Tables II and III into the equations (9.9) to (9.14).

EXAMPLE 9.4: CONCERNING WET STEAM AND THE CRITICAL POINT

Problem. 1 kg of wet steam is contained in a rigid container at 16 °C. What should be (a) the volume of the container, (b) the initial dryness, (c) the proportion of the volume initially occupied by water, if the meniscus is neither to rise to the top nor to descend to the bottom when the system is heated to 400 °C.

Solution. (a) The requirement is that the system should pass through the critical point. Since the volume of the vessel is fixed and contains 1 kg, its volume must equal 1 times the specific volume at the critical point (374·15 °C, 22·12 × 10⁶ N/m²). From Table II (or III), $v_g = v_f = 0.003\ 17$ m³ kg at the critical point. Therefore

$$\text{volume of container} = 0.00317 \text{ m}^3 \qquad \textit{Answer (a)}$$

Solution. (b) If the initial dryness is x, and the initial specific volumes of liquid and vapour are v_f and v_g, we have from eq (9.11)

$$0.003\ 17 = (1 - x)v_f + xv_g$$

v_f and v_g are found from Table II. The equation then runs .

$$0.003\ 17 = (1 - x) \times 0.001\ 001 + x \times 73.384$$

$$\therefore \qquad x = \frac{0.003\ 17 - 0.001\ 001}{73.384 - 0.001\ 001}$$

$$= 29.6 \times 10^{-6} \qquad \textit{Answer (b)}$$

Thus the initial dryness is very small indeed. In this case it is certainly not permissible to neglect the volume of the water in comparison with that of the steam.

Solution. (c) The volume of the steam at 16 °C is

$$xv_g = 29.6 \times 10^{-6} \times 73.384$$

$$= 2.17 \times 10^{-3} \text{ m}^3$$

The volume of the water at 16 °C,

$$(1 - x)v_f = (1 - 29.6 \times 10^{-6}) \times 0.001\ 001 = 1.00 \times 10^{-3} \text{ m}^3.$$

giving of course a total volume of 0·003 17 m³ as in answer (a).

The proportion of the total volume occupied by water is

$$\frac{0.001\ 00}{0.003\ 17} = 0.315 \qquad \textit{Answer (c)}$$

Therefore the vessel should be slightly less than one third full of water.

Table IV: Superheated steam

This table gives data for the pure vapour. h, v and s (i.e. entropy, Chapter 13) are contained in the body of the table, for various values of pressure and temperature.

In general, data will be required for combinations of pressure and temperature which do not appear explicitly in the tables; then interpolation is necessary. Linear interpolation is sufficiently accurate for most purposes.

EXAMPLE 9.5: ILLUSTRATING INTERPOLATION WHERE THE PRESSURE APPEARS IN THE TABLE BUT THE TEMPERATURE DOES NOT

Problem. What are h, v and u for $p = 1 \cdot 5 \times 10^6$ N/m², $t = 429$ °C?

Solution. At $p = 1 \cdot 5 \times 10^6$ N/m², $t = 400$ °C, we have $h = 3256 \cdot 6 \times 10^3$ J/kg (from Table IV)

At $p = 1 \cdot 5 \times 10^6$ N/m², $t = 500$ °C, we have $h = 3472 \cdot 8 \times 10^3$ J/kg (from Table IV)

Therefore at $p = 1 \cdot 5 \times 10^6$ N/m², $t = 429$ °C, we have

$$h = 3256 \cdot 6 \times 10^3 + \frac{429 - 400}{500 - 400} \times (3472 \cdot 8 - 3256 \cdot 6) \times 10^3$$

$$= 3256 \cdot 6 \times 10^3 + \frac{29 \times 216 \cdot 2}{100} \times 10^3$$

$$= 3256 \cdot 6 \times 10^3 + 62 \cdot 7 \times 10^3 = 3319 \cdot 3 \times 10^3 \text{ J/kg} \qquad \textit{Answer}$$

Similarly for v we have:

At $p = 1 \cdot 5 \times 10^6$ N/m², $t = 400$ °C, $v = 0 \cdot 202 \, 92$ m³/kg (from Table IV)

At $p = 1 \cdot 5 \times 10^6$ N/m², $t = 500$ °C, $v = 0 \cdot 235 \, 03$ m³/kg (from Table IV)

Therefore at $p = 1 \cdot 5 \times 10^6$ N/m², $t = 429$ °C, we have:

$$v = 0 \cdot 202 \, 92 + \frac{429 - 400}{500 - 400}(0 \cdot 235 \, 03 - 0 \cdot 202 \, 92)$$

$$= 0 \cdot 202 \, 92 + \frac{29}{100} \times 0 \cdot 032 \, 11$$

$$= 0 \cdot 202 \, 92 + 0 \cdot 009 \, 31 = 0 \cdot 212 \, 23 \text{ m³/kg} \qquad \textit{Answer}$$

Now $u = h - pv$

$$= 3319 \cdot 3 \times 10^3 - 1 \cdot 5 \times 10^6 \times 0 \cdot 212 \, 23$$

$$= 3319 \cdot 3 \times 10^3 - 318 \cdot 3 \times 10^3 = 3001 \cdot 0 \times 10^3 \text{ J/kg} \qquad \textit{Answer}$$

EXAMPLE 9.6: ILLUSTRATING INTERPOLATION WHERE NEITHER THE PRESSURE NOR THE TEMPERATURE APPEARS IN THE TABLE

Problem. What is h for $p = 1 \cdot 68 \times 10^6$ N/m², $t = 429$ °C?

Solution. We first obtain the values of h at $t = 429$ °C for $p = 1 \cdot 5 \times 10^6$ N/m² and for $p = 2 \cdot 0 \times 10^6$ N/m², by use of the method given in the previous problem,

At $t = 429$ °C, $p = 1 \cdot 5 \times 10^6$ N/m²: $h = 3319 \cdot 3 \times 10^3$ J/kg (see example 9.5)

At $t = 429$ °C, $p = 2 \cdot 0 \times 10^6$ N/m²:

$$h = 3248 \cdot 7 \times 10^3 + \frac{429 - 400}{500 - 400}(3457 \cdot 3 - 3248 \cdot 7) \times 10^3$$

$$= 3248 \cdot 7 \times 10^3 + \frac{29}{100} \times 218 \cdot 6 \times 10^3$$

$$= 3248 \cdot 7 \times 10^3 + 63 \cdot 4 \times 10^3 = 3312 \cdot 1 \times 10^3 \text{ J/kg}$$

We now interpolate linearly between the values of h at 1.5×10^6 N/m^2 and 2.0×10^6 N/m^2 to obtain the value of h corresponding to 1.68×10^6 N/m^2, as follows:

At $t = 429$ °C, $p = 1.68 \times 10^6$ N/m^2:

$$h = 3319.3 \times 10^3 - \frac{1.68 - 1.5}{2.0 - 1.5} (3319.3 - 3312.1) \times 10^3$$

$$= 3319.3 \times 10^3 - \frac{0.18}{0.5} \times 7.2 \times 10^3$$

$$= 3319.3 \times 10^3 - 2.6 \times 10^3 = 3316.7 \times 10^3 \text{ J/kg} \qquad Answer$$

Remarks about Table IV. 1. The saturation temperatures corresponding to the various pressures are shown in brackets. They are necessary for finding the properties corresponding to states just to the left of those tabulated in Table IV, i.e. of only slightly superheated steam. They are also needed when the temperature quoted is not the true temperature but the "degrees superheat" (see p. 180).

EXAMPLE 9.7

Problem. A boiler, fitted with a superheater, generates steam at 1.0×10^6 N/m^2 with 10 K superheat. What is the enthalpy of the steam?

Solution. At 1.0×10^6 N/m^2, the saturation temperature of steam $t_g = 179.88$ °C (from Table III).

Therefore the temperature, t, of the steam produced by the boiler is given by

$$t = (179.88 + 10) \text{ °C}$$

$$= 189.88 \text{ °C}$$

This temperature is less than the lowest tabulated temperature at $p = 1.0 \times 10^6$ N/m^2 in Table IV. Therefore to obtain the required value of h we have to interpolate between Tables III and IV as follows:

From Table III: At $p = 1.0 \times 10^6$ N/m^2 and $t_g = 176.88$ °C, h_g equals 2776.2×10^3 J/k

From Table IV: At $p = 1.0 \times 10^6$ N/m^2 and $t = 200$° C, $h = 2826.8 \times 10^3$ J/kg

At $p = 1.0 \times 10^6$ N/m^2 and $t = 189.88$ °C we have

$$h = 2776.2 \times 10^3 + \frac{189.88 - 179.88}{200 - 179.88} (2826.8 - 2776.2) \times 10^3$$

$$= 2776.2 \times 10^3 + \frac{10}{20.12} \times 50.6 \times 10^3$$

$$= 2776.2 \times 10^3 + 25.1 \times 10^3 = 2801.3 \times 10^3 \text{ J/kg} \qquad Answ$$

2. At high temperatures and low pressures it will be seen from Table I that h varies only slightly with p at a fixed temperature. For example $t = 800$ °C, $h = 4158.7 \times 10^3$ J/kg when $p = 2.0 \times 10^3$ N/m^2 and 4144 $\times 10^3$ J/kg when $p = 3.0 \times 10^6$ N/m^2.

3. The internal energy is not tabulated. When it is needed, it must b derived from eq. (9.23) as before.

4. Linear interpolation may become rather inaccurate when applied t specific volume at low pressures. One may use more precise (i.e. non-linea

interpolation formulae; often however it is preferable to turn to more detailed tables, for example the UK Steam Tables in SI Units 1970 (see bibliography), from which the tables in Appendix B are extracted.

EXAMPLE 9.8: THROTTLING CALORIMETER

Problem. A throttling calorimeter is used to measure the dryness fraction of the steam in a steam main, Fig. 9.14. The calorimeter readings are: pressure 100×10^3 N/m^2; temperature, 120 °C. The main pressure is 600×10^3 N/m^2. Calculate the dryness fraction of the steam in the main.

Solution. It has been shown that the enthalpies of the steam, at entry to and exit from a throttling calorimeter, are equal (see p. 188, eq. (9.22)). The downstream (state 2) measurements enable h_2, and hence h_1, to be evaluated as follows:

At state 2: $p_2 = 100 \times 10^3$ N/m^2; $t_2 = 120$ °C.

Since t_2 is greater than 99·63 °C, the saturation temperature corresponding to $p_2 = 100 \times 10^3$ N/m^2, the steam is superheated downstream of the throttle; therefore we have to use Table IV to evaluate h_2.

From Table IV: at $p = 100 \times 10^3$ N/m^2 and $t = 100$ °C, $h = 2676·2 \times 10^3$ J/kg

at $p = 100 \times 10^3$ N/m^2 and $t = 150$ °C, h equals $2776·3 \times 10^3$ J/kg

at $p_2 = 100 \times 10^3$ N/m^2 and $t_2 = 120$ °C

$$h_2 = 2676·2 \times 10^3 + \frac{120 - 100}{150 - 100}(2776·3 - 2676·2) \times 10^3$$

$$= 2676·2 \times 10^3 + \frac{20}{50} \times 101·1 \times 10^3 = 2716·6 \times 10^3 \text{ J/kg}$$

$$\therefore \qquad h_1 = 2716·6 \times 10^3 \text{ J/kg}$$

Now this value of enthalpy is less than h_g at 600×10^3 N/m^2, namely, $2755·5 \times 10^3$ J/kg (from Table III); the steam in the main is wet (Fig. 9.15). We use eq. (9.12) to evaluate the dryness fraction, for we have

$$h_1 = h_g - (1 - x_1)h_{fg} \text{ at } 600 \times 10^3 \text{ N/m}^2$$

$$2716·6 = 2755·5 - (1 - x_1) 2085·0 \text{ from Table III}$$

whence $x_1 = 0·981$ (N.B. < 1)   *Answer*

Compressed-liquid states

Symmetry demands that, having dealt first with the saturated phases and so with the L + V region, and secondly with the superheat (V) region, we should now turn to the compressed-liquid (L) states. Nevertheless, no table for these states is provided in Appendix B because the dependence of the properties of liquid water on pressure is quite small, at least when far from the critical point. It is sufficiently accurate for the examples in this book, and for many practical problems also, to take the saturated-water data for the *temperature* in question as valid even though the pressure exceeds the saturation pressure for that temperature. For the h and u

of liquid H_2O, it is often satisfactory to use the expression given in note 5, p. 191.

In professional work of course it is never permissible to make this assumption without examination of the inaccuracy involved. Data for compressed liquid water will be found in the Steam Tables listed in the bibliography.

BIBLIOGRAPHY

ANON. *International Critical Tables*. National Research Council. McGraw-Hill, New York, 1926–30.

ANON. *UK Steam Tables in SI Units*, 1970. United Kingdom Committee on the Properties of Steam. Edward Arnold (Publishers) Ltd. London, 1970.

BAIN, R. A. *Steam Tables*, National Engineering Laboratory. H.M.S.O., London, 1964.

KEENAN, J. H., and KEYES, F. G., *Thermodynamic Properties of Steam*. Wiley & Sons, New York, 1936.

PERRY, J. H., CHILTON, C. H., KIRKPATRICK, S. D. *Chemical Engineers Handbook*. 4th. Ed. McGraw-Hill, New York and Maidenhead, 1963.

SCHMIDT, E. *VDI~Wasserdampftafeln*. 6th Ed. Verein Deutscher Ingenieure. Springer-Verlag, 1963.

CHAPTER 9—PROBLEMS

9.1 Using the Steam Tables given in Appendix B, write down the symbol, the magnitude and the units for each of the following:

(a) Specific volume of saturated water at a pressure of $6\cdot0 \times 10^3$ N/m².

(b) Specific volume of dry-saturated steam at a temperature of 200 °C.

(c) Temperature of saturated water at a pressure of $1\cdot013\ 25 \times 10^5$ N/m².

(d) Temperature of dry-saturated steam at a pressure of $1\cdot013\ 25 \times 10^5$ N/m².

(e) Temperature of wet steam, dryness 0·9 at a pressure of $1\cdot013\ 25 \times 10^5$ N/m².

(f) Enthalpy (latent heat) of vaporization at a pressure of 630×10^3 N/m².

(g) Enthalpy of dry-saturated vapour at a temperature of 100 °C.

(h) Internal energy of saturated water at a pressure of $3\cdot75 \times 10^6$ N/m².

(i) Specific volume, enthalpy and internal energy of steam at a pressure of 400×10^3 N/m² and a temperature of 400 °C.

(j) Temperature and specific volume of steam at a pressure of $7\cdot0 \times 10^6$ N/m² and an enthalpy of $2903\cdot0 \times 10^3$ J/kg.

(k) Pressure and internal energy of steam at a temperature of 350 °C and a specific volume at 0·30 m³/kg.

9.2 Using the Steam Tables given in Appendix B, sketch, approximately to scale, the boundaries of the liquid- and vapour-phase regions of H_2O on diagrams with the following coordinates:

(a) $t \sim p$; (b) $h \sim p$; (c) $h \sim t$; (d) $p \sim v$; (e) $u \sim v$. On sketches (b) and (d) show a few lines of constant t and on (c) a few lines of constant p. Mark the triple and critical points in each case.

9.3 Using the Steam Tables given in Appendix B, determine the value of the enthalpy, the internal energy and the specific volume of a mixture of 0·1 kg of saturated water and 0·9 kg of saturated steam at (a) a pressure of $1\cdot4 \times 10^6$ N/m² and (b) a temperature of 250 °C.

9.4 A vessel of volume 0·04 m³ contains a mixture of saturated water and saturated steam at a temperature of 240 °C; the mass of liquid present is 8 kg. For the mixture, evaluate the pressure, the mass, the specific volume, the enthalpy and the internal energy.

9.5 (a) Use Table II of the Steam Tables to find the mass, the enthalpy and the internal energy of a system comprising 0·1 m³ of wet steam at a temperature 160 °C and a dryness fraction of 0·94.

(b) The system undergoes a fully-resisted process at constant pressure to a temperature of 250 °C. During the process there are work and heat interactions between the system and its surroundings. Show the path of this process on a sketch of the $p{\sim}v$ diagram and evaluate the increase in enthalpy, the increase in internal energy, the work done and the heat transfer.

9.6 (a) A rigid vessel of volume 0·58 m³ contains 1 kg of steam at a pressure of 300 × 10³ N/m². Evaluate the specific volume, the temperature, the dryness fraction, the internal energy and the enthalpy of the steam.

(b) Heat transfer to the steam causes its temperature to rise to 160 °C. Show the path of this process on a sketch of the $p{\sim}v$ diagram and evaluate the pressure, the increase in enthalpy, the increase in internal energy of the steam and the heat transfer. Evaluate also the pressure at which the steam becomes dry-saturated.

9.7 A rigid vessel contains 1 kg of a mixture of saturated water and saturated steam at a pressure of 140 × 10³. When the mixture is heated the state passes through the critical point.

Evaluate:

(a) the volume of the vessel;

(b) the mass of liquid and of vapour in the vessel initially;

(c) the temperature of the contents of the vessel when the pressure has risen to 30 × 10⁶ N/m²;

(d) the heat transfer required to produce the final state (c).

9.8 A vertical cylinder, fitted with a frictionless leak-proof piston, contains 0·03 kg of dry-saturated steam. The upper face of the piston is exposed to the atmosphere; the weight of the piston is such that the steam pressure is 300 × 10³ N/m². A quantity of saturated water at a pressure of 300 × 10³ N/m² is introduced into the cylinder and mixes thoroughly with the steam. When the mixture is heated subsequently, with the piston held stationary, its state passes through the critical point. Find the mass of water introduced.

9.9 A throttling calorimeter sampling the steam generated by a boiler at a pressure 800 × 10³ N/m² gives readings of pressure, 100 × 10³ N/m² and temperature 116 °C. Evaluate the dryness of the boiler steam.

9.10 A combined separating and throttling calorimeter is often used to measure the dryness of very wet steam. The separator is a rigid vessel fitted in the steam-sampling pipe between the steam main and the throttling calorimeter; the steam sample flows through the separator before entering the calorimeter. The purpose of the separator is to remove some of the suspended water particles from the sample, which otherwise would not become superheated in passing through the throttle. A test is carried out by measuring, over a specified period of time, the mass of water drained from the separator and the mass of steam flowing through the throttle. In addition the main pressure and the pressure and temperature after the throttle are recorded.

A particular test gave the following results:

Main pressure: 3·4 × 10⁶ N/m² gauge.

Mass of water drained from the separator: 0·33 kg.

Mass of steam condensed after passing through the throttle: 4·66 kg.

Throttling calorimeter readings:

pressure, 51 mm water gauge
temperature, 145 °C.

Barometer reading: 746 mm mercury.

Evaluate the dryness of the steam in the main and state, with reasons, whether the throttling calorimeter alone could have been used for this test.

9.11 A sample of wet steam from a steam main flows steadily through a partially open valve into a pipe-line in which is fitted an electric coil; the valve and the pipe-line are well-insulated. The steam mass flow rate is 0·007 kg/s while the coil takes 3·78 amperes at

230 volts. The main pressure is 400×10^3 N/m^2 and the pressure and temperature of the steam downstream of the coil are 200×10^3 N/m^2 and 155 °C respectively. Steam velocities may be assumed to be negligible.

(a) Evaluate the dryness of the steam in the main.

(b) State, with reasons, whether an insulated throttling calorimeter could have been used for this test.

9.12 A 280-mm diameter cylinder fitted with a frictionless leakproof piston contains 0·02 kg of steam at a pressure of 600×10^3 N/m^2 and a temperature of 200 °C. As the piston moves slowly outwards through a distance of 305 mm, the steam undergoes a fully-resisted expansion during which the steam pressure p and the steam volume V are related by $pV^n =$ constant, where n is a constant. The final pressure of the steam is 120×10^3 N/m^2. Determine:

(a) the value of n

(b) the work done by the steam

(c) the magnitude and sign of the heat transfer.

9.13 A kilogram of steam at a pressure of $1·0 \times 10^6$ N/m^2 and a temperature of 250 °C is contained in a cylinder closed by a piston. The steam is compressed by the motion of the piston inwards until the pressure is $2·0 \times 10^6$ N/m^2. The work done on the steam is 610×10^3 N m and the heat transfer from the steam is 890×10^3 J.

Find the final steam temperature if the steam is superheated, or the dryness if it is wet.

9.14 Water flows steadily into a domestic radiator at a temperature of 80 °C and leaves at a temperature of 65 °C.

Evaluate the heat transfer from the water per kilogram of water flowing, assuming:

(a) the specific heat of water at constant pressure to be constant and equal to $4·19 \times 10^3$ J/kg K;

(b) the enthalpy of water at a given temperature to be equal to the enthalpy of saturated water at that temperature.

Neglect changes in kinetic and gravitational-potential energies.

9.15 (a) Steam enters a convergent-divergent nozzle with a velocity of 60 m/s, with a pressure 800×10^3 N/m^2 and with 79·6 K superheat. The steam leaves the exit section of the nozzle at a pressure of 160×10^3 N/m^2 with a dryness of 0·96. The cross-sectional area of the exit section is 12 cm^2. The flow is adiabatic.

Determine the steam velocity at the exit section and the steam mass flow rate.

(b) The exhaust steam from the nozzle flows into a condenser and flows out as water at a temperature of 95 °C and negligible velocity. The cooling water enters the condenser at a temperature of 10 °C and leaves at a temperature of 25 °C; determine its mass flow rate.

9.16 Steam at a pressure of 2×10^6 N/m^2 and a temperature of 250 °C flows steadily into a turbine with negligible velocity. The steam leaves the turbine at a pressure of 15×10^3 N/m^2 with a velocity of 200 m/s. The heat-transfer rate from the turbine casing to the atmosphere is 160×10^3 W. The power developed by the turbine is 3430×10^3 W. The mass flow rate is 6·1 kg/s.

Determine the dryness of the steam leaving the turbine and the cross-sectional area of the exit section.

9.17 A well-insulated rigid vessel of volume 1 m^3 contains 28 kg of a mixture of liquid water and steam at 95 °C. A thin-walled coiled tube within the vessel, immersed in the mixture, is supplied with a steady flow of steam at a pressure of 400×10^3 N/m^2 and a dryness of 0·9. Some of the inflowing steam condenses and collects in the coil and the remainder flows out at a pressure of 400×10^3 N/m^2; the pressure within the coil is uniform at 400×10^3 N/m^2. Changes in velocity and elevation are negligible.

(a) Evaluate the maximum pressure reached by the mixture in the vessel.

(b) Calculate the mass of water which collects in the coil to achieve this maximum pressure.

9.18 A well-insulated pressure vessel of volume 6 m³ is connected, via a valve, to a steam main in which the pressure and temperature are maintained at 3·0 × 10⁶ N/m² and 250 °C respectively. Initially the vessel contains water and steam at a pressure of 300 × 10³ N/m² in the proportion 1:2 by volume. The valve is opened, so allowing steam from the main to flow into the vessel. Determine the increase in the mass of the contents of the vessel and the proportions by volume of liquid and vapour finally in the vessel when the pressure has risen to 3·0 × 10⁶ N/m². (The arrangement described is known as a *steam accumulator*.)

9.19 Two rigid vessels are connected by a short pipe and a valve, initially shut. One vessel has a volume of 0·0495 m³ and is evacuated. The other vessel has a volume of 0·0085 m³ and contains dry-saturated ammonia at a temperature of 50 °C.

The valve is opened; when conditions throughout the two vessels have become uniform, it is found that the pressure and temperature of the ammonia are 355 × 10³ N/m² and 40 °C respectively.

Determine the heat transfer to the ammonia, using the data shown below for ammonia.

Pressure kN/m²	Saturation temperature ° C	Specific volume m³/kg		Specific enthalpy kJ/kg		
		Saturated liquid	Saturated vapour	Saturated liquid	Saturated vapour	Vapour superheated by 50 K
355	−5	0·001 55	0·3472	158·2	1438·9	1559·7
2033	50	0·001 78	0·0635	421·9	1474·6	1633·1

10 The Second Law of Thermodynamics

Introduction

We now return to the subject of producing mechanical power. The First Law of Thermodynamics shows that there is a fixed rate of exchange between heat and work. Since, by burning fuel, either chemical or nuclear, high temperatures can be obtained and heat transfer caused, all that the engineer has to do, it might be thought, is to convert the heat into the corresponding amount of work.

In this chapter some of the means for producing work from heat will be examined, and it will be seen that they do not by any means effect a complete conversion. Is this because so far engineers have been insufficiently skilful? The answer to this question confronts us with the Second Law of Thermodynamics.

Symbols

C_{hp}	Coefficient of performance of a heat pump.	t_2	Temperature of a cold body (system).
C_{ref}	Coefficient of performance of a refrigerator.	w	Net work done by (on) a system.
q	Heat transfer.	w_{aux}	Work to auxiliaries (in an internal-combustion engine).
q_1	Heat transfer between a heat engine and a hot body (system).	w_p	Shaft work input to a feed pump.
q_2	Heat transfer between a heat engine and a cold body (system).	w_{shaft}	Shaft work.
q_{atm}	Heat transfer to the atmosphere.	w_t	Shaft work delivered by a turbine.
		w_x	External (shaft) work.
q_c	Heat transfer to the coolant (in an internal-combustion engine).	w_{100}	Net work done by a one-hundred-per-cent-efficient heat engine.
t_1	Temperature of a hot body (system).	η	Efficiency of a heat engine, thermal efficiency.

Engines

The steam power plant

A large proportion of the world's power is produced by steam power plants. The four elements of such plants are illustrated in Fig. 10.1:

boiler, turbine, condenser and feed pump. The working fluid, H_2O (vapour and liquid), flows steadily through these components in turn, so executing a cyclic process. Entering the boiler as water, the working fluid is vaporized and passes into the turbine; emerging from there still as steam but at a lower temperature and pressure, the working fluid enters the condenser; there it becomes water once again, and, after being compressed to boiler pressure by the feed pump, is returned to the boiler ready to begin the circuit once more.

A system boundary has been drawn around the power plant in Fig. 10.1. Four interactions occur at it: in unit time, or per unit quantity of steam

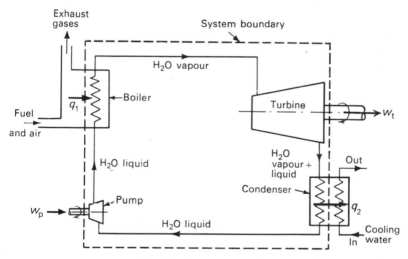

Fig. 10.1 Steam power plant.

flow, q_1 is the heat transfer from the hot combustion products to the steam in the boiler; w_t is the external work done by the turbine; q_2 is the heat transfer from the condensing steam to the cooling water, drawn perhaps from a river; while w_p is the external work driving the feed pump. The small letters, q and w, used for heat and work, indicate that the earlier sign convention has been dropped; here q and w are positive in the directions of the arrows and hence $Q_1 \equiv q_1$, $Q_2 \equiv -q_2$.

Application of the First Law, eq. (6.1), to this system, for steady conditions, leads to an expression for the net work, namely

$$w_t - w_p = q_1 - q_2 \tag{10.1}$$

Normally w_p is very much smaller than the turbine output w_t; q_2 on the other hand may amount to more than two-thirds of the magnitude of q_1. This means that the work obtained is less than one-third of that which the First Law would permit if the entire heat input, q_1, were converted. Why do engineers tolerate this?

An immediate answer to the question is that, if there were no heat transfer q_2, the steam would not be condensed. If it were not condensed, the pump would have to raise the pressure of the exhaust steam to that of the boiler in order that the steam should work in a cycle. To do this the pump would have to work harder than when pumping condensed steam, i.e. water, back into the boiler; so much more, in fact, that w_p would exceed w_t and the net work output of the plant would be negative.

It might be thought that this is an idiosyncrasy of steam which could perhaps be avoided by using a different working fluid; alternatively one might look for a different way of using steam in order to eliminate the

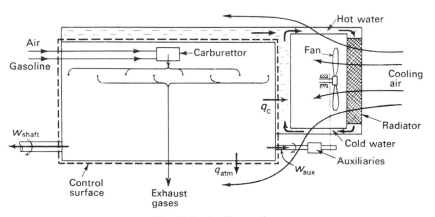

Fig. 10.2 Gasoline engine.

heat transfer q_2. The reader is invited to try to invent such a power-plant arrangement before finishing this chapter.

In the subsequent discussion it will be important to note a difference between plants like the steam power plant, in which combustion occurs, if at all, externally to the working fluid, and other power plants in which the working fluid itself takes part in combustion. We therefore now discuss the internal-combustion engine.

The internal-combustion engine

The important features of these engines will be described by reference to Fig. 10.2, which represents a gasoline engine for a road vehicle. Gasoline and air pass, via the carburettor, into the engine cylinders, where, after combustion and expansion, the resulting exhaust gases (i.e. a mixture of carbon dioxide, carbon monoxide, nitrogen, steam and other gases) flow out to atmosphere. The cylinder block of the engine is maintained at a reasonable working temperature by the circulation of cooling water between it and the engine "radiator". A fan, driven by the engine, draws atmospheric air through the radiator, and so cools the water circulating through it.

The dotted boundary of Fig. 10.2, though drawn as a rectangle, should be thought of as fitting over the metal surfaces. It encloses a control volume, not a system, for its contents change continuously as air and gasoline flow in and exhaust gases flow out; the cooling air, on the other hand, flows *past* the surface, not *through* it. In addition to the material flows across the boundary, there are the two interactions, at the boundary, between the fuel-air-product stream and its surroundings: the net external work W_x and the net heat transfer Q. In this case W_x is the algebraic sum of the engine crankshaft work, w_{shaft}, and the work delivered to the cooling fan and other auxiliaries (fuel pump, dynamo, etc.), w_{aux}: thus $W_x = w_{shaft} + w_{aux}$. Further, Q is the algebraic sum of the heat transfers to the coolant q_c, and to the atmosphere (from exposed hot surfaces), q_{atm}: hence $Q = - q_c - q_{atm}$.

Application of the First Law to the control volume, with neglect of kinetic energy and gravitational potential energy, eq. (8.45), yields

$$(-q_c - q_{atm}) - (w_{shaft} + w_{aux}) = \text{enthalpy increase of the} \qquad (10.2)$$
$$\text{fuel-air-products stream;}$$

so this time the algebraic sum of the heat and work interactions is not equal to zero. A further difference from the steam power plant is that there is no inward heat transfer; heat is transferred only *from* the reacting mixture. The flow of gasoline into the engine is not a heat transfer, nor is the outflow of hot exhaust gases; for neither is an interaction between systems by reason of a temperature difference. It should be noted also that although the engine works cyclically, i.e. it executes the same sequence of events periodically, the air and gasoline do not execute a cycle—the exhaust gases leaving the engine are never returned to it as air and gasoline to repeat the process.

Analysis of the performance of internal combustion engines, and comparison of their work output with the theoretical maximum, form important parts of thermodynamics; however, they are not treated until the end of the present book. For the moment we merely note the differences between internal-combustion engines and systems of the steam-power-plant type, and restrict discussion to the latter, to which we will give the name *heat engines*.

The Heat Engine

A definition which is logically sufficient is:

A heat engine is a continuously operating thermodynamic system at the boundary of which there are heat and work interactions.

Explanatory remarks. 1. "Continuously operating" means that the state of the system exhibits only periodic changes; the phrase covers both rotary and reciprocating machines. If a working fluid is present it must undergo cyclic processes.

2. Since a system is in question, the internal-combustion engine cannot be classed as a heat engine. Matter flows continuously into and out of an internal-combustion engine.

3. The steam power plant, as described above, is covered by the definition; the contents of the boundary in Fig. 10.1 comprise a system. If, however, the boundary were enlarged to include the combustion space of the boiler, it would enclose a control volume, not a system; for, in addition to the heat and work interactions at the boundary, the air, fuel and flue gases would

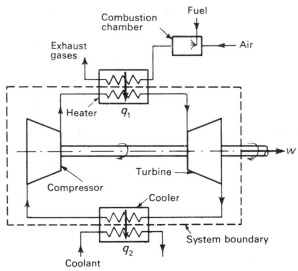

Fig. 10.3 "Closed-cycle" gas-turbine engine.

flow across it. The plant enclosed by this enlarged boundary would cease to be a heat engine in the sense defined above.

4. The turbine by itself is not a heat engine, for steam flows both in and out. A reciprocating steam engine, if understood to comprise merely cylinder, piston, and valve gear, is also not a heat engine, for the same reason.

5. The distinction which is being emphasized is neatly illustrated by the two types of gas-turbine engine.

The "closed-cycle" gas-turbine engine: a heat engine

This power plant is illustrated in Fig. 10.3. Its four elements correspond to those of the steam plant: heater, turbine, cooler and compressor. The working fluid, a gas such as air, flows steadily into the heater, from which it emerges at a higher temperature. After expansion in the turbine, it passes to the cooler where its temperature is reduced before re-compression in the compressor. The heating and cooling are effected in order that the specific volume of the gas in the turbine should exceed that in the compressor; then the power output of the turbine can exceed the power input to the compressor and allow a net output of work.

Examination of the dotted boundary in Fig. 10.3 shows that the net work, w, and the two heat transfers, q_1 and q_2, occur at the boundary, but no material crosses it; therefore the boundary encloses a system and hence the "closed-cycle" gas turbine is a heat engine.

We noted above (pp. 201, 202) that the steam power plant required an outward heat transfer q_2, in the condenser, for continuous operation. Despite being more modern than steam plants, the gas-turbine power plant

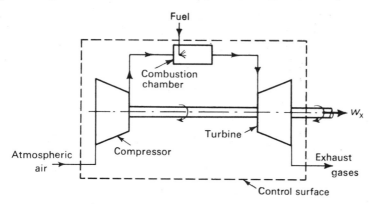

Fig. 10.4 "Open-cycle" gas-turbine engine.

has not managed to dispense with the outward heat transfer q_2, which may amount to 75 per cent of q_1 in practice.

The "open cycle" gas turbine engine: not a heat engine

A plant which is similar in many ways to that last described is the so-called "open-cycle" gas-turbine engine, which is illustrated in Fig. 10.4. This has a steady-flow combustion chamber in place of a heater; instead of the cooler, the exhaust from the turbine is passed directly to the atmosphere, while the compressor takes in fresh cold air.

Examination of the dotted boundary in Fig. 10.4 shows that this plant is not a heat engine: fresh air and fuel oil flow inwards; hot combustion products flow outwards; the working fluid does not perform a cycle, since nowhere are the combustion products turned back into oil and air. There are no heat transfers across the boundary, apart from minor ones to the atmosphere from exposed hot surfaces ("radiation losses"); but there is, of course, a net work output.

It may be remarked that the features which cause this plant not to be a heat engine are responsible for its popularity as a prime mover. For a combustion chamber is much cheaper to make than the gas-heater of the "closed-cycle" plant, and it is a considerably simpler matter to allow the gases flowing out of the turbine to blow into the atmosphere than to provide a cooler and a supply of coolant.

The reversed heat engine: refrigerators and heat pumps

Nothing was said in the definition of *heat engine* about the directions of the heat and work interactions. Nevertheless the phrase, when unqualified, usually implies that there is a net work output and correspondingly a net inward heat transfer; often such engines are called *direct*

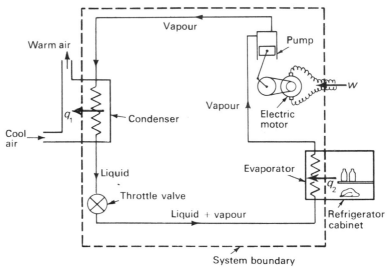

Fig. 10.5 Domestic refrigeration plant.

heat engines. However, there are important engineering applications in which the directions are reversed. These will now be discussed.

Fig. 10.5 illustrates a refrigerator of a type found in many homes. It has four elements: a condenser, a vapour compressor, an evaporator, and a throttle. Comparison of Fig. 10.1 with Fig. 10.5 shows that, to form the refrigeration plant, we may imagine the role of each component of the steam plant to be *reversed*: the boiler (an evaporator) becomes a condenser; the expanding device, the turbine, operates backwards and becomes a compressor; the condenser becomes an evaporator; and the compressing device, the pump, is replaced by an expander, the throttle valve. The working fluid, which may be carbon dioxide or ammonia but is more probably Freon,* executes a cycle but in an opposite sense to that executed by the steam. The condenser is at the higher temperature and pressure; from it, liquid flows through the throttle valve to the evaporator which is at

---

* A commercial refrigerant, e.g. refrigerant -12, dichlorodifluormethane, CCL_2F_2. The refrigerant number is usually preceded by the trade name of the manufacturer, e.g. Freon-12, Arcton-12, etc.

a lower pressure and temperature; thence the vapour passes to the compressor which pumps it into the condenser.

Examination of the dotted boundary in Fig. 10.5 shows that it encloses a system engaging continuously in heat and work interactions with its surroundings: the plant is a heat engine. However, the work, w, has an inward direction, from the electric supply to the motor; the heat transfer q_1, at the higher temperature, is *from* the working fluid; the heat transfer q_2, at the lower temperature, is *to* the working fluid.

If the purpose of the plant is to cause heat transfer *from* the low-temperature region, it is called a *refrigerator*. If its purpose is to provide heat transfer *to* the higher temperature region, it is called a *heat pump*.

Both refrigerators and heat pumps are *reversed* heat engines. The adjective "reversed", which here means "operating backwards", must not be

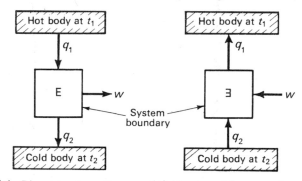

(a) Direct heat engine   (b) Reversed heat engine
Fig. 10.6 Diagrammatic representation of heat engines.

confused with another, which will be introduced in the next chapter, namely "reversible". Curiously, thermodynamic usage permits an engine to be reversed without being reversible.

It is customary to represent heat engines diagrammatically as shown in Fig. 10.6. The system boundary encloses the engine; q_1, q_2 and w represent respectively the heat and the work interactions, and their directions are indicated by the arrows at the system boundary. Fig. 10.6a shows a direct heat engine, E; Fig. 10.6b shows a reversed heat engine, ⌿ (i.e. E backward), which may be described as either a refrigerator or a heat pump.

The thermocouple as a heat engine

In order to demonstrate that the definition of *heat engine* covers devices which are not ordinarily regarded as engines, we consider the thermocouple system illustrated in Fig. 10.7. Two metal wires of dissimilar composition, for example copper and the copper-nickel alloy, constantan, are connected so that one junction communicates with a hot body and the other one with a cold body. A voltage difference develops and an electric current flows. If a motor is in the circuit, this may be used to perform work. The voltage difference increases with the temperature difference; but it is always very small (millivolts), and permits only tiny power outputs.

As will have been expected, the work output, if maintained, is accompanied by heat transfer from the surroundings to the system. Experimentally it is found that there are two heat transfers: an inward one, q_1, from the hot body, and a very slightly smaller outward one, q_2, to the cold body.

Examination of the dotted system boundary of Fig. 10.7 shows that the thermocouple system should be classed as a direct heat engine. Indeed it has the feature noted in every other example, though not specified in the

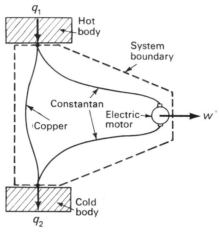

Fig. 10.7 The thermocouple as a heat engine

definition, namely the presence of an outward heat transfer to the cold body which amounts to a large proportion of the inward heat transfer; this feature appears to be wasteful.

Heat-Engine Performance

Heat-engine efficiency

Since the "wastage" of heat by engines appears to be so common, it is convenient to measure it by defining the *efficiency* of a heat engine, η, by

$$\eta \equiv \frac{w}{q_1} \tag{10.3}$$

where w is the net work output, while q_1 is the heat input at the higher temperature. η is a measure of the excellence of the heat engine in that it compares the work, w, which we want and obtain, with the heat, q_1, which we have to provide. Application of the First Law, eq. (6.1), to the engine gives: $w = q_1 - q_2$; hence eq. (10.3) may be written:

$$\eta = \frac{q_1 - q_2}{q_1} = 1 - \frac{q_2}{q_1} \tag{10.4}$$

Throughout this chapter, q and w will be expressed in identical units, in accordance with the special convention described on page 26; η is therefore dimensionless.

Actual values of the efficiency of heat engines are given in Table 10.1. It is seen that values above 35 per cent are rare. A significant trend is that η seems to increase with the maximum temperature employed in the cycle.

TABLE 10.1 *Efficiencies of Power Plants*

Plant	Date of installation	Working fluid	Maximum temperature of the working fluid	Efficiency %
Great Britain: central power stations.	1936–40	Steam	425 °C	28
Great Britain: West Thurrock (Conventional power station).	1962	Steam	565 °C	36
Great Britain: Calder Hall (Nuclear).	1956	Steam	310 °C	19
Great Britain: Dungeness "A" (Nuclear).	1965	Steam	390 °C	33
United States.	1956	Steam	650 °C	40
United States.	1949	Steam + mercury	510 °C	34
Switzerland: closed-cycle gas turbine.	1944	Air	690 °C	32
France: closed-cycle gas turbine.	1950	Air	680 °C	34

No internal-combustion plants are included in Table 10.1 because the definition, eq. (10.3), applies only to direct *heat engines*. Although it *is* possible to define an efficiency for an internal-combustion engine, this is a more sophisticated concept; it is discussed in Chapter 16.

Often the efficiency of eq. (10.3) is called the *thermal efficiency*, to distinguish it from other "efficiencies".

Eq. (10.3) is used only for heat engines operating "directly", that is to say producing net power. It is convenient at this point to introduce the corresponding quantities which are used for reversed heat engines.

Coefficients of performance of refrigerators and heat pumps

Eq. (10.3) defines efficiency as "what we want" divided by "what we have to pay for". The same principle is adopted for measuring the performance of reversed heat engines.

With refrigerator plants, "what we want" is heat transfer q_2 from the low-temperature region; "what we have to pay for" is the work w. Consequently the *coefficient of performance of a refrigerator*, C_{ref}, is defined as:

$$C_{ref} \equiv \frac{q_2}{w} \qquad (10.5)$$

TABLE 10.2 *Coefficients of Performance*

Location	Date	Type of plant	Working fluid	Upper temperature °C	Lower temperature °C	Coefficient of performance
Portsmouth, Ohio, U.S.	1940	Heat pump	Freon	29 36	−3 −5	3·6 3·2
Coshocton, Ohio, U.S.	1940	Heat pump	Freon	30 32 42	11 11 11	5·0 3·9 3·9
Norwich, Great Britain	1945	Heat pump	Sulphur dioxide	49	7	3·5
Zurich, Switzerland	—	Heat pump	—	50 25	River water River water	3·5 7·0
London (Festival Hall), Great Britain	1951	Heat pump	Freon	54	18	2·8
Great Britain	1951	Domestic refrigerator	Freon	32 32 32	−1 −7 −29	4·8 3·8 1·3
Great Britain	1945	Refrigerated rail car	Ammonia	49	−21	1·8

In a heat-pump plant, the interesting product is heat transfer q_1 to the high-temperature region; again w is the commodity which must be supplied. The *coefficient of performance of a heat pump*, C_{hp}, is therefore defined as

$$C_{hp} \equiv \frac{q_1}{w} \tag{10.6}$$

Since the application of the First Law gives: $w = q_1 - q_2$; and since the same plant may be regarded as either a refrigerator or a heat pump, we have the relations

$$C_{hp} = \frac{w + q_2}{w}$$

$$= 1 + C_{ref} \tag{10.7}$$

Typical practically-obtained values of C_{ref} and C_{hp} are contained in Table 10.2. It may be seen that the coefficients of performance are greatest when the temperature differences are least.

The 100%-Efficient Engine, E.100

To return now to "direct" heat engines, it is clear that all we have to do is to design one like that illustrated in Fig. 10.8. This transfers no heat to a

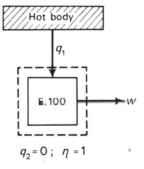

Fig. 10.8   The 100%-efficient heat engine: E.100.

cold body and therefore has an efficiency of 100 per cent. We will designate this engine the E.100.

Before discussing the possibility of attaining this ideal, it should be pointed out that the E.100 is more remarkable than perhaps appears at first sight: *it will be unnecessary to buy any fuel for it at all.* To show how this comes about, let us suppose that E.100 will work satisfactorily if it communicates with a hot body at a temperature of, say 500 °C. Rather than burn fuel to provide hot gases at this temperature, it will be cheaper in running costs to employ the arrangement of Fig. 10.9. This shows the E.100 driving a reversed heat engine Ǝ which is so arranged that it transfers just as much heat to the 500 °C body as is transferred to E.100. The reversed engine Ǝ does not require the whole work output of the E.100 however, because we have,

for Ǝ: $w = q_1 - q_2$ where both q_1 and q_2, stand for
positive numbers   (10.8)

and for E.100: $w_{E100} = q_1$ (10.9)

It follows that there is a net work output from the system within the dotted boundary of Fig. 10.9, given by:

$$w_{E100} - w = q_2 \qquad (10.10)$$

Now examination of the dotted boundary shows that the only heat transfer is that from the river. There is no heat transfer to the 500 °C body

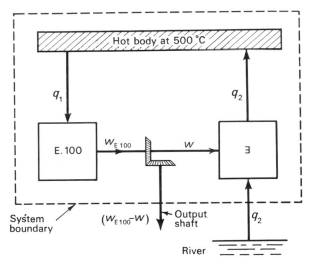

Fig. 10.9 How to make E.100 operate without using fuel.

from outside the boundary, and therefore no fuel has to be burned. All that is needed to keep the engine running is a plentiful supply of low-temperature material, e.g. the oceans, the atmosphere or the earth. A ship powered by an E.100 coupled to a heat pump could sail the seas for ever.

The Second Law of Thermodynamics

All attempts to build an E.100 have failed, and it is now accepted that the task is an impossible one. This fact of nature is embodied in the Second Law of Thermodynamics, which may be expressed as:

It is impossible to cause an engine to operate in a (thermodynamic) cycle, in which the only interactions are *positive* work done on the surroundings and heat transfer from a system which remains at constant temperature.

In symbols, the Law may be expressed as

$$\oint dW \leqslant 0 \qquad\qquad (10.11)$$

Single constant-
temperature system

Remarks. 1. The above form of the Second Law is due to Max Planck (1897). The essential idea of the Second Law was first conceived by Sadi Carnot in 1824, that is to say well before the First Law was established. Although other forms will be discussed later, this one is given prominence here because it most clearly shows the mechanical engineer the obstacle he faces.

The Second Law must not be taken merely as an expression of faint-heartedness on the part of engine builders. The Law has many consequences, some of which will be discussed below. To disprove the Second Law, it would suffice to disprove just one of its corollaries. No single instance of disproof is on record; as a result, the Law is now regarded as among the most firmly established of all the laws of nature.

2. In practice, the "constant-temperature system" has one or other of two main forms. It may consist of a body of material which is sufficiently large for its temperature to be uninfluenced by the heat interaction with the engine; a large river, the atmosphere, or a well-stirred mixture of ice and water are examples. This kind of constant-temperature system is sometimes called a "reservoir"; this word is a metaphor, and represents a legacy from the caloric theory which taught that heat could be stored or contained in a body. Alternatively, a constant-temperature system may be limited in extent but in communication with other systems which maintain the constancy of temperature; for example, the metal tubes of a boiler, though thin, are kept at constant temperature by radiation from the furnace.

3. Whereas the First Law states that heat and work are interchangeable, the Second Law states that complete conversion is possible only in one direction, namely from work into heat. In financial terms, work is a "freely-convertible currency"; heat is not. It is for this reason that such care has been necessary to distinguish heat from work in earlier chapters; only with heat and work defined as above is the Second Law true as stated.

The *directional* implication of the Second Law just mentioned is illustrated in Fig. 10.10; the entirely possible process on the left, which might for example represent Rumford's cannon-boring experiment, cannot be reversed. The concept of *reversibility* will prove to be of great importance below.

4. The 100-per-cent-efficient engine, E.100, is sometimes known as a *perpetual-motion machine of the second kind* (abbreviated to PMM 2). A perpetual-motion machine of the first kind (PMM 1) would be a contravener of the First Law: the system continuously produces work while isolated, as regards heat transfer, from its surroundings. Fig. 10.11 makes the point. Both are impossible.

5. The Second Law is also the reason for excluding internal-combustion engines from the heat-engine class, instead of regarding them as heat engines in which "the heat is in chemical form". For an engine *can* be

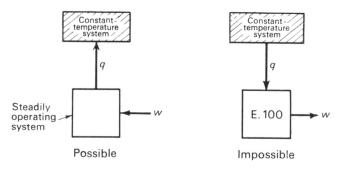

Fig. 10.10 A directional implication of the Second Law.

constructed which converts into work the whole of the "chemical energy", i.e. enthalpy change on complete isothermal reaction, of the fuel (p. 380).

6. The Second Law is a blank statement that the dearest wish of the power-plant engineer is unattainable. Of course we cannot leave the matter there, but must go on to ask "Well, how near can we get?" It is in answering this question, and looking for paths leading as closely as possible to the

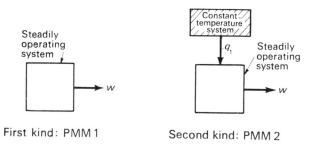

Fig. 10.11 Perpetual-motion machines.

unattainable, that the thermodynamicist has had to invent and use the concepts of reversibility, absolute temperature, and entropy. These matters will therefore occupy our attention for the next three chapters.

CHAPTER 10—PROBLEMS

10.1 A heat engine performs many cycles while doing work equal to $21 \cdot 5 \times 10^3$ N m and receiving a heat transfer of 90×10^3 J. Evaluate the efficiency of the engine and the heat transfer from the working fluid.

10.2 A heat engine working at the rate of 100 kW has an efficiency of 20 per cent. Evaluate the magnitudes of the heat-transfer rate to and from the working fluid.

10.3 The efficiency η of a certain type of ideal closed-cycle gas-turbine engine (Fig. 10.3) is given by

$$\eta = 1 - \frac{1}{(r)^{\gamma - 1}}$$

where r is the "pressure ratio", i.e. the ratio of the pressures at outlet from and at inlet to the compressor and γ equals c_p/c_v, the ratio of the specific heats of the working fluid.

(a) Evaluate η for $r = 7$ assuming air (for which $\gamma = 1\cdot4$) to be the working fluid.

(b) The heat-transfer rate from the air in the cooler is $3\cdot2 \times 10^6$ J/s. Evaluate the heat-transfer rate to the air in the heater, and the power developed, assuming all other heat transfers to be negligible.

10.4 A steam power plant (Fig. 10.1), incorporating a reciprocating steam engine, gave the following results during a test under steady conditions.

Boiler.
   Steam-outlet conditions: pressure, 700×10^3 N/m²; temperature, 200 °C.
   Feed water: temperature, 55 °C; mass flow rate, 0·0262 kg/s.

Engine.
   Shaft power: 6·5 kW.
   Steam-inlet conditions: as for boiler outlet.

Condenser.
   Cooling water: flow rate, 0·73 kg/s; temperature rise, 19 K.

There are unmeasured heat transfers to the atmosphere from the exposed hot surfaces of the plant. Fluid velocities may be assumed to be negligible.

Evaluate, per unit mass of working fluid (H_2O):

(a) the heat transfer to the H_2O in the boiler;
(b) the external work done by the H_2O in the engine;
(c) the heat transfer from the condensing steam in the condenser, assuming the heat transfer from the condenser casing to the atmosphere to be zero;
(d) the heat transfer to the atmosphere assuming the feed-pump work to be zero;
(e) the efficiency of the plant.

10.5 In a steady-flow closed-cycle gas-turbine engine (Fig. 10.3), the temperature and velocity of the working fluid (air) between the various components are as follows:

Cooler and compressor: 30 °C,   90 m/s.
Compressor and heater: 230 °C,   30 m/s.
Turbine and cooler: 520 °C,   100 m/s.

The velocity of the air entering the turbine is 300 m/s and the heat transfer to the air in the heater is 642×10^3 J/kg. All heat transfers to the atmosphere may be assumed to be negligible.

Calculate: (a) the temperature of the air at entry to the turbine;
(b) the heat transfer from the air in the cooler;
(c) the net work output of the plant;
(d) the efficiency.

Assume that the enthalpy of air is a function of temperature only and that the specific heat at constant pressure c_p equals $1\cdot005 \times 10^3$ J/kg K (see problem 7.7).

10.6 In a reversed heat engine, the work done on the engine is 75×10^3 N m and the heat transfer to the engine from the low-temperature region is 220×10^3 J. Evaluate the heat transfer to the high-temperature region and the coefficient of performance as a refrigerator and as a heat pump.

10.7 The coefficient of performance of a heat pump is 5 when the power supplied to it is 50 kW.

(a) Evaluate the magnitudes of the heat-transfer rates to and from the working fluid.

(b) The heat transfer from the heat pump is used to heat the water flowing through the radiators of a building. Evaluate the mass flow rate of the heated water given that its temperature increases from 50 °C to 70 °C in flowing through the heat pump. Assume the water velocity to be negligible.

10.8 This problem is stated in terms of the Imperial system of units and illustrates concepts commonly used in refrigeration engineering. Find the coefficient of performance and the heat transfer in the condenser of a refrigerator (in Btu/hp h) which has a capacity of 0·9 tons of refrigeration per hp. (The "ton of refrigeration" is a unit peculiar to the refrigeration industry; it is equivalent to a heat-transfer rate of 200 Btu/min. The unit was originally intended to represent the heat-transfer rate required to produce 1 U.S. short ton of ice, of "latent heat" 144 Btu/lb$_m$, in 24 hours from water at the same temperature.)

10.9 In a refrigerating plant (Fig. 10.5), the states of the working fluid (Freon-12) between the various components are as follows:

Evaporator and compressor: wet vapour at a temperature of −15 °C.

Compressor and condenser: dry-saturated vapour at a temperature of 30 °C.

Condenser and expansion valve: saturated liquid at a temperature of 30 °C.

Expansion valve and evaporator: wet vapour at a temperature of −15 °C.

The heat-transfer rate from the Freon in the condenser is 1·5 × 10³ J/s and the power required to compress the Freon is 310 W. All heat transfers to the atmosphere, and also fluid velocities, may be assumed to be negligible.

Sketch the cycle on an enthalpy∼pressure diagram and, using the data given below, calculate:

(a) the mass flow rate of the Freon in kg/s;

(b) the heat-transfer rate to the Freon from the cold region;

(c) the enthalpy of the Freon after the expansion valve;

(d) the enthalpy and dryness fraction of the Freon at entry to the compressor;

(e) the coefficient of performance.

Extract from table of properties of Freon-12 (Datum for enthalpy: saturated liquid at −40 °C):

Pressure 10⁶ N/m²	Saturation temperature °C	Specific enthalpy 10³ J/kg	
		Saturated liquid	Saturated vapour
0·1825	−15	22·3	181·0
0·745	30	64·6	199·6

10.10 A heat engine is used to drive a heat pump. The heat transfers from the heat engine and from the heat pump are used to heat the water circulating through the radiators of a building. The efficiency of the heat engine is 27 per cent and the coefficient of performance of the heat pump is 4. Evaluate the ratio of the heat transfer to the circulating water to the heat transfer to the heat engine.

10.11 Show that if invention 2, page 15, functioned successfully, a perpetual motion machine of the first kind would have been devised

11 Reversibility

Introduction

In this chapter we introduce the idea of the *reversible engine* and show that its efficiency in given circumstances is the highest possible. This prompts the desire to see how *reversibility* can be achieved, and how its opposite, *irreversibility* can be identified and eliminated. A number of practical processes will be shown, by reference to the Second Law, to involve irreversibility; they include friction, unresisted expansion, heat transfer with a finite temperature difference, and combustion. As far as possible these processes must be avoided in engines. This prescription will be illustrated by discussion of the engine cycle suggested by Carnot.

Symbols

C_{hp}	Coefficient of performance of a heat pump.		t	Temperature.
F	Force.		w	Net work done by (on) a heat engine.
g	Gravitational acceleration.		w_x	External work.
g_c	Constant in Newton's Second Law.		η	Efficiency of a heat engine, thermal efficiency.
m	Mass.			
p	Pressure.		*Subscripts*	
Q	Heat transfer.		C	Cold constant-temperature system.
q	Heat transfer.			
q_1	Heat transfer between heat engine and hot constant-temperature system		H	Hot constant-temperature system.
			I	Irreversible.
q_2	Heat transfer between heat engine and cold constant-temperature system.		R	Reversible.
			1, 2, ...	States of the working fluid.

Reversibility and Engine Efficiency

The reversible engine

Suppose that a reciprocating steam "engine" is driven backwards. It will act like the reciprocating compressor of a refrigerator. Similarly a turbine, when driven backwards, will act like a rotary compressor, albeit inefficiently. "Inefficiently" in this connexion means that much more work will be necessary, to drive one kilogram of steam from the low pressure to the high pressure, than could be done by the same kilogram of steam when flowing in the normal direction.

It is, however, possible to *imagine* a heat engine, i.e. a complete power plant, all components of which would work *just as well* backwards as forwards. Such a heat engine will be called a *reversible engine*. By way of definition, we state:

A heat engine which engages in heat transfer with two systems of fixed, but different, temperatures, is reversible if its efficiency when operating directly is equal to the reciprocal of its coefficient of performance when operating as a heat pump.

i.e.
$$\eta_R = \frac{1}{C_{hp,R}} \qquad (11.1)$$

Fig. 11.1 illustrates the definition. E_R is the engine when producing power.

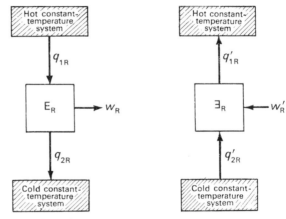

Reversible engine operating directly.

The same reversible engine operating as a heat pump between the same constant-temperature systems.

Fig. 11.1.

$\exists_R$ is the engine working as a heat pump. The definition states therefore that:

$$\frac{w_R}{q_{1R}} = \frac{w'_R}{q'_{1R}} \qquad (11.2)$$

An important consequence of the definition of a reversible engine

We now show that the efficiency of a reversible engine is independent of the natures of the engine and of the constant-temperature systems.

Proof. Consider the reversible engines and constant-temperature systems made of differing materials, shown in Fig. 11.2. Let us suppose that the efficiency of engine $E_{R,A}$ exceeds that of $E_{R,B}$. Then reverse $E_{R,B}$ (i.e. set $\exists_{R,B}$ in operation) and cause $E_{R,A}$ to drive it so that the same heat q_1 is returned to the hot system H_B as is withdrawn from H_A. H_A and H_B are in good thermal contact so that no temperature differences arise between them. Then since η_A exceeds η_B, we have

$$\frac{w_A}{q_1} > \frac{w_B}{q_1} \tag{11.3}$$

so

$$w_A - w_B > 0 \tag{11.4}$$

Now C_A and C_B, which will also be supposed in good thermal contact, are at the temperature t_2 and so form a single cold constant-temperature

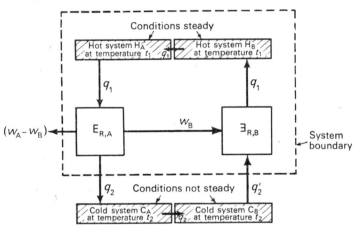

Fig. 11.2.

system. The combination of $E_{R,A}$ and $\exists_{R,B}$, together with H_A and H_B, form a system operating continuously, doing positive work, and engaging in heat transfer with a constant-temperature system.* Since this is impossible according to the Second Law, our original supposition must have been incorrect. **Therefore $E_{R,A}$ cannot be more efficient than $E_{R,B}$.**

In the same way it can be shown that $E_{R,B}$ cannot be more efficient than $E_{R,A}$. It follows that the efficiencies of $E_{R,A}$ and $E_{R,B}$ must be equal; so, in general, all reversible engines have the same efficiency when operating between two systems of fixed but different temperatures. The efficiency does not depend on the nature of either the constant-temperature systems or the reversible engines.

Corollaries. (i) The result just proved, together with eq. (11.1) which defines the reversible engine, implies that *all reversible engines acting as heat*

---

* N.B. The boundary of the combined system must exclude the cold systems, since they are not in a steady state; the net heat transfer from them is $q_2' - q_2$.

pumps, operating between systems at the same two temperatures, have equa
coefficients of performance.

(ii) If only engine B is supposed reversible, the above argument must stop
at the point at which it is shown that engine A cannot be more efficient
than engine B. We deduce that *the efficiency of a reversible engine* (and so of
all reversible engines) *is the greatest that can be exhibited by any engine
operating between systems having the same two temperatures.*

(iii) If only engine A is supposed reversible, the heat pump $\exists_B$ (N.B. we
have dropped the subscript R, for an obvious reason) has, it may be shown
by a simple modification of the argument, a coefficient of performance
which cannot exceed that of engine A when *it* is reversed. We deduce that
the coefficient of performance of a reversible heat pump (and so of all reversible heat pumps) *is the greatest that can be exhibited by any engine operating
between systems having the same two temperatures.*

Definition of irreversible engines and heat pumps. In practice, real engine
have efficiencies which are lower than the maximum possible, for reasons
that we shall discuss; similarly, real heat pumps have coefficients of performance that are lower than the maximum. We distinguish these from reversible machines, for which η and C_{hp} always *equal* the maximum values, b
calling them *irreversible*. In symbols, with the subscript I denoting irreversibility, we have:

$$\eta_I < \eta_R \tag{11.5}$$

and $$C_{hp,I} < C_{hp,R} \tag{11.6}$$

Reversible Processes

So far the term "reversible" has been applied only to heat engines and, b
inference, to the cyclic processes which they perform. If a heat engine
known to be irreversible, it is important to examine which features render
so. For this, an extension of meaning is made to non-cyclic processes a
follows:

**A process is reversible if, after the process has been completed, means ca
be found to restore the system and all elements of its surroundings to the
respective initial states.**

Discussion. 1. This definition clearly includes the processes carried o
by reversible heat engines: heat engines operate cyclically in any case, a
the system is easily restored to its initial state; and, if the engine is reve
sible, the effects on the surroundings of the heat interactions q_1 and q_2 ar
the work interaction w_R are exactly cancelled when the same transfers tak
place in the opposite direction. An irreversible engine on the other hand
requires, for example, a larger work interaction to restore q_1 to the h

system when operating reversed than is developed when q_1 is supplied to the engine when operating directly; as a result there is a net heat transfer to the low-temperature system which causes permanent changes in the surroundings.

2. We require only that "means can be found" to restore the initial states, but do not demand that these states shall be restored automatically.

3. Any process which is not reversible is *irreversible*. This is an extension of the definition. The way to decide whether a process is reversible or irreversible is to apply the only natural law at our disposal which concerns direction: the Second Law of Thermodynamics. This will be done below.

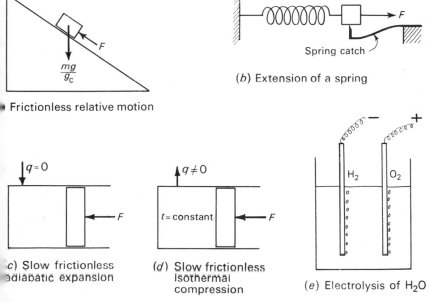

(b) Extension of a spring

Frictionless relative motion

(c) Slow frictionless adiabatic expansion

(d) Slow frictionless Isothermal compression

(e) Electrolysis of H_2O

Fig. 11.3 Reversible processes.

Examples of reversible processes

We now consider a number of processes which are important in engineering and which may be regarded as reversible. It will be seen that in practice departures from reversibility always occur to a limited extent; but often these imperfections are small. The examples are illustrated in Fig. 11.3.

(a) *Frictionless motion of solids.* A block accelerates down a smooth plane under the influence of the gravitational force. The process is reversible because, if at the base of the plane the block were led along a smooth runway and up another inclined plane, it would eventually come to rest, reverse direction, and return to its original state. Then the system (block) and all elements of its surroundings would be restored.

(b) *Extension of a spring.* The appropriately-varying force F slowly extends a spring until a weak catch is engaged. Then relaxation of F does not cause the process to be reversed, for the catch prevents this. Nevertheless the process is reversible because *means can be found* to restore the initial state, namely lowering the catch and infinitesimally relaxing of F. The work involved in moving the catch is neglected.

(c) *Slow adiabatic expansion of a gas.* The gas in the cylinder expands without heat transfer against a frictionless piston which is acted on by the external force Γ. The process is reversible because an infinitesimal increase in F will suffice to cause the system and its surroundings to trace out each step in the process in the reversed order.

(d) *Slow isothermal compression of steam.* The force F slowly pushes the leak-proof frictionless piston inward, thus causing the steam in the cylinder to condense. The steam transfers heat to the walls of the cylinder. Since the process is very slow, this heat transfer necessitates only an infinitesimally higher temperature in the steam than in its surroundings. The process is reversible because, when the force F is reduced by a negligible amount, the direction of movement reverses, the steam temperature drops infinitesimally below its previous value, the direction of heat transfer reverses, and the condensate re-evaporates. Thus the process is traced out in reverse and the initial states of system and surroundings are restored.

(e) *Electrolysis of water.* A potential difference of 1·47 volts exists between the plates of a cell containing water rendered electrically conducting by dissolved potassium hydroxide. A small current passes and bubbles of hydrogen appear at the negative electrode and bubbles of oxygen at the positive electrode. If the potential difference is slightly reduced, the process stops but does not go backwards: the bubbles of gas do not start travelling downwards to the plates and disappear there. Nevertheless the process is reversible (apart from the motion of the bubbles) because *means can be devised* by which the hydrogen and oxygen are recombined to form water and produce electrical power. Such a device is known as a *fuel cell*, and requires special electrodes. The principle is the same as that of the lead-acid accumulator but the technique is more difficult. So far it has proved possible to recover in re-combination only about 65 per cent of the work expended in electrolysis (see bibliography at the end of this chapter).

Irreversible Processes

In this section the main types of irreversibility which occur in engines will be discussed; their irreversible nature will be proved by reference to the Second Law. The procedure of proof will be the same in each case, namely:

(i) Describe the process, P, noting in particular its end-states and the heat and the work.

(ii) Assume that the process is reversible, i.e., that a reverse process, exists, in which the *magnitudes* of the heat and work, and of the changes of properties, are the same, but their *directions* are reversed.

(iii) Devise a mechanism making use of ꟼ as part of a cyclic process which amounts to a heat engine of 100 per cent efficiency.

(iv) Invoke the Second Law to prove that the hypothesis (ii) was incorrect.

The purpose of the proofs is two-fold. First, they demonstrate that the irreversible natures of the various processes are not a miscellany of unconnected facts, but are all manifestations of a single law. Secondly, they show

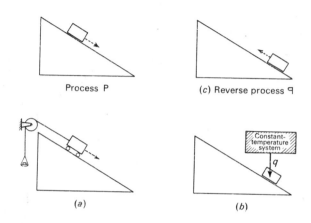

Fig. 11.4 Processes used in proving that solid friction is irreversible.

not merely that the processes fail, of themselves, to go backwards, but that nothing that the wit of man can devise can succeed in undoing their effects. This means that it is no good leaving friction in a machine, for example, and then hoping to nullify its effect by a friction-cancelling device: the friction *must* be eliminated if the machine is to give the highest performance.

Solid friction

An example of friction between solids in relative motion will be considered. Fig. 11.4 illustrates this.

(i) In the process P, the block slides slowly down the rough inclined plane. Both are good conductors of heat but there is no heat or other interaction with their surroundings. Both block and plane increase in temperature during the process.

(ii) Assume that a mechanism can be devised to execute the reverse process ꟼ, the end-states of which are such that the block rises up the plane to its previous starting position, that both block and plane decrease in temperature, and that there are no net heat or work interactions with the surroundings; i.e., the effects of ꟼ completely cancel those of P in the block, in the plane and in their surroundings.

(iii) Cause the system comprising the block and the plane to execute the following cycle:

(a) The block starts at the top of the incline. It is placed on rollers and attached by a rope and pulley to a suitably chosen weight (Fig. 11.4a). The block then rolls down, so raising the weight.

(b) Heat transfer occurs between the system and a constant-temperature system ("reservoir") at a higher temperature; it raises the temperature of the system to that which would have been reached if the block had slid down under friction (Fig. 11.4b).

(c) The process Я is then carried out (Fig. 11.4c), with the result that the block returns to its starting point and the temperatures fall to their initial value. The cycle is thus completed.

(iv) The result of this cyclic process is that the work done in raising the weight is exactly equal to the heat transfer from the "reservoir". The system is therefore a heat engine of 100 per cent efficiency; but this is impossible according to the Second Law of Thermodynamics. Now processes (a) and (b) can certainly be carried out; it follows that it is the reverse process Я which is impossible. Therefore the process P is irreversible.

Q.E.D.

Discussion. The conclusion is valid for all forms of solid friction such as that between brake block and drum, between piston rings and cylinder wall, and between a shaft and its bearing. All types of solid friction must be reduced to the practical minimum if an engine is to achieve maximum efficiency. This is done by the use of lubricants and by keeping bearing loads and rubbing velocities as low as possible.

Fluid friction

The existence and causes of fluid friction have been mentioned in Chapter 3 (p. 69). As an example, we now consider the long thermally-insulated steam pipe shown in Fig. 11.5.

(i) The steam flows steadily from point 1 to point 2. The enthalpy remains constant but the pressure falls. There are no heat or external-work interactions with the surroundings, although there is in general a small net flow work. Let the process be called P.

(ii) Assume that a mechanism can be devised to execute the reverse process Я, such that the steam flows steadily from state 2 to state 1 without external work or heat interactions with the surroundings.

(iii) Imagine the steam to execute a cyclic process by means of the following steps:

(a) Instead of the process P, the steam flows through an adiabatic turbine performing shaft work. Its enthalpy decreases, but the initial and final pressures and velocities are as in P.

(b) The steam then passes through a heater in communication with a "reservoir" to restore its temperature, and therefore its enthalpy, to the values prevailing at state 2.

(c) Process $\mathfrak{P}$ is applied, to return the steam to its initial state 1. This completes the cycle, which may be repeated indefinitely.

(iv) In this cyclic process, all the heat transfer from the "reservoir" is converted into shaft work; the circuit therefore comprises a heat engine

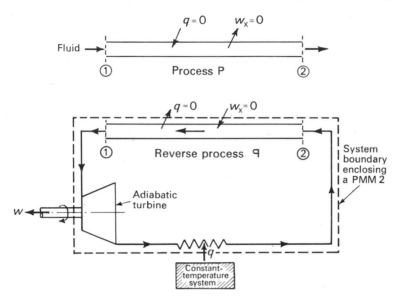

Fig. 11.5 Demonstrating that fluid friction is irreversible.

of 100 per cent efficiency. But this is impossible according to the Second Law. We conclude that the reverse process $\mathfrak{P}$ is impossible, and, hence, that P is irreversible.   Q.E.D.

Discussion. It follows that the frictional flow of fluids must be reduced to the practicable minimum in engines. This is accomplished in practice by keeping flow velocities low, by using smooth pipes of the largest permissible diameter, and by avoiding any unnecessary length of pipe.

Unresisted expansion

At the end of the working stroke of a gasoline or diesel engine, the gases in the cylinder are at a pressure greater than atmospheric (e.g. 3 atm). The exhaust valve then opens and the gas rushes into the exhaust pipe, eventually emerging into the atmosphere at atmospheric pressure and relatively low velocity. This is illustrated in Fig. 11.6; it will be used as an example of an unresisted-expansion process.

(i) In the process P, gas expands from state 1 in the cylinder to state 2 where it is partly outside the cylinder. Work is done on the atmosphere; the internal energy, pressure and temperature of the gas are reduced. Heat transfer and mixing of gas and air will be neglected.

(ii) Assume that a process q can be devised, the net effect of which is to restore the gas to its initial state and position without heat transfer and with

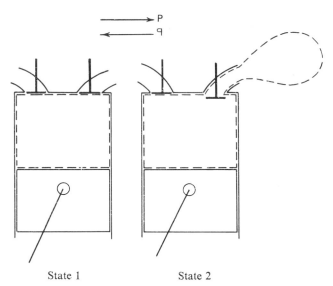

State 1   State 2

Fig. 11.6 Diagram used in the proof that unresisted
expansion is irreversible.

an amount of external work equal exactly to that done on the atmosphere in P.

(iii) Imagine the gas to execute a cyclic process by the following steps:

(a) A piston fitted in the (lengthened) exhaust pipe allows the gas to expand slowly from state 1 to the atmospheric pressure. The piston is connected to a mechanism so that its movement causes weights to be raised. The internal energy and temperature of the gas meanwhile fall below those corresponding to state 2.

(b) Heat is transferred to the gas in the exhaust pipe from a sufficiently hot "reservoir" so that its temperature and internal energy rise to the values of state 2, meanwhile keeping the pressure constant. The gas expands still further in this process and so raises additional weights. It will be found that the work done on the atmosphere during (a) and (b) is the same as in process P.

(c) Process q is applied to restore the gases from the exhaust pipe to the cylinder at state 1. The cycle has now been completed.

(iv) As in the previous examples we have designed an E.100. But this is an impossibility. Therefore P is irreversible.   Q.E.D.

Discussion. A practical implication is that, as far as possible, all gas expansions should be "resisted"; this partly explains the popularity of exhaust-gas turbines in diesel-engine practice. Another device for reducing the irreversibility of the sudden opening of exhaust valves is to keep the flow fast but concentrated into orderly pressure waves which, with proper exhaust-pipe tuning, assist in scavenging and so reduce or eliminate the power input to the scavenging blower.

Heat transfer with a finite temperature difference

When high-pressure steam first enters the cylinder of a reciprocating steam engine, it is at a higher temperature than the cylinder walls, which

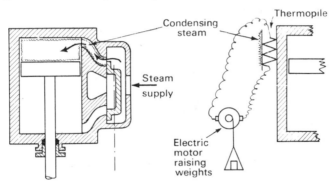

Fig. 11.7 Diagram used in proving that heat transfer with a finite temperature difference is irreversible.

have just been in contact with expanded steam at a lower temperature. Some of the entering steam therefore condenses on the walls, to which heat is transferred. This will serve as an example of the irreversibility of heat transfer with a finite temperature difference, Fig. 11.7.

(i) In the process P, a mass m of steam condenses on the walls of the cylinder, the temperature of which rises from t_1 to t_2; let the associated heat transfer, from the steam to the walls, be q. The steam pressure falls from p_1 to p_2. The accompanying motion of the piston is immaterial to the problem and will be ignored for simplicity.

(ii) Assume the process ꟼ exists, which has as its only net effects the fall in wall temperature from t_2 to t_1, a heat transfer from wall to steam of q, and the consequent re-evaporation of a mass m of steam, the pressure of which rises from p_2 to p_1.

(iii) (*a*) Replace the process P by another one in which the steam con-
denses at constant volume on the surface of a thermopile (assembly of thermocouples), the cold junctions of which are in contact with the

cylinder walls. The thermopile drives a small electric motor which does work by raising weights. When the mass m of steam has condensed, the circuit is disconnected. It is found that the cylinder walls have risen from t_1 to t_2', where $t_2' < t_2$.

(b) Suppose that heat is transferred to the walls from a "reservoir" to cause their temperature to rise from t_2' to t_2.

(c) Then process ꟼ is applied to re-evaporate the condensate and lower the cylinder-wall temperature from t_2 to t_1. This completes the cycle, steam and walls being in their initial states.

(iv) The result is that the system (mass m) executes a cycle in which positive work is done and heat transfer occurs from a "reservoir"; according to the Second Law this is impossible. The only doubtfully possible step in (iii) was ꟼ; we conclude that it is impossible. It follows that P is irreversible.   Q.E.D.

Discussion. 1. In the reciprocating steam engine, the ill-effects of the process just considered are sometimes mitigated by adopting the *Uniflow* principle, in which the steam exhausts through ports near the end of the piston stroke but enters at the ends of the (double-acting) cylinder. The entry ends of the cylinder therefore tend to stay at a higher temperature than the middle (exhaust) part of the cylinder. One of the advantages of the steam turbine is that its blades stay in regions of constant temperature so that alternate condensation and re-evaporation does not occur.

2. The necessity to avoid, as far as possible, heat transfer with finite temperature differences has far-reaching effects in power-plant engineering. It governs the design of boiler plant and dictates the use of feed-water heating by bled steam; it is one reason for the use of the counter-flow arrangement in heat-exchange plant; and it plays a part in the tendency to use re heating and inter-cooling in gas-turbine prime movers.

3. Of course it is impracticable to avoid this form of irreversibility entirely, because the smaller the temperature difference across, say, a boiler tube, the lower the *rate* of heat transfer in J/m² s and so the larger the boiler *size* for a given duty. Design has to be a compromise in which economic factors play a large part: large size entails high capital costs; large temperature differences cause low efficiency and so high running costs; these have to be balanced, and the optimum arrangement will depend on the purpose and required life of the plant.

4. The irreversibility of heat transfer with finite temperature differences forms the core of an alternative statement of the Second Law due to Clausius:

It is impossible for a system working in a cycle to have as its sole effect the transfer of heat from a system at a constant low temperature to a system at a constant higher temperature.

We have just proved this postulate by reference to Planck's statement of the Second Law. It is no less easy to derive the latter starting from Clausius's statement.

5. In general there are at least as many possible statements of the Second Law as there are irreversible processes. We could, if we wished, state the law as "Friction is irreversible"; then all the rest would follow.

Combustion

Nearly all power plants employ the combustion of fuel in some way. True heat engines, in practice, often use the combustion products of

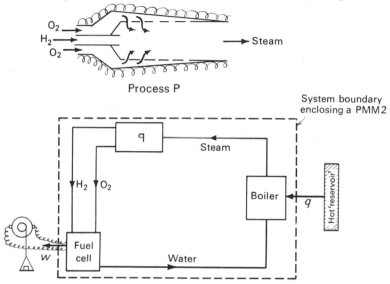

Fig. 11.8 Demonstrating that combustion is irreversible.

fuel and air as their hot constant-temperature system; internal-combustion engines, on the other hand, use the product gases as a working fluid. As an example which will serve to show the irreversibility of these processes, we will consider the steady-flow adiabatic gas burner shown in Fig. 11.8.

(i) In the process P, streams of hydrogen and oxygen at atmospheric temperature and pressure enter the combustion chamber, where they react in a flame. A steady stream of steam (the combustion product) leaves the chamber at high temperature but still at atmospheric pressure. For simplicity, we suppose there is no excess hydrogen or oxygen. The combustion chamber is thermally insulated so there is no heat transfer to the surroundings. There is no shaft work.

(ii) Assume that a device exists which can execute the reverse process P; i.e. assume that, without heat transfer or shaft work, the hot products can be steadily reconverted into the cold oxygen and hydrogen streams.

(iii) Then imagine the following cyclic process:

(a) Atmospheric streams of hydrogen and oxygen are supplied to a fuel cell. There they re-combine to form water, while electrical power is produced which can be used for raising weights. Probably the water produced will be at a higher temperature than the gases supplied.

(b) The water is now passed to a boiler and engages in heat transfer with a hot "reservoir", emerging as steam at atmospheric pressure and at the temperature of the steam leaving the combustion chamber in P.

(c) Then ꟼ is applied so as to turn the steam back into hydrogen and oxygen at the initial temperature. This completes the cycle.

(iv) Once again a perpetual-motion machine of the second kind has been devised. The Second Law states this to be impossible; hence P must be irreversible.   Q.E.D.

Discussion. 1. Generalizing the above result, we may conclude that all combustion processes are irreversible, whatever the fuel and whether carried out in steady flow or at constant volume. They should therefore be avoided if it is desired to extract the maximum work from limited fuel supplies.

2. Unfortunately this is easier said than done. However, the hydrogen-oxygen reaction has been carried out in fuel cells for several years. More recently (1964–65) fuel cells employing methyl alcohol and air have been operated successfully, with efficiencies of from 20 to 40 per cent; the reader is referred to the bibliography given at the end of the chapter for further details on these advances in technology.

Although substantial engineering problems have had to be solved to bring the fuel cell to its present stage of development, immensely greater ones will require solution before the coal-air reaction, for example, can also be carried out in a fuel cell. Nevertheless it should never be forgotten that the use of combustion to obtain power is a second-best, which yields only about one third of the power from each kilogram of fuel which the Second Law would permit us to obtain. The same is true of the nuclear reactions currently used in atomic power stations; indeed the fraction there still smaller.

Other irreversible processes

The above examples by no means exhaust the irreversible processes of importance in engineering. A list of other such processes is appended. The reader is invited to test his grasp of the foregoing method of argument by demonstrating their irreversibility by reference to the Second Law. They are:

(a) Flow of electric current through a resistor. (Flows through inductors and capacitors on the other hand are reversible.)

(b) The flow of a river over a waterfall.

(c) Unresisted expansion of an elastic structure.

(d) Plastic deformation of a material.

(e) Magnetization of a material exhibiting hysteresis.

(f) "Hydraulic jump" in the flow of water in a channel.

(g) The mixing of two unlike fluids. (To prove the irreversibility of this it is necessary to postulate membranes which are permeable to one fluid but not to the other; these membranes are then used as pistons. Such membranes exist for some pairs of fluids.)

The Carnot Cycle

Now that we have seen what must be avoided in heat-engine designs, an ideal engine will be considered which embodies these lessons. It was devised

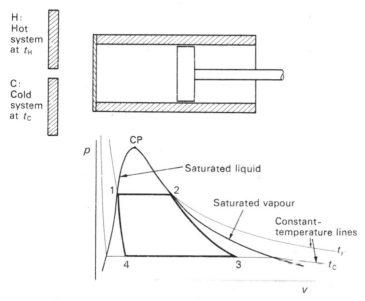

Fig. 11.9 Carnot cycle with H_2O as the working fluid.

by Carnot at the early date of 1824. Although never realized completely, modern steam and closed-cycle gas turbine power plants approach it as nearly as is considered economically practicable.

Description of the Carnot cycle and of a machine for performing it

Figure 11.9 illustrates a cylinder closed by a piston. There are no valves. The cylinder barrel and the piston are thermally non-conducting, but the end opposite the piston is thin and a good conductor. Vertically below the sketch of the cylinder is an indicator diagram, which, in this case, is simultaneously a $p \sim v$ property diagram for the working fluid in the cylinder. This we shall suppose to be steam, although Carnot thought of a permanent gas. Two constant-temperature systems H and C can be caused to communicate with the cylinder head.

Starting with the piston in the inmost position 1, let the contents of the cylinder be saturated water at the temperature of the hotter system, t_H. H is caused to communicate with the cylinder head, and the piston is allowed to move outwards slowly. The latter is connected to a mechanism which raises weights. Friction is supposed to be absent. Since the motion is slow, heat transfer, q_1, from H to the fluid occurs with only an infinitesimal temperature difference, so the fluid executes an *isothermal expansion* to point 2, say. In this case, though not with the permanent gas used by Carnot, the expansion is also at constant pressure. We will suppose the expansion to end when all the water has just evaporated. State point 2 is therefore on the saturated-steam line of the $p{\sim}v$ diagram.

Now H is removed from the cylinder head, but the piston continues to move slowly outwards; the steam performs an *adiabatic and reversible expansion* until the steam temperature falls to the temperature of the cooler system t_C. The pressure will also be lower, and some of the steam will, as it happens, have condensed (point 3).

At this stage, C is caused to communicate with the cylinder head and the direction of the piston movement is reversed. Heat, q_2, is transferred from the steam to C with an infinitesimal temperature difference, and the steam slowly condenses *isothermally*.

Before condensation is completed, however, C is removed. The steam state is now represented by point 4, which has been so chosen that the continuation inwards of the piston motion causes a *reversible adiabatic compression* which returns the steam exactly to state 1.

The steam has therefore performed a cyclic process. All the parts of the process are reversible; indeed they can in this case be simply reversed. The whole cycle is therefore reversible. We note that heat is transferred to the steam by H at t_H, and from the steam by C at t_C. In addition the finite area of the indicator diagram and the sense of rotation show that finite work w, has been done.

The Carnot engine is therefore a reversible heat engine which can in principle be used equally well as a direct heat engine to produce power or in reverse to act as a heat pump. Since it is reversible, its efficiency when when working directly as a heat engine $\left(\eta_R = \dfrac{w}{q_1}\right)$ will be equal to the reciprocal of its coefficient of performance when working reversed as a heat pump $\left(C_{hp,R} \equiv \dfrac{q_1}{w}\right)$.

Remarks on the practicability of the Carnot cycle. 1. The realization of a Carnot engine depends partly on keeping the motion slow, so that at any instant the pressure of the working fluid should be uniform (this is not difficult), and so that the working fluid temperature should be uniform. This would require a very large engine of very small power; for the power depends not only on the work per stroke but also on the strokes per unit time. Such an engine would be excessively expensive to build. However the ideal is approached in practical plant by carrying out the functions of heating and cooling in separate organs, the boiler and condenser, where the working fluid can remain in contact with the heat-transfer surfaces for much longer times than it spends in the cylinder. It is significant that the sizes of the boiler and of the condenser in a steam plant are very much greater than that of the engine cylinder.

2. It has also been postulated that friction between the piston, or rather its rings, and the cylinder barrel, is zero. All that can be done in practice is to reduce the irreversibility due to this cause, when expressed in terms of work, to a small fraction of the useful power output.

3. No materials have been discovered which fail to conduct heat entirely; moreover those which are known to be poor conductors will not stand high stress. If therefore heat transfers to cylinder and piston are to be eliminated, the time spent by the working fluid in the cylinder must be kept small. This is another reason for performing the necessary heat transfers in separate organs, as was realized by the early steam-engine builders (Savory, 1698, and Newcomen, 1705, in the case of the boiler; and Watt, 1763, in the case of the condenser), well before Carnot was born (1796).

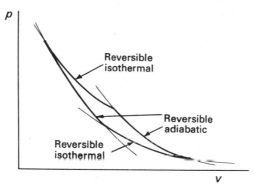

Fig. 11.10 Carnot cycle with a gas as the working fluid.

4. An additional engineering requirement for realizing Carnot's cycle is correct timing; for example, if C is removed too late, the adiabatic compression will cause the steam to arrive at the inner dead centre at a temperature lower than t_H; then an irreversible transfer of heat will occur. In an engine with a separate boiler and condenser, this difficulty appears in a different form: the timing is carried out by means of valves which open and shut to control conditions in the cylinder. This is true of internal-combustion engines also; it is not the easiest part of the engine-designer's task. Usually the design must be a compromise; for an engine must operate under varying loads: the valve timing that is correct under one condition will not give the best performance under another.

5. Steam has been employed as the working fluid in the above example,

so that the isothermal processes should be at constant pressure; the control of the process is easy to visualize. The condensable nature of the fluid is however in no way essential to the argument. With a permanent gas as the working fluid, as Carnot suggested, the indicator (and $p{\sim}v$) diagram would be as shown in Fig. 11.10. Even with steam, it may be added, there is no need for any of the "corners" of the diagram to be situated on the saturation line. Indeed no engineer would care to contemplate what might actually happen at the end of the compression stroke 4–1 when the piston is designed to come to rest, leaving just sufficient room in the cylinder for a cushion of water.

Internal and External Reversibility

Carnot's conception of a reversible engine cycle is a most important one, which has had a profound effect on power-plant practice. It is necessary however to emphasize that the realization of this ideal, difficult though it may be, is not the end even of the thermodynamic aspects of power-plant design. At least two points need to be made here.

1. High-temperature systems are rarely found in nature. They usually have to be made by burning fuel. The combination of a Carnot engine with a fuel-burning appliance is not, *treated as a whole*, a reversible plant. We should always prefer therefore a reliable fuel cell to such a Carnot engine.

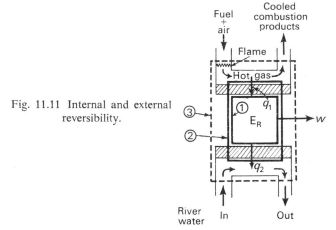

Fig. 11.11 Internal and external reversibility.

2. In practical circumstances it is often impermissible to raise the temperature of the working fluid to that of the source (e.g. the combustion products in a boiler furnace), usually because the containing walls would burst or melt. Even though the working fluid may pass through a reversible cycle therefore, we are still left with the irreversibility of heat transfer with finite temperature differences at the contact between the fluid and the hot source. The same sort of irreversibility may arise at the boundary between the fluid and the cold system (e.g. river water), from the need to restrict the area of the surface at which heat transfer takes place.

These points are illustrated by Fig. 11.11, which shows a reversible engine enclosed by boundary 1; if the system boundary 1 is considered, we have a reversible power plant. If the system boundary 2 is considered however, the system is no longer reversible since irreversibility occurs at both heat transfer surfaces. Usually the hot surface is the worse culprit, for the combustion gases may be at 1650 °C, whereas the mean wall temperature has to be restricted to, say, 650 °C; as a result the fluid temperature of E_R may not exceed 600 °C.

If finally the boundary 3 is considered, firstly we no longer have a system, because matter crosses the boundary; the contents of the boundary can not now be called a heat engine. Secondly, the additional irreversibility of combustion is present.

These distinctions are sometimes expressed by use of the terms *internal and external reversibility*. The engine E_R would then be called internally reversible but externally irreversible.

Sometimes it is useful to make other distinctions. Thus friction and sudden expansion are termed *mechanical irreversibility*; heat transfer with a finite temperature gradient is called *thermal irreversibility*; while combustion is called *chemical irreversibility*.

BIBLIOGRAPHY

ANON. Hydrocarbon Fuel Cell. *The Engineer*, **215**, No. 5063, 1963.
ANON. Shell Portable Fuel Cell. *Engineering*, **198**, No. 5148, 1964.
ANON. Esso Direct-Conversion Fuel Cell. *Engineering*, **199**, No. 5162, 1965.
JUSTI, E. W. Fuel Cell Research in Europe. *Proc. Inst. Elect. Electron. Engrs*, **51**, No. 5, 1963.
KIRKLAND, T. J. and JASINSKI, R. Fuel Cells—A State of the Art. *Inst. Elect. Electron. Engrs Trans.*, Vol. 1E-10, No. 1, 1963.
MITCHELL, W. *Fuel Cells*. Academic Press, New York, 1963.
Recent Research in Great Britain on Fuel Cells. Fifth World Power Conference. Section ref. 119 K/4. 1956.
LORD ROTHSCHILD. Fuel Cells. *Science Journal*, **1**, No. 1, 1965.
BARAK, M. Fuel Cells—Present Position and Outstanding Problems. *Adv. Energy Conversion*, **6**, No. 1, 1966.

CHAPTER 11—PROBLEMS

11.1 (*a*) A reversible heat engine operating between hot and cold constant-temperature systems delivers a work output of 54×10^3 N m. The heat transfer from the engine is 66 kJ. Evaluate the efficiency of the heat engine.

(*b*) The engine in (*a*) is reversed and operates as a heat pump between the same systems. Evaluate (i) the coefficient of performance of the heat pump and (ii) the power input to the pump when the heat-transfer rate to the hot system is 7·5 kW.

(*c*) If the reversed engine in (*b*) were considered to be a refrigerator what would be its coefficient of performance?

11.2 An engine operating on the Carnot cycle employs 1 kg of H_2O as the working fluid. The (constant) temperatures of the hot and of the cold bodies are 200 °C and 40 °C respectively. Fig. 11.9 illustrates the cycle; state 1 corresponds to saturated liquid, state 2 to dry-saturated vapour, state 3 to wet vapour, dryness 0·762, and state 4 to wet vapour, dryness 0·229.

Evaluate the work done and the heat transfer in each process, the net work done during the cycle, and the efficiency.

11.3 Instead of considering the Carnot cycle to be executed by a system in a piston-cylinder mechanism (see p. 231), it is possible to conceive of the four processes taking place in sequence in a steady-flow heat engine. In this case the working fluid flows steadily in turn through a heater (state 1 to state 2), a reversible turbine (2 to 3), a cooler (3 to 4) and a reversible compressor (4 to 1).

In such a Carnot engine, with H_2O as the working fluid, the states of the H_2O are as given in problem **11.2**. Evaluate, per unit mass of H_2O flowing: the external work done and the heat transfer during each process, the net work done by the engine, and the efficiency.

11.4 An inventor claims to have constructed a device which will produce shaft work continuously at a steady rate when it is supplied with a steady stream of steam.

The device consists of a well-insulated box, through the side of which projects a shaft; an essential requirement, it is stated, is that the insulation be such that it reduces the heat transfer from the device to the surroundings to negligible proportions. Steam flows steadily into the box at one point and flows steadily out at another point. The only fact divulged by the inventor is that the steam merely condenses in a coiled tube inside the box.

The claim is that when dry and saturated steam at a pressure of 1·4 bar is supplied steadily at the rate of 0·058 kg/s it will leave at a pressure of 1 bar with a dryness of 0·98, while the shaft power developed will be 3·5 kW.

Examine the feasibility of these claims by reference to (a) the First Law and (b) the Second Law.

11.5 By applying the Second Law of Thermodynamics, demonstrate that the processes listed on page 230 are irreversible.

12 The Absolute Temperature Scale

Introduction

We have seen that a heat engine of 100 per cent efficiency is an impossibility. Once this is accepted, the natural next question is, "How high an efficiency *can* be reached?" The answer is that the highest efficiency depends only on the temperatures of the two constant-temperature systems with which the engine must interact. Generally speaking, the greater the temperature difference, the higher is the attainable efficiency; but it is not the temperature *difference* alone that matters, as will be seen. Further, we have shown that the best engines are reversible ones; we therefore now consider the relationship between the efficiency of a reversible engine and the temperatures of the hot and cold systems between which it operates.

We might establish the relation between engine efficiency and temperature by experiment; measurements of heat and work could be made on a reversible engine for many pairs of constant-temperature systems. However, as was stressed in Chapter 4, all our temperature scales are arbitrary in any case; the temperature~efficiency relation would therefore differ according to whether we were using mercury-in-glass or platinum-resistance thermometers, for example.

In this situation therefore, why not take the bold step of *stating* what the efficiency~temperature relation shall be, that is to say of *defining* a temperature scale in terms of the efficiency of reversible engines? Then the above-mentioned experiments can be used to *calibrate the thermometers*. This move was made by Kelvin (1851); it has the advantage of providing a temperature scale which is independent of the thermometric substance.

Kelvin's innovation of the *Absolute (Thermodynamic) Temperature Scale* has been universally accepted, and in this chapter we discuss some of its implications. The attentive reader will however have perceived a difficulty: in our discussion of reversible engines, it was made clear that a completely reversible one is only a theoretical possibility; yet now we are proposing to use reversible engines for the task of precise temperature measurement. The way in which this difficulty can be surmounted is indicated in the section on the Clausius-Clapeyron relation (p. 242).

Finally, some examples will be given of the use of the Absolute Temperature in determining the maximum efficiencies and coefficients of performance of heat engines and heat pumps.

Symbols

C_{hp} Coefficient of performance of a heat pump.

h Specific enthalpy of a pure substance.

p Pressure.

q Heat transfer.

T Absolute (thermodynamic) temperature (Kelvin's proposal).

t Temperature.

v Specific volume.

w Net work done by a system (heat engine).

η Efficiency of a heat engine, thermal efficiency.

θ Absolute temperature (Alternative to Kelvin's proposal).

Subscripts

R Reversible.

f Saturated liquid (fluid).

g Saturated vapour (gas).

fg Saturated liquid to saturated vapour.

1, 2 Hot, cold constant-temperature system.

The Absolute Temperature Scale

Definition. **The absolute temperature* scale is defined by**

$$\eta_R \equiv \frac{T_1 - T_2}{T_1} \tag{12.1}$$

where η_R = efficiency of a reversible heat engine,

T_1 = absolute temperature of the hotter constant-temperature system,

T_2 = absolute temperature of the colder constant-temperature system.

Discussion

Alternative forms. It is often convenient to express the definition in terms of the magnitudes of the heat and work interactions. Referring to Fig. 12.1,

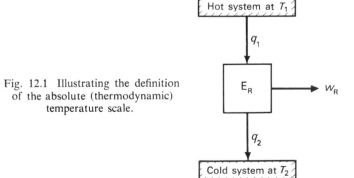

Fig. 12.1 Illustrating the definition of the absolute (thermodynamic) temperature scale.

we have, from eq. (12.1) and the definition of efficiency,

$$\frac{w_R}{q_1} = \frac{T_1 - T_2}{T_1} \tag{12.2}$$

* Frequently absolute temperature is called thermodynamic temperature.

and, from the First Law, eq. (6.1)

$$w_R = q_1 - q_2 \qquad (12.3)$$

Equations (12.2) and (12.3) can be re-arranged to give

$$\frac{w_R}{T_1 - T_2} = \frac{q_1}{T_1} = \frac{q_2}{T_2} \qquad (12.4)$$

The symmetrical form of this expression renders it convenient to use and easy to remember.

The size of the temperature unit. The expression in eq. (12.1) is a ratio. This implies that, given two constant-temperature systems, *any* number can be ascribed to the temperature of one of them; then a measurement on a reversible engine will fix the number which *must* be ascribed to the second one. There is therefore a choice to be made; this expresses itself as a choice of the *size of the temperature unit* in order that we may have either 100 or 180 degrees between the absolute temperatures corresponding to the freezing point and boiling point of water at atmospheric pressure. An important difference should be noted between the procedure here and that given in Chapter 4 for the construction of "conventional" thermometers. There, in addition to the specification of the size of the degree, numbers had to be assigned to *both* the ice-point and the steam-point temperatures. Here, the size of the degree and only a *single number*, the *ratio* of the temperatures, are required. This reduction in specification arises from the use of a reversible engine as a "thermometer".

Experiment shows that the value of the ratio is 1·3661 (see pages 82, 240); hence, when 100 divisions are selected, the temperatures 273·15 and 373·15 must be ascribed to the two fixed points. This fact is conventionally expressed as: "At atmospheric pressure, water freezes at 273 °C abs* and boils at 373 °C abs".* Here " °C abs" means "degrees on the Celsius absolute scale". An alternative description which we shall use from now on, is "K" which stands for "degrees on the Kelvin scale"; "°C abs" and "K" have identical meanings.

When the number of divisions between the freezing and boiling points is chosen as 180, the corresponding temperatures are respectively 491·67 °R or °F abs and 671·67 °R† or °F abs, where °R stands for "degrees Rankine", and °F abs for "degrees on the Fahrenheit absolute scale". The relations between the four scales are probably seen most clearly by reference to Fig. 12.2. We also have the relations:

$$T \text{ in K} = t \text{ in } °C + 273 \qquad (12.5)$$

$$T \text{ in } °R = t \text{ in } °F + 460 \qquad (12.6)$$

---

* These rounded figures usually suffice.

† Usually rounded to 492 °R and 672 °R respectively.

Relation of absolute temperature scale to scales based on particular thermo-metric substances. When the relative expansion of mercury in a glass container is plotted against absolute temperature, the curve is found to be slightly non-linear. This means that the markings of a mercury-in-glass thermometer should be somewhat unequally spaced if the corresponding absolute-temperature intervals are to be equal. However this is rarely done in practice; for the maximum error in a mercury-in-glass thermometer with equally spaced markings between 0 °C and 100 °C is about 0·1 K; it occurs near 40 °C.

Similar non-linearities are exhibited by all other thermometric substances. Those however which are most nearly linear are thermometers using the expansion at constant pressure, or pressure rise at constant volume, of

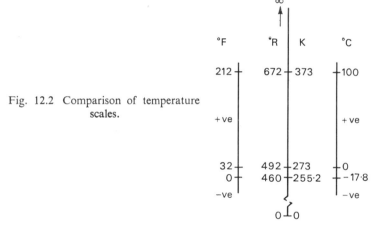

Fig. 12.2 Comparison of temperature scales.

"permanent gases" such as air or hydrogen (p. 82). It is by the use of such gas thermometers that the *ratio* of the ice-point and steam-point temperatures, mentioned above, may be obtained. This subject will be returned to in Chapter 14.

The International Scale of Temperature discussed earlier (page 81) is an attempt to provide a range of thermometric substances which, by means of piecewise scales, provide easy experimental techniques for measuring the temperature of a system on the absolute scale.

It will be noticed that negative temperatures have not been indicated on the absolute scales of Fig. 12.2. This is because no finite (macroscopic) system can possibly exist at such temperatures.

On the possibility of other definitions of the absolute temperature. Emboldened by the apparent success of Kelvin's innovation, the reader may inquire whether other definitions would do just as well. The answer is that other definitions *could* be used, and have been suggested. Eq. (12.1) has found favour however because of its simple form and because of its close relation to that of the gas thermometers. Not all definitions will do however. The permissibility of eq. (12.1) will now be demonstrated.

Consider the constant-temperature systems and reversible heat engines shown in Fig. 12.3. The combined work of the two engines is given from eq. (12.2) by

$$
\begin{aligned}
w_{12} + w_{23} &= \frac{T_1 - T_2}{T_1} q_1 + \frac{T_2 - T_3}{T_2} q_2 \\
&= \frac{T_1 - T_2}{T_1} q_1 + \frac{T_2 - T_3}{T_2} \cdot \frac{T_2}{T_1} q_1 \\
&= \frac{T_1 - T_3}{T_1} q_1
\end{aligned}
\tag{12.7}
$$

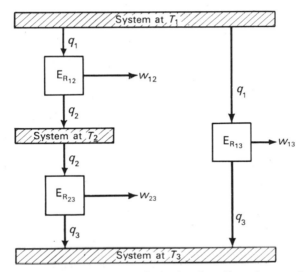

Fig. 12.3 Engine arrangements for testing the self-consistency of the definition of the absolute temperature scale.

The last expression is that which would be given by eq. (12.2) for a reversible engine operating between systems at temperatures T_1 and T_3, as it should be. The definition (12.1) is therefore self-consistent.

Consider by contrast the alternative definition

$$
\eta_R \equiv T_1 - T_2
\tag{12.8}
$$

This would lead to

$$
\begin{aligned}
w_{12} - w_{23} &= (T_1 - T_2)q_1 + (T_2 - T_3)q_2 \\
&= (T_1 - T_2)q_1 + (T_2 - T_3)(q_1 - w_{12}) \\
&= \{(T_1 - T_2) + (T_2 - T_3)(1 - T_1 - T_2)\}q_1
\end{aligned}
\tag{12.9}
$$

from which T_2 cannot be made to disappear as it ought. Definition (12.8) is therefore self-contradictory and would not be satisfactory.

The reader may care to check that the definition of an absolute temperature θ by way of:

$$\eta_{\mathrm{R}} \equiv \frac{\ln^{-1}\theta_1 - \ln^{-1}\theta_2}{\ln^{-1}\theta_1} \tag{12.10}$$

would be self-consistent. It is not used however.

Measuring the Absolute Temperature: the Clausius-Clapeyron Relation

In this section it is intended to show that, even though reversible engines cannot be constructed in practice, it is possible to use the definition of

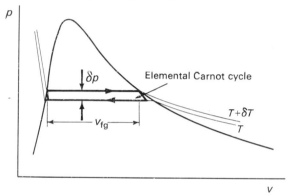

Fig. 12.4 Illustrating the proof of the Clausius-Clapeyron relation.

eq. (12.1) to determine the absolute temperature, by means of measurements of pressure, volume and heat transfer.

We may imagine a Carnot engine, using steam contained in a cylinder, operating between two systems which differ in temperature by a small magnitude δT. We are free to ascribe to this *difference* any value we like. The $p{\sim}v$ diagram for the steam will be as in Fig. 12.4. The work done in a single cycle by unit mass of steam is given by the area of the diagram and is clearly*

$$w_{\mathrm{R}} = \delta p \,.\, v_{\mathrm{fg}} \tag{12.11}$$

The heat transfer during the isothermal (and constant-pressure) heating is given for unit mass by

$$q_1 = h_{\mathrm{fg}} \tag{12.12}$$

the efficiency is therefore given by

$$\eta_{\mathrm{R}} = \frac{w_{\mathrm{R}}}{q_1} = \frac{v_{\mathrm{fg}} \,.\, \delta p}{h_{\mathrm{fg}}} \tag{12.13}$$

* Provided that the pressure$\sim$volume relationships of steam in the reversible-adiabatic parts of the cycle are not represented by horizontal lines, which may be established by quite crude experiments.

But the efficiency is also given by the definition (12.1). We therefore deduce

$$\frac{\delta T}{T} = \frac{v_{\text{fg}} \cdot \delta p}{h_{\text{fg}}} \qquad (12.14)$$

or

$$T = \frac{h_{\text{fg}}}{v_{\text{fg}}} \cdot \frac{\delta T}{\delta p} \qquad (12.15)$$

Now h_{fg} can be determined by a heat-transfer measurement; v_{fg} is measured by way of the dimensions of the container; and the difference of saturation pressure corresponding to (what we agree shall be called) a unit difference of absolute temperature, i.e. $\delta p/\delta T$, can be established by conventional techniques. Thus all the quantities on the right-hand side of eq. (12.15) can be measured; so the absolute temperature can be determined without recourse, except in imagination, to a reversible engine.

Eq. (12.15) is known as the *Clausius-Clapeyron Relation*, after its formulators.

EXAMPLE 12.1

Problem. Use eq. (12.15) to check the self-consistency of the Steam Tables given in Appendix B.

Solution. In the saturated-steam tables we find:

t °C	p kN/m²	v_f m³/kg	v_g m³/kg	v_{fg} m³/kg	h_{fg} kJ/kg
195	1398·7	0·001 148 9	0·140 84	0·139 69	1957·9
200	1554·9	0·001 156 5	0·127 16	0·126 00	1938·6

Taking mean values of v_{fg} and h_{fg} over the temperature interval between 195 °C and 200 °C, we deduce, from eq. (12.18)

$$T = \frac{1948\cdot3 \times 10^3}{0\cdot137\ 85} \times \frac{5}{156\cdot2 \times 10^3}$$

$$= 469\cdot4 = 273 + 196\cdot4$$

which is approximately the arithmetic mean of the two temperatures considered. If a sufficiently small temperature interval is used, good precision can be achieved. We conclude that the table is self-consistent in this region.

Determining the absolute temperature by means of the Clausius-Clapeyron Relation requires extremely accurate instrumentation. In Chapter 14, where the Ideal Gases are discussed, an experimentally easier way of determining T will be indicated. (See pages 292, 293.)

Examples Relating to Attainable Efficiency and Coefficient of Performance

We now use the above results to establish the limits of performance of heat engines and heat pumps in cases of engineering interest. When

temperatures are mentioned, these are supposed to be measured on thermometers calibrated with reference to the absolute scale of temperature, since
these are the standard instruments now universally used. Between the first
definition of the absolute temperature scale, however, and the existence of
these thermometers, an immense amount of experimental work has had to
be performed by physicists; present-day engineers owe them much.

EXAMPLE 12.2

Problem. A steam power plant operates between a hot system at 550 °C (a typical
value for the temperature of the boiler tubes) and a cold system at 12 °C (the mean
temperature of the condenser cooling water taken from a river). What is the maximum
attainable efficiency of the plant?

Solution. The maximum efficiency attainable will be that of a reversible engine operating between systems at the given temperatures.

$$T_1 = 550 + 273 = 823 \text{ K}$$

$$T_2 = 12 + 273 = 285 \text{ K}$$

$$\therefore \quad \eta_R = \frac{823 - 285}{823}$$

$$= 0 \cdot 654 \text{ or } 65 \cdot 4\% \qquad \qquad Answer$$

Note. For various reasons associated with mechanical and fluid friction, with heat
transfer to the surroundings, and with the finite temperature differences between the steam
and the tubes of the boiler and the water of the river, the actual plant efficiency will be
less than two-thirds of this. When evaluating the fuel consumption, it is also necessary
to remember that the boiler usually transfers only about 80% of the heat to the steam
which could be obtained by "isothermal" combustion of the fuel, i.e. steady-flow combustion in which the products leave at the temperature of the ingoing fuel and air.

EXAMPLE 12.3

Problem. A heat pump for domestic heating works between a cold system (the contents
of a refrigerator cabinet) at 0 °C and the water in the radiator system at 80 °C. What is
the minimum electrical power consumption to provide a heat output of 90 000 kJ/h?

Solution. The minimum power consumption will be achieved when the heat pump is
reversible.

$$T_1 = 80 + 273 = 353 \text{ K}$$

$$T_2 = 0 + 273 = 273 \text{ K}$$

$$\therefore \quad C_{hp} = \frac{1}{\eta_R} = \frac{353}{353 - 273} = 4 \cdot 41$$

$$\therefore \quad \text{Minimum power input} = \frac{90\ 000}{4 \cdot 41} \text{ kJ/h}$$

$$= \frac{90\ 000}{4 \cdot 41 \times 3600} \text{ kJ/s}$$

$$= 5 \cdot 67 \text{ kW} \qquad \qquad Answer$$

Note. This is of course only 1/4·41 times the electricity requirement which would be
required by the use of heating by electrical resistors. That heat pumps are still little used
is due firstly to the fact that considerably lower coefficients of performance are obtained

in practice (e.g. about 2 for the above case) and secondly to the large capital expense that is required, not only in providing the heat pump itself, but also in providing thermal contact between the working fluid and a suitable low-temperature system. (Rivers run through few back-gardens; the earth is a poor conductor; air deposits ice crystals on the heat-exchange surfaces, thus reducing heat-transfer rates.)

It should not be forgotten that the electricity often has to be obtained by burning fuel in a power station, perhaps at less than 30% efficiency. The next example examines a combined installation.

Problem. A heat pump for the temperature range of example 12.3 is to be driven by a steam power plant with boiler tubes at a temperature of 550 °C as its hot system and the domestic radiator plant at 80 °C as its cold system. What is the maximum possible value of the ratio: (combined heat transfer to the radiator plant)/(heat transfer from the boiler tubes)?

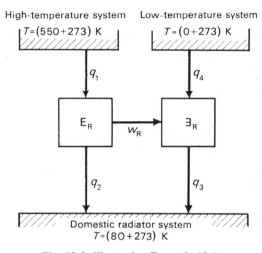

Fig. 12.5 Illustrating Example 12.4.

Solution (see Fig. 12.5). We suppose engine and heat pump to be reversible. From eq. (12.4) we have:

For the engine, E_R,

$$\frac{q_2}{353} = \frac{q_1}{823} = \frac{w_R}{470}$$

For the heat pump, $\exists_R$,

$$\frac{q_3}{353} = \frac{w_R}{80}$$

therefore

$$\frac{q_2 + q_3}{q_1} = \frac{353}{823} + \frac{470}{823} \times \frac{353}{80}$$

$$= 0 \cdot 429 + 2 \cdot 520$$

$$= 2 \cdot 949 \qquad\qquad\qquad Answer$$

Note. This means that nearly three times as much heat transfer to the building can be obtained from the boiler fuel as is possible by direct combustion. This is an attractive possibility, but examination of the capital cost of the plant may reduce its attractiveness.

A careful study of the economics of the particular proposal is always needed before it can be decided whether a heat pump will be advantageous.

The above three-fold increase in output from the fuel is quite distinct from that mentioned in Chapter 11 as obtainable from replacing combustion by use of a fuel cell. Combining a fuel cell with a heat pump promises even greater rewards. But the difficulties will remain formidable for many years.

CHAPTER 12—PROBLEMS

12.1 (a) A reversible heat engine operates between systems at constant temperatures of 150 °C and 10 °C. Evaluate the efficiency of the engine.

(b) The work output from the engine is 2000 N m. Evaluate the heat transfer from the system at 150 °C and the heat transfer to the system at 10 °C.

(c) The engine in (a) is reversed and operates as a heat pump between the same systems. Evaluate the coefficient of performance of the heat pump and the power input required when the heat-transfer rate from the system at 10 °C is 40 kW.

12.2 For an engine operating on the Carnot cycle, the work output is 0·2 times the heat transfer to the cold body. The difference in the working-fluid temperature when in contact with the hot and the cold body, is 60 K. Evaluate the efficiency of the engine and the maximum and minimum temperatures of the working fluid.

12.3 A refrigerator operates on the reversed Carnot cycle. The temperature of the working fluid as it evaporates in the evaporator is −20 °C while its temperature in the condenser is 25 °C. The heat-transfer rate from the cold region to the working fluid ("refrigerating effect") is 80 kJ/s. Evaluate the power required to drive the refrigerator.

12.4 In a steam power plant, the temperature of the walls of the boiler tubes is 510 °C. The cooling water circulating through the condenser is supplied from a river at a temperature of 12 °C. What is the maximum possible efficiency of the plant?

12.5 An ice-making plant produces ice at atmospheric pressure and at 0 °C from water at 0 °C. The mean temperature of the cooling water circulating through the condenser of the refrigerating machine is 18 °C. Evaluate the minimum electrical work in kWh required to produce one tonne of ice. (The enthalpy of fusion of ice at atmospheric pressure is 333·5 kJ/kg.)

12.6 An office block is heated by means of a heat pump. The mean air temperature within the building is 20 °C when the outside mean air temperature is −4 °C. The heat-transfer rate to the heat pump is 108 000 kJ/h and the power required to drive the pump is 10 kW.

(a) Evaluate the heat-transfer rate to the building and the coefficient of performance of the heat pump.

(b) Evaluate the maximum possible coefficient of performance of a heat pump for these conditions, and the minimum power input to the heat pump to satisfy the given heating requirements.

12.7 Exhaust steam from a process plant is to be used as the hot system for a heat engine. The steam is available at a pressure of 140×10^3 N/m² with a dryness fraction of 0·6. Heat transfer from the steam to the engine causes the steam to condense to saturated liquid at 140×10^3 N/m². River water is available at a mean temperature of 14 °C.

Evaluate the maximum possible power which could be developed by the engine when the exhaust-steam mass flow rate is 1800 kg/h.

12.8 An inventor claims to have designed a heat engine which has an efficiency of 38 % when using the exhaust gas from an engine, at a temperature of 145 °C, as the high temperature system.

Examine the validity of this claim.

12.9 A reversible heat engine operates between two systems at constant temperatures of 600 °C and 40 °C. The engine drives a reversible refrigerator which operates between systems at constant temperatures of 40 °C and −20 °C. The heat transfer to the heat engine is 2000 kJ and the net work output of the combined engine-refrigerator plant is 350×10^3 N m.

(a) Evaluate the heat transfer to the refrigerant and the net heat transfer to the system at 40 °C.

(b) Reconsider (a), given that the efficiency of the heat engine and the coefficient of performance of the refrigerator are each 40% of their respective maximum possible values.

13 Entropy

Introduction

We now know how to calculate the maximum efficiency of a system which works cyclically, i.e. of a heat engine. But cycles are made up of individual non-cyclic processes, each of which may require quantitative analysis; moreover many important engineering processes are not cyclic at all, for example those undergone by the air and the fuel in the "open-cycle" gas turbine. It is therefore necessary to see whether the definition of the absolute temperature scale can be used to answer such questions as, "What is the maximum work obtainable from a given process?" A new quantitative test for reversibility will be developed from our study of this matter.

Symbols

g	Gravitational acceleration.	w_p	Shaft work to a feed pump.
g_c	Constant in Newton's Second Law.	w_x	Shaft work delivered by a turbine.
h	Specific enthalpy of a pure substance.	X, Y	Arbitrary independent properties.
m	Mass.	x	Dryness, dryness fraction.
p	Pressure.	z	Elevation above an arbitrary datum.
Q	Heat transfer.		
q	Heat transfer.	η	Efficiency of a heat engine, thermal efficiency.
S	Entropy of a system.		
s	Specific entropy of a pure substance.	η_{isen}	Isentropic efficiency.
		$\eta_{Rankine}$	Efficiency of a heat engine working on the Rankine cycle.
T	Absolute temperature.		
u	Specific internal energy of a pure substance.		
		η_{ratio}	Efficiency ratio.
V	Velocity.	ρ	Density.
v	Specific volume.	*Subscripts*	
W	Net work done by a system (heat engine).	f	Saturated liquid (fluid).
		g	Saturated vapour (gas).
W_x	External work. Shaft work delivered by a turbine.	fg	Saturated liquid to saturated vapour.
W_p	Shaft work to a feed pump.	R	Reversible.
		1, 2, . . .	States of the working fluid

A turbine example

To make the task clear, the steam turbine illustrated in Fig. 13.1a will be considered. This receives a steady supply of steam of fixed state 1, and exhausts the steam at a state 2. Sufficient (two) conditions at state 1 are specified, but at state 2 only the pressure p_2 is known. The turbine is

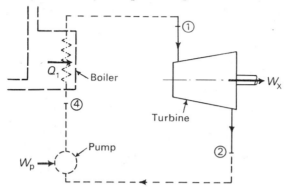

Fig. 13.1 (a) Steam turbine with imaginary boiler and pump making a cyclic process possible, if the turbine exhaust is water.

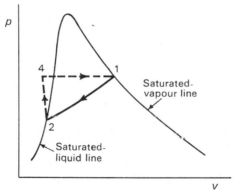

Fig. 13.1 (b) Illustrating the cyclic process undergone by the steam, if the turbine exhaust is water.

adiabatic and the kinetic and gravitational energy terms are negligible. How great a shaft-work output W_x is obtainable per unit mass of steam? The Steady-Flow Energy Equation, eq. (8.11), leads immediately to

$$0 - W_x = h_2 - h_1 \qquad (13.1)$$

Here h_1 is known; but we need another piece of information about state 2 before h_2, and so W_x, can be evaluated. Can this information be deduced from the Second Law?

The following argument makes it certain that the Second Law has at least something to say about state 2. Suppose this state is that of saturated water at p_2; then we could connect a feed pump at the turbine outlet, pump the water into a boiler, supply heat there, and so provide the turbine with its steady supply of steam at state 1. The $p \sim v$ diagram of Fig. 13.1b illustrates the cyclic path followed by the steam if this is done.

Examination of the boiler-turbine-pump system shows however that we have constructed a heat engine of 100 per cent efficiency: for the feed-pump

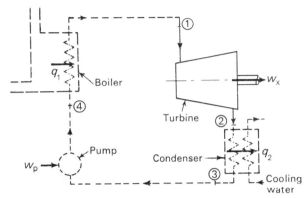

Fig. 13.2 (a) Steam turbine with imaginary boiler, pump and condenser making a cyclic process possible; the turbine discharges wet steam.

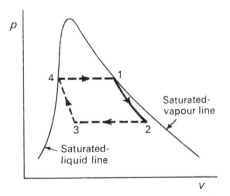

Fig. 13.2 (b) Illustrating the cyclic process undergone by the steam; the turbine discharges wet steam.

power is certainly very small, so the net work transfer is outward. The Second Law therefore implies that our supposition is incapable of realization: the dryness fraction of the exhausted steam cannot be as low as zero. How low *can* it be?

This question can be answered by re-constructing the imaginary heat engine: a condenser is placed between the turbine exhaust and the feed-pump inlet, Fig. 13.2a; in this, the steam leaving the turbine is almost completely condensed before being pumped back into the boiler as saturated water. The cyclic path followed by the steam is shown in Fig. 13.2b. Now it is reasonable to suppose that the turbine work output is greatest when the turbine, and all other engine components, work reversibly (this point is established on p. 258 below). The Steady-Flow Energy Equation applied to the reversible adiabatic turbine gives*

$$-w_{x,R} = h_2 - h_1 \tag{13.2}$$

Further, from the definition of the absolute-temperature scale, we know that the heat interactions q_1 and q_2 are related to the temperatures T_1 and T_2 prevailing in the boiler and condenser respectively by

$$\frac{q_1}{T_1} = \frac{q_2}{T_2} \tag{13.3}$$

Now q_2 is given by the S.F.E.E., applied to the condenser plus feed pump, as

$$-q_2 + w_{p,R} = h_4 - h_2 \tag{13.4}$$

The quantity $w_{p,R}$ is the reversible feed-pump work; like h_1 and h_4 (on the saturated-water line) $w_{p,R}$ may be evaluated: it is $(p_1 - p_2)v_{f4}$, since the specific volume does not change appreciably in the pump.

Applying the S.F.E.E. to the boiler, we obtain

$$q_1 = h_1 - h_4$$

Our problem can now be solved; for, by combining this equation with eq. (13.3) and eq. (13.4), we obtain the unknown h_2 as

$$h_2 = h_4 + \frac{T_2}{T_1}(h_1 - h_4) - w_{p,R} \tag{13.5}$$

When this is inserted in eq. (13.2), the maximum possible turbine-shaft work, when receiving steam of state 1 and exhausting at pressure p_2, becomes

$$w_{x,R} = (h_1 - h_4)\left(1 - \frac{T_2}{T_1}\right) + w_{p,R} \tag{13.6}$$

This expression provides us with a standard of comparison for the actual performance of the turbine.

Discussion

The above argument was rather clumsy; for the very simple case of steady flow through a turbine, it was necessary to construct a complete imaginary steam power plant and to argue directly from the definition of the absolute-temperature scale. Calculation was fortunately facilitated by the fact that the reversible-feed-pump term was easily evaluated with sufficient accuracy because the specific volume of the fluid remains almost constant during the pumping process. Most engineering problems of this type are more difficult; we therefore need a simpler method of calculation.

* Here we write the turbine work as $w_{x,R}$ (and not as $W_{x,R}$) to conform with the convention used earlier in Chapter 10 when discussing engines.

An analogy with the First Law. It will be remembered that, in Chapter 6, the First Law of Thermodynamics was stated in terms of the heat and the work in a cyclic process. In order to apply the law to non-cyclic processes, we had to invent a new property: energy. This could then be tabulated for particular substances, with great advantage in ease and speed of calculation. Is it possible to use the same approach in the present case? Can we invent a new property, enabling the Second Law to be applied to non-cyclic processes?

A possible line of attack. One characteristic of a property is that its change is zero in a cyclic process (p. 101); so we might look for an entity the change of which is zero in the steam-turbine example. Eq. (13.3) is suggestive: if we return to the earlier convention regarding the sign of heat, indicated by putting $Q_1 \equiv q_1$ and $Q_2 \equiv -q_2$, eq. (13.3) becomes

$$\frac{Q_1}{T_1} + \frac{Q_2}{T_2} = 0 \tag{13.7}$$

Since Q_1 and Q_2 are the only heat transfers to the steam in the imaginary cyclic process, which, it will be remembered, was assumed to be *reversible*, we can write for this case

$$\oint \frac{\mathrm{d}Q_R}{T} = 0$$

where $\oint$ means the summation around the cycle, $\mathrm{d}Q_R$ stands for an element of the reversible heat transfer to the steam, and T denotes the absolute temperature of the steam as it engages in the heat transfer $\mathrm{d}Q_R$.

Here and below we especially emphasize that the heat transfer occurs during a *reversible* process by means of the subscript R. Previously, when using the symbol Q for heat transfer, it was not necessary to distinguish between heat transfers in reversible and irreversible processes; in this chapter, however, the distinction is important; Q will represent the heat transfer in a general process (reversible or irreversible) and Q_R the heat transfer in a reversible process.

Eq. (13.7) shows that $\int [(\mathrm{d}Q_R)/T]$ has a zero sum *for this particular case*, and so might serve in general as our required property. Clearly however the situation is somewhat more difficult than in the case of energy and the First Law, because it will be necessary to make sure whether reversible processes are in question or not.

Summary of the remainder of the chapter. It will be shown below, in general terms, that the magnitude $\int [(\mathrm{d}Q_R)/T]$ it indeed *always* a property; this will be given the name *entropy*. The argument is somewhat abstract; but it is important and not difficult; it should be studied carefully. Thereafter we shall show how entropy can be used in engineering calculations, and how its relation to the other properties of systems can be established.

Entropy

Definition. Entropy, S, is a property of a system, such that its increase, $S_2 - S_1$, as the system changes from state 1 to state 2, is given by

$$S_2 - S_1 \equiv \int_1^2 \frac{dQ_R}{T} \qquad (13.8)$$

Notes. 1. That the quantity so defined really *is* a property will be proved below. The argument is in three steps, and the intermediate results are also important.

2. Because S is a property, the value of the integral is independent of the path of the change of state, and depends only on the end-states 1 and 2. It is the latter feature which will be demonstrated.

3. When the word "entropy" is pronounced, the first syllable carries the stress.

4. In differential form, eq. (13.8) runs

$$dS \equiv \frac{dQ_R}{T} \qquad (13.9)$$

Proof that entropy is a property. Step 1: The Clausius Inequality

Statement. When any system undergoes a cyclic process, the integral around the cycle of $(dQ)/T$ is less than or equal to zero. In symbols

$$\oint \frac{dQ}{T} \leqslant 0 \qquad (13.10)$$

Here dQ stands for an infinitesimal heat transfer, while T represents the absolute temperature *of the part of the system to which the heat transfer dQ occurs.*

Proof

Re-constructing the surroundings of the system. Consider the system illustrated in Fig. 13.3. Replace the actual surroundings of the system by a single constant-temperature system, at absolute temperature T_0, and a reversible heat engine E_R; this substitution can make no difference as far as the system is concerned. The system will, in general, do work on the surroundings; let the element of this work simultaneous with dQ be dW; and let the corresponding element of work done by the heat engine be dW_R. The heat engine is small and performs many cycles while the system performs one.

Argument. Applying the First Law to the system as it completes its cyclic process, we have

$$\oint (dQ - dW) = 0 \qquad (13.11)$$

Application of the Second Law to the cyclic process of the combined system within the dotted boundary of Fig. 13.3, from eq. (10.11), yields

$$\oint (dW + dW_R) \leqslant 0 \tag{13.12}$$

for otherwise this combined system would be a perpetual-motion machine of the second kind.

Invoking the definition of the absolute temperature scale, (eq. 12.1), we have, for the reversible engine,

$$\frac{dW_R}{T_0 - T} = \frac{dQ}{T} \tag{13.13}$$

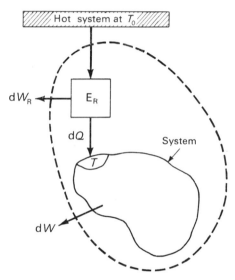

Fig. 13.3 Illustrating the proof of the Clausius inequality.

Conclusion. Elimination of dW and dW_R from equations (13.11), (13.12) and (13.13), now leads to

$$\oint \left\{ dQ + \frac{(T_0 - T)}{T} dQ \right\} \leqslant 0$$

i.e.

$$T_0 \oint \frac{dQ}{T} \leqslant 0$$

i.e.

$$\oint \frac{dQ}{T} \leqslant 0 \qquad \text{Q.E.D.} \tag{13.14}$$

Statement (13.14) is known as the *Clausius Inequality*.

Proof that entropy is a property. Step 2

Statement. For any system which undergoes an internally reversible cycle, the integral of $(\mathrm{d}Q_R)/T$ is zero; in symbols:

$$\oint \frac{\mathrm{d}Q_R}{T} = 0 \qquad (13.15)$$

Argument. Let the system execute a cyclic process, starting at state 1, proceeding to state 2 via the reversible path A, and returning to state 1 via

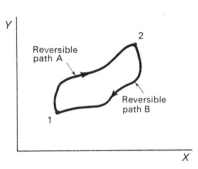

Fig. 13.4 An arbitrary reversible cyclic process.

the different reversible path B. Fig. 13.4 represents this process as a plot of any property Y against any other independent property X.

Then, from the Clausius Inequality, eq. (13.10), we have:

$$\text{along 1A2B1,} \qquad \oint \frac{\mathrm{d}Q_R}{T} \leqslant 0 \qquad (13.16)$$

Since the process is reversible, we may reverse it and thus cause the system to retrace its path precisely. Let the element of heat transfer corresponding to the system boundary at temperature T be $\mathrm{d}Q'_R$ for this reversed process. Then from eq. (13.10) we have:

$$\text{along 1B2A1,} \qquad \oint \frac{\mathrm{d}Q'_R}{T} \leqslant 0 \qquad (13.17)$$

But, since the second cycle is simply the first one with the direction reversed, we have

$$\mathrm{d}Q'_R = -\mathrm{d}Q_R \qquad (13.18)$$

Therefore statement (13.17) becomes

$$\text{along 1B2A1} \qquad -\oint \frac{\mathrm{d}Q_R}{T} \leqslant 0$$

$$\text{or} \qquad \oint \frac{\mathrm{d}Q_R}{T} \geqslant 0 \qquad (13.19)$$

Conclusion. Comparing statements (13.16) and (13.19), we see that they can both be true simultaneously only if

$$\oint \frac{dQ_R}{T} = 0 \qquad\qquad \text{Q.E.D.} \quad (13.20)$$

Proof that entropy is a property. Step 3

Statement. The integral of $(dQ_R)/T$, when a system executes any reversible process between fixed end-states, is independent of the path of the process. In symbols,

$$\text{for arbitrary paths A and B,} \qquad \int_{1_A}^{2} \frac{dQ_R}{T} = \int_{1_B}^{2} \frac{dQ_R}{T}$$

Argument. Let the system execute the reversible cyclic process, illustrated in Fig. 13.5, from 1 via path A to 2, and back via path C to 1. Then from eq. (13.20) we have

$$\int_{1_A}^{2} \frac{dQ_R}{T} + \int_{2_C}^{1} \frac{dQ_R}{T} = 0 \qquad\qquad (13.21)$$

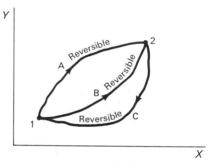

Fig. 13.5 Arbitrary reversible cyclic
processes.

Similarly, for the reversible cyclic process 1B2C1, we have from eq. (13.20)

$$\int_{1_B}^{2} \frac{dQ_R}{T} + \int_{2_C}^{1} \frac{dQ_R}{T} = 0 \qquad\qquad (13.22)$$

Conclusion. Subtracting eq. (13.22) from eq. (13.21), we obtain

$$\int_{1_A}^{2} \frac{dQ_R}{T} = \int_{1_B}^{2} \frac{dQ_R}{T} \qquad\qquad \text{Q.E.D.} \quad (13.23)$$

Final remarks. We have now proved the result foreshadowed on p. 253 for in eq. (13.23) the paths A and B are arbitrary. This means that

$\int_{1}^{2} \dfrac{dQ_R}{T}$ has the same value for *any* reversible path between 1 and 2.

$$(13.24$$

Hence from the definition of entropy, eq. (13.8), we may write statement (13.24) as:

$(S_2 - S_1)$ has the same value for *any* reversible path between 1 and 2. It follows therefore that *entropy is a property*.

Discussion of entropy

Tabulation of entropy. Since entropy is a property, it can, for pure substances, be tabulated as a function of two other independent properties. How its magnitude is established will be described later. For the moment it suffices to state that values of the property are now tabulated in the scientific literature for a large number of important substances. Appendix B, for example, contains entropy values for water and steam tabulated in the same way as those of enthalpy and specific volume.

Like energy, entropy has been defined only as a difference; it is therefore necessary to specify its value arbitrarily in some state. In the tables of Appendix B, entropy is taken as zero for saturated water at the triple point. Often however tables will be found for which entropy is zero at the absolute zero of temperature in the crystalline solid state.

Entropy is an *extensive* property (see p. 184); for, if the mass of the system is doubled, twice the amount of heat is necessary to bring about the same change in the system state. We therefore introduce the symbol s, standing for the entropy of unit mass of material, i.e. the specific entropy.

The units of S used in this book are J/K, and those of s are J/kg K in conformity with SI. In the Imperial system, the units of s are Btu/lb_m °R or $ft lbf/lb_m$ °R; such a change of unit system would cause changes in the numerical value of s.

Because entropy is extensive, its value in a region comprising two phases can be adequately described by tabulating its values for the saturated states. Thus, for the substance H_2O, s is given in Table III of Appendix B for the saturated-water and saturated-steam states, viz. s_f and s_g. The entropy s of wet steam of dryness x is then evaluated, for a given pressure, from

$$s = (1 - x)s_f + xs_g \qquad (13.25)$$

Reversible adiabatic processes. Many engineering processes occur with negligible heat transfer: the flow of steam or gas through a turbine is an example. The standard process with which these adiabatic processes may be conveniently compared is the reversible adiabatic process. Since by definition dQ equals 0 here, and the suffix R is appropriate, the entropy change is zero in such a process. Thus, from eq. (13.8), for a reversible adiabatic process

$$dQ_R = 0: \qquad \Delta S = 0 \qquad (13.26)$$

A process in which the entropy remains constant is termed an *isentropic* process. All reversible adiabatic processes are therefore isentropic. However it does not follow that all isentropic processes are either reversible or adiabatic.

The Use of Entropy

Entropy has been defined and tabulated in order to make engineering calculations easier. This Chapter began with a discussion of a steam-turbine example: we will shortly show how entropy can be used in this case. First however it is desirable to prove rigorously that a reversible adiabatic expansion is the one that gives the greatest work of all adiabatic expansions from a fixed initial state to a fixed final pressure. As usual, the proof is based on the Second Law and so involves the construction of an imaginary heat engine.

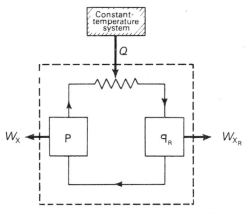

Fig. 13.6 Demonstrating that reversible adiabatic steady-flow processes do more work than irreversible adiabatic ones.

Proof about the superiority of reversible processes

Statement. Of all the possible *adiabatic* processes which can be executed by a system starting at a fixed initial state and ending at a fixed final *pressure*, the reversible one gives the maximum external work.

Proof. Suppose that a process P exists which gives an external work, W_x, greater than the $W_{x,R}$ given by the reversible process P_R. Then allow the system to execute a cyclic process, comprising:

(a) The process P.

(b) A heat interaction at constant pressure bringing the system to the state at the end of process P_R.

(c) The process P_R reversed, i.e. q_R.

This is illustrated by Fig. 13.6, for the case in which the processes occur sequentially in steady flow.

Examination of the dotted boundary shows that finite work, $W_x - W_{x,R}$, is done in the cycle, and that heat transfer occurs from a constant-temperature system. But this is impossible according to the Second Law. It follows that W_x cannot exceed $W_{x,R}$.

It is easy to see that, if P is reversible, W_x equals $W_{x,R}$; if, however, P is irreversible, W_x is less than $W_{x,R}$.

Comments. This proof was necessary because, previously, we had demonstrated the superiority of reversible *cycles* only; a similar proof can be devised for the case where it is the final *volume* that is specified: then the heat interaction in (*b*) must take place at constant volume. It has been assumed that the kinetic and gravitational potential energies are negligible; modifications to account for these should be obvious.

EXAMPLE 13.1: ON THE MAXIMUM WORK FROM AN ADIABATIC TURBINE

Problem. Dry saturated steam at 600×10^3 N/m² flows steadily into an adiabatic turbine which exhausts to a condenser at 30×10^3 N/m². What is the maximum possible shaft work per unit mass of steam, if the kinetic energies at inlet and exit can be neglected?

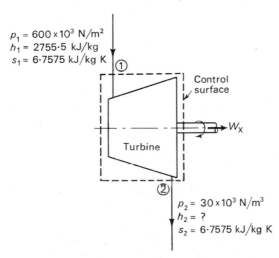

Fig. 13.7 Calculating the maximum shaft work of an adiabatic turbine.

Solution. Fig. 13.7 illustrates the situation. On it is written the information obtained by inserting the data in Steam Tables, and also the unknowns. The first task is to evaluate h_2, from which W_X can be evaluated via eq. (13.1). We do this by noting that the maximum-work process is a reversible one. Then, since the flow is also adiabatic, the maximum-work process must be isentropic. So:

$$s_1 = 6.7575 \text{ kJ/kg K}$$

$$s_1 = s_2 = (1 - x_2)s_{f_2} + x s_{g_2}$$

$$6.7575 = (1 - x_2)0.9441 + x_2(7.7695)$$

Here the numerical values are obtained from Steam Table III (Appendix B) for a pressure of 30×10^3 N/m², on the assumption that the steam is wet, a condition which must be checked later (see below).

Hence the unknown dryness at state 2, x_2, is given by

$$x^2 = \frac{(6\cdot7575 - 0\cdot9441)}{(7\cdot7695 - 0\cdot9441)}$$

$$= 0\cdot852$$

The fact that this value lies between 0 and 1 confirms our guess that the exhaust steam would be wet (it often is). Had we obtained a value greater than 1, it would have been necessary to go back and establish state 2 by looking in the Superheated-Steam Tables (Table IV, Appendix B) to find, by interpolation, a state with $s = 6\cdot761$ kJ/kg K and $p = 30 \times 10^3$ N/m². Had we obtained a value of x less than zero on the other hand, it would have meant that the arithmetic was wrong!

Looking up h_f and h_g from Steam Table III at 30×10^3 N/m² as $289\cdot3$ and $2625\cdot4$ J/kg, we now obtain h_2 from eq. (9.9) as

$$h_2 = (1 - 0\cdot852)\, 289\cdot3 + 0\cdot852 \times 2625\cdot4$$

$$= 2279\cdot7 \text{ kJ/kg}$$

Hence, from eq. (13.1),

$$W_{x,R} = h_1 - h_2 = 2755\cdot5 - 2279\cdot7$$

$$= 475\cdot8 \text{ kJ/kg} \hspace{3cm} \textit{Answer}$$

Comment. This procedure is frequently necessary in steam-power-plant calculations. It is important both theoretically and practically. It should be practised until its execution becomes mechanical.

Isentropic processes as standards of comparison

It is convenient to assess the performance of an actual adiabatic machine by comparing it with that of its reversible-adiabatic counterpart. The result of the comparison is usually expressed by means of an *isentropic efficiency* or an *efficiency ratio*.

Definition. The isentropic efficiency, η_{isen}, of an adiabatic work-producing machine is the ratio of the actual work output to the work output of a reversible adiabatic machine taking in fluid at the same initial state and exhausting it at the same final pressure (or, more rarely, volume). In symbols

$$\eta_{isen} \equiv \frac{W_x}{W_{x,R}} \hspace{3cm} (13.27$$

Similar measures of performance are used for adiabatic machines in which work is done *on* the fluid, e.g. rotary compressors. In these cases isentropic efficiency is defined by:

$$\eta_{isen} \equiv \frac{W_{x,R}}{W_x}$$

EXAMPLE 13.2

Problem. A steam turbine, taking in dry saturated steam at 600×10^3 N/m² and exhausting it at 30×10^3 N/m², produces shaft work per unit mass of steam equal to 380·6 kJ/kg. What is its isentropic efficiency?

Solution. Since the maximum possible shaft work is 475·8 J/kg (see example 13.1), we have

$$\eta_{\text{isen}} = \frac{380\cdot6}{475\cdot8} = 0\cdot80 = 80\% \qquad\qquad Answer$$

This is a typical value for rotary machinery. A reciprocating steam engine would give in practice an isentropic efficiency nearer 50%.

The Rankine cycle. In steam-power-plant practice, it is common to make the same comparison in a somewhat less direct fashion as a ratio of efficiencies. First the (thermal) efficiency is calculated, of a complete hypothetical heat engine employing the work-producing machine from the formula

$$\eta = \frac{W_x}{h_1 - h_{f_2}} \qquad\qquad (13.28)$$

where h_1 is the enthalpy of the steam entering the machine, and h_{f_2} is the enthalpy of saturated water at the pressure of the exhaust steam. It will be seen, by applying the Steady-Flow Energy Equation to the boiler and feed pump, and neglecting the small feed-pump work term, that the denominator of eq. (13.28) is equal to the heat transfer to the steam in the boiler. (See Fig. 13.2; for this case, h_3 equals h_{f_2}.) This heat transfer is necessary to complete the cycle after the steam has been condensed.

Then the same expression is evaluated for the reversible work-producing machine.

The *Rankine cycle efficiency,* η_{Rankine}, is defined as

$$\eta_{\text{Rankine}} \equiv \frac{W_{x,R}}{h_1 - h_{f_2}} \qquad\qquad (13.29)$$

The *efficiency ratio* is then defined as

$$\eta_{\text{ratio}} \equiv \frac{\eta}{\eta_{\text{Rankine}}} \qquad\qquad (13.30)$$

Comparison of eq. (13.30) with eq. (13.27) shows that in this case the efficiency ratio of the cycle is identical with the isentropic efficiency of the turbine (or reciprocating engine). This is not so for heat engines in general, however; it happens to be true in this case because the small feed-pump work has been neglected.

Strictly the feed-pump term should be taken account of in defining η and η_{Rankine}. There is little advantage in this, except in the case of high-pressure plant.

A quantitative test for irreversibility in adiabatic systems

Returning to more abstract matters, we now prove (i) that *the entropy increase in an irreversible process is always greater than the integral of* $(dQ)/T$, and (ii) *that the entropy of an adiabatic system can only increase.* Symbolically

(i) $$S_2 - S_1 > \int_1^2 \frac{dQ}{T} \text{ for any irreversible process,}$$

and

(ii) $S_2 - S_1 > 0$ for an adiabatic irreversible process.

These propositions can be used as somewhat more convenient tests of irreversibility, or, for that matter, tests of possibility, than those employed earlier. These tests are also quantitative.

Proof. (i) Suppose that a system executes the cyclic process represented by Fig. 13.8 where the outward path A is irreversible with heat transfer Q,

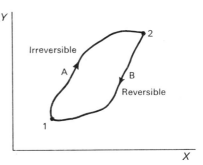

Fig. 13.8 An irreversible cyclic process.

and the return path B is reversible with heat transfer Q_R. For example, a mass of water might be evaporated by stirring ($Q = 0$), and then returned to its original state slowly by cooling ($Q < 0$).

From the Clausius Inequality, we have

$$\int_1^2 \frac{dQ}{T} + \int_2^1 \frac{dQ_R}{T} < 0 \qquad (13.31)$$

The $<$ sign is appropriate because the cycle is irreversible, involving as it does the irreversible process A.

Introducing entropy by way of its definition, eq. (13.8), namely:

$$S_2 - S_1 \equiv \int_1^2 \frac{dQ_R}{T} \quad \text{independently of the path,}$$

we have, for the reversible process B between states 2 and 1,

$$S_1 - S_2 = \int_2^1 \frac{dQ_R}{T}.$$

Eq. (13.31) now reduces to

$$S_2 - S_1 > \int_{1}^{2} \frac{dQ}{T}$$

and hence, since the path A is arbitrary,

$$S_2 - S_1 > \int_{1}^{2} \frac{dQ}{T} \text{ for } \textit{any irreversible} \text{ path between 1 and 2.}$$

$$(13.32)$$

Proof. (ii) For any *adiabatic* process, dQ equals 0. Therefore

$$\int_{1}^{2} \frac{dQ}{T} = 0$$

Hence, from statement (13.32), we deduce that for any irreversible *adiabatic* process:

$$S_2 - S_1 > 0 \qquad (13.33)$$

This means that, if the adiabatic system undergoes an irreversible process, its entropy increases. If the process is reversible, we already know that the entropy remains constant. Hence only increases are possible in adiabatic processes.   Q.E.D.

The differential forms of statement (13.32) and of the definition of entropy, eq. (13.9), are often required (see Chapter 16). By combining them we may write

$$dS \geqslant \frac{dQ}{T} \qquad (13.34)$$

or

$$dQ \leqslant T \, dS \qquad (13.35)$$

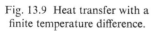
Fig. 13.9 Heat transfer with a finite temperature difference.

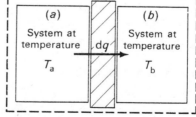

Heat transfer with a finite temperature difference. Fig. 13.9 illustrates heat conduction between two systems a and b at temperatures of T_a and T_b. The dotted boundary encloses the composite system (a + b) which is assumed to be isolated, as regards heat transfer, from its surroundings. We can apply statement (13.35), to the component systems a and b, in turn, while the small heat interaction dq occurs between them.* Since each component system is at a uniform temperature, and because there is an infinitesimal temperature difference at each system boundary, both heat interactions are reversible. The equality sign in statement (13.34) is therefore appropriate; thus we have:

For system a,
$$dS_a = -\frac{dq}{T_a}$$

For system b,
$$dS_b = \frac{dq}{T_b}$$

Hence the entropy increase, dS, for the combined system (a + b) is given by

$$dS = dS_a + dS_b$$

$$= dq\left(\frac{1}{T_b} - \frac{1}{T_a}\right)$$

$$= dq\left(\frac{T_a - T_b}{T_a T_b}\right) \tag{13.36}$$

The right-hand side of eq. (13.36) is clearly positive if T_a exceeds T_b; so heat transfer between systems at different temperatures is irreversible. If T_a equals T_b, the entropy increase is zero; this shows that the process is then reversible. If T_b exceeds T_a, dS is still positive, because then dq changes sign; the process is again irreversible.

Of course the heat transfer dq in general causes changes in the temperatures of both systems. Eq. (13.36) is therefore valid only for heat interactions of negligible magnitude; for larger heat transfers, the dependences of T_a and T_b on dq must be inserted and the total entropy increase evaluated by *integration*.

The conclusion that heat transfer with a finite temperature difference is irreversible was reached in a more roundabout fashion in Chapter 11; the test for entropy increase therefore speeds analysis; moreover it may be used in much more complex situations than the above, for example in examining the claims of the inventor of a plant which acts simultaneously as engine, heat pump, and chemical process.

Determination of Entropy

Although eq. (13.8) suffices as a definition of entropy, its restriction to reversible processes appears to leave us without a way of calculating the magnitude of the entropy change of an irreversible process: in the present section a general procedure for doing this will be provided. By using this procedure, we shall show that there exist quantitative relations between the

* We use q rather than Q to avoid breaking the sign convention adopted for the latter throughout the book.

entropy of a system and other properties such as enthalpy, pressure and volume.

General method of calculating entropy change in an irreversible process

The problem. Given a system, its initial state 1, its final state 2, and an irreversible process from 1 to 2, the entropy increase $S_2 - S_1$ is to be found.

Argument. We know two things about entropy:—(*a*) that it is a property; (*b*) its definition, eq. (13.8), in terms of a *reversible* process. We use the fact (*a*) to argue that $S_2 - S_1$ is independent of the path of the process from 1 to 2. This being so, we recognize as immaterial the fact that the process connecting the two states happened to be irreversible; the entropy difference between the two states is the same *whatever* the path joining them.

In order to evaluate the entropy difference, we must employ equation (13.8), the only equation (as distinct from inequality) in which entropy has appeared so far. We must find a reversible path from 1 to 2 (any one will do); the difference $S_2 - S_1$ can then be found by integrating $\int_1^2 [(\mathrm{d}Q_R)/T]$ along the path selected.

Conclusion. It is therefore always possible to calculate the entropy change, provided sufficient information about the other properties of the system is available to enable an alternative, reversible, path to be imagined.

The procedure will now be exemplified, first for a "mechanical" system, and secondly for a pure substance.

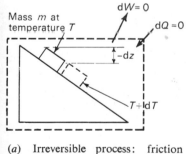

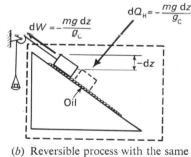

(*a*) Irreversible process: friction between block and plane.

(*b*) Reversible process with the same end states.

Fig. 13.10

Entropy increase of a block and inclined plane with friction

Problem. The first example is illustrated in Fig. 13.10. The thermally-conducting block of mass m descends through a vertical distance $-\,\mathrm{d}z$ in a gravitational field of acceleration g, while sliding down the rough thermally-conducting plane. There is no change in kinetic energy, but the block and plane rise in (uniform) temperature from T to $T + dT$. What is the entropy increase of the system comprising block plus plane?

Argument. Replace the actual process by a reversible one between the same end-states, as follows:

(*a*) Lubricate the plane and allow the block to fall $-dz$ vertically without friction, simultaneously raising an external weight. The system does work given by

$$dW = -\frac{mg}{g_c} dz$$

(*b*) Restore the internal energy of the system to the value corresponding to that at the end of the actual process, reversibly, by a heat transfer dQ_R from an external system. Since in the actual process there is no heat transfer, no work, and therefore no increase in internal energy, the combined effect of the reversible steps (*a*) and (*b*) must, by the First Law, eq. (6.5), give

$$dQ_R - \left(-\frac{mg}{g_c} dz\right) = 0$$

or

$$dQ_R = -\frac{mg}{g_c} dz$$

Conclusion. Now compute the entropy increase for the reversible process (a) + (b) from eq. (13.8). In (a) it is zero; in (b) it is

$$dS = \frac{dQ_R}{T} = -\frac{mg}{g_c T} dz \qquad (13.37)$$

in which it has to be remembered that dz is a negative quantity, so that dS is positive.

It should be noted that the entropy increase, dS, is *not* simply $\frac{dQ}{T}$ for the actual process; for in this adiabatic process $\frac{dQ}{T}$ equals 0. *The entropy increase can be calculated for the reversible process only.*

The entropy increase in a large fall can be computed by integrating eq. (13.37). If however the temperature changes are very small, it is sufficient to write

$$\Delta S = -\frac{mg}{g_c T} \Delta z \qquad (13.38)$$

where $-\Delta z$ is the large vertical fall of the block, and T is the mean block temperature during the process.

The entropy increase of a pure substance

We now shift the emphasis from the particular process undergone by the system, which we have seen is important only in that it specifies end-states, to the relation of the entropy of a pure substance to the other properties. We will derive two relations, which are of fundamental importance in the study of the thermodynamics of substances.

The T ds relations. For any infinitesimal process undergone by unit mass of a pure substance in the absence of gravity, motion, electricity, magnetism and capillarity, the First Law of Thermodynamics eq. (7.5) gives

$$\mathrm{d}Q = \mathrm{d}u + \mathrm{d}W \tag{13.39}$$

If in addition the process is reversible, we have

$$\mathrm{d}Q_\mathrm{R} = \mathrm{d}u + p\,\mathrm{d}v \tag{13.40}$$

since the only reversible work which such a system is capable of doing is fully-resisted displacement work (p. 50).

From the definition of entropy, we have

$$\mathrm{d}Q_\mathrm{R} = T\,\mathrm{d}s \tag{13.41}$$

Combining eq. (13.40) and eq. (13.41), we have

$$\boldsymbol{T\,\mathrm{d}s = \mathrm{d}u + p\,\mathrm{d}v} \tag{13.42}$$

A companion relation is obtained by eliminating $\mathrm{d}u$ from eq. (13.42) by means of the definition of enthalpy, $h \equiv u + pv$. In differential form the definition becomes:

$$\mathrm{d}h = \mathrm{d}u + p\,\mathrm{d}v + v\,\mathrm{d}p \tag{13.43}$$

which, combined with eq. (13.42), gives

$$\boldsymbol{T\,\mathrm{d}s = \mathrm{d}h - v\,\mathrm{d}p} \tag{13.44}$$

We shall refer to eqs. (13.42) and (13.44), which are sufficiently important to be committed to memory, as "the $T\,\mathrm{d}s$ relations".

Remarks on the T ds relations. 1. Although proved by considering reversible processes in the absence of gravity, etc., these relations are concerned merely with properties. They may therefore be used to calculate the entropy change of a pure substance in any process, whether reversible or irreversible. Gravity and motion may also be present.

2. Since internal energy, pressure, volume and absolute temperature are comparatively easy to measure, equations (13.42) and (13.44) afford convenient means of deducing the value of entropy without contriving measurements in reversible processes. Steam Tables are constructed in this way.

3. Combination of eq. (13.44) with the Steady-Flow Energy Equation leads to an interesting comparison with equations derived in textbooks on fluid mechanics. Reference should also be made to the discussion in Chapter 8 on p. 160. *If the flow is reversible* we deduce, from the Steady-Flow Energy Equation and eq. (13.41), for unit mass of fluid:

$$T\,\mathrm{d}s - \mathrm{d}W_x = \mathrm{d}\left[h + \frac{V^2}{2g_c} + \frac{gz}{g_c}\right] \qquad (13.45)$$

Subtraction of eq. (13.44) now yields:

$$\mathrm{d}\left[\frac{V^2}{2g_c} + \frac{gz}{g_c}\right] + \mathrm{d}W_x = -v\,\mathrm{d}p \qquad (13.46)$$

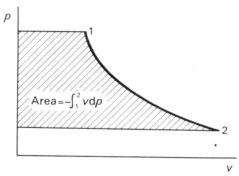

Fig. 13.11   Area representing $W_x + \Delta\left(\dfrac{V^2}{2g_c} + \dfrac{gz}{g_c}\right)$ in a *reversible* flow process.

For the case of zero external work ($\mathrm{d}W_x = 0$), this is the *Euler Equation* of fluid mechanics. It represents the application of Newton's Second Law of Motion to steady frictionless flow.

In integrated form, eq. (13.46) becomes

$$\Delta\left[\frac{V^2}{2g_c} + \frac{gz}{g_c}\right] + W_x = -\int v\,\mathrm{d}p \qquad (13.47)$$

The integral on the right-hand side is an area on the $p\sim v$ diagram representing the process (Fig. 13.11). If the process is adiabatic as well as reversible, that is to say if the process is isentropic ($\mathrm{d}s = 0$), this area is also equal to the enthalpy *decrease* in the process; for the integrated form of eq. (13.44), with $\mathrm{d}s = 0$, is

$$h_1 - h_2 = -\int_1^2 v\,\mathrm{d}p \qquad (13.48)$$

If the fluid is *incompressible*, so that v is constant, eq. (13.47) may be written without the integral sign. With v replaced by $1/\rho$, where ρ is the density of the fluid, and with zero shaft work, eq. (13.47) becomes the *Bernoulli* equation of fluid mechanics, namely

$$\frac{p_1}{\rho} + \frac{V_1^2}{2g_c} + \frac{gz_1}{g_c} = \frac{p_2}{\rho} + \frac{V_2^2}{2g_c} + \frac{gz_2}{g_c} \qquad (13.49)$$

Here there is no necessity for the flow to be adiabatic, but it must be frictionless: shear stresses must be absent from material boundaries across which there is relative motion.

Property Diagrams with Entropy as a Co-ordinate

Since entropy is just as much a property as pressure or internal energy, it may, like them, be used as one co-ordinate of the property diagrams of a

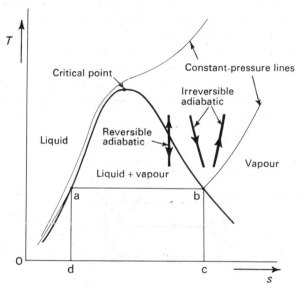

Fig. 13.12 Temperature~entropy diagram.

pure substance. Two such diagrams are in common use, the other co-ordinates being respectively absolute temperature and enthalpy.

The temperature~entropy *(T~s)* diagram. Fig. 13.12 shows, to scale, the liquid- and vapour-phase regions of the temperature~entropy diagram of steam. Some lines of constant pressure are also shown.
   This diagram is useful because it possesses the following features:
   1. Areas on the diagram have the *dimensions* of heat (or work).
   2. In the mixed-phase region, the constant-pressure lines are horizontal.
   3. In a *reversible* process, the area under a curve is equal to the heat

transfer in the corresponding process, since, from equation (13.9) for a unit-mass system:

$$\int dQ_R = \int T\,ds.$$

For example, the heat transfer, Q_R, in a *reversible* isothermal process, ab, Fig. 13.12, is given by

$$Q_R = \int_a^b T\,ds \tag{13.50}$$

$$= T_a \int_a^b ds = T_a(s_b - s_a) \qquad \text{in this case.}$$

The quantity $T_a(s_b - s_a)$ is, of course, the area abcd on the $T{\sim}s$ diagram (Fig. 13.12).

4. *Isentropic* processes are vertical straight lines.

5. In a *reversible cycle*, the area enclosed by the curve representing the process is equal to the net heat transfer to the fluid and so, from the First Law, is also equal to the net work.

6. The *efficiency of the reversible cycle* is given, if the cycle is clockwise and so represents a work-producing process, by the ratio of the *area enclosed by the curve* to the *area beneath the upper part of the curve*; for the first area represents the work done by the system, while the latter represents the heat transfer to the system.

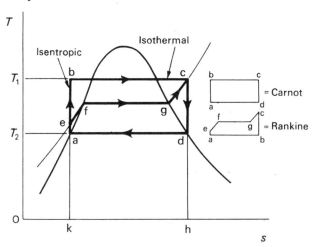

Fig. 13.13 Comparison of Rankine and Carnot cycles.

7. If the substance executes a *Carnot cycle*, its path on the $T{\sim}s$ diagram is a rectangle. For example, Fig. 13.13 shows a Carnot cycle abcd with steam as the working fluid. The cycle consists of two isothermal processes (horizontal lines, bc, da) and two isentropic processes (vertical lines ab, cd). (Compare Chapter 11, p. 231.) In this case, in which all processes are

reversible, the area ratio mentioned in note 6 is easily evaluated, giving the efficiency:

$$\eta = \frac{\text{area abcda}}{\text{area kbchk}} \qquad (13.51)$$

From geometry, the area ratio in eq. (13.51) is equal to the ratio of the heights of these two rectangles, i.e. to $(T_1 - T_2)/T_1$. This is in accordance with the definition of the Absolute Scale of Temperature, as it should be.

8. An *irreversible adiabatic* process is represented by a line tending to the right (Fig. 13.12); for the entropy always increases in such a process.

The Rankine cycle on the T~s diagram. Fig. 13.13 also shows a second reversible cycle (aefgcda) with the same temperature limits as the Carnot cycle; this is the Rankine cycle, as can be confirmed by comparison with the earlier mention of the cycle on p. 261. Here cd is the expansion in the turbine; da is the condensation in the condenser; ae is the compression of the condensate in the feed pump; and efgc is the constant-pressure heating in the boiler. In the case illustrated, the steam becomes superheated.

It is evident that, though the same amount of heat is transferred to the cold "reservoir" as in the Carnot cycle (equal areas beneath ad), the work done by the Rankine engine is less (smaller area enclosed by the curve). The Rankine-cycle efficiency for given temperature limits is therefore less than that of the Carnot cycle, even though both are reversible as far as the steam is concerned. How does this apparent contradiction of the result on p. 219 come about?

The answer is that, in the Rankine-cycle engine, the whole of the heat input does not occur at the highest temperature: the cycle is therefore *internally* reversible, but involves *external, thermal irreversibility*; for the temperature of the H_2O during the heating process efgc is less than T_1, the supposed temperature of the hot constant-temperature system from which the heat transfer occurs (see p. 233).

The Rankine cycle is regarded as a more satisfactory standard of comparison for steam-power machinery than the Carnot cycle (see p. 270), just because of this feature. It is "fairer" to the engine or turbine not to attribute to it the irreversibility associated with the way the boiler and feed arrangements are operated.

Consideration of Fig. 13.13 shows that the efficiency of the Rankine cycle will approach that of the Carnot cycle more nearly if the superheat temperature rise is reduced. There are several reasons why superheat is nevertheless regarded with favour; one is that, *with fixed upper temperature*, the pressure of the steam will rise to values which are difficult to deal with structurally if the superheat is reduced; another is that, without superheat, the steam is wet throughout the expansion, which leads to excessive wall-condensation in a reciprocating engine and to erosion by water droplets of the blades in a turbine.

Further discussion of these matters, together with measures to raise the efficiency of practical steam-power plant to that of the Carnot cycle, may be found in advanced texts on the subject.

The enthalpy∼entropy (h∼s) diagram

Fig. 13.14 shows, to scale, a diagram of the properties of the steam plotted with enthalpy as ordinate and entropy as abscissa. This is known as the *Mollier Diagram* for steam.* Large-scale charts are available and

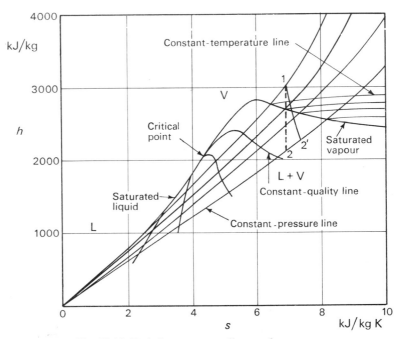

Fig. 13.14 Enthalpy∼entropy diagram for steam.

are widely used because of the appearance of enthalpy in calculations on steady-flow and because of the significance of the isentropic process as a standard of comparison for practical adiabatic expansions and compressions.

Some lines of constant pressure and temperature are plotted on Fig. 13.14, together with the lines of constant dryness fraction. The pressure and temperature lines naturally coincide in the two-phase region.

The *slope* of a line of constant pressure p may be calculated from eq. (13.44) by putting $\mathrm{d}p = 0$. We obtain

$$\left(\frac{\mathrm{d}h}{\mathrm{d}s}\right)_p = T \tag{13.52}$$

* After R. Mollier (1863 to 1935), a German scientist who introduced this and other diagrams for the representation of thermodynamic properties and processes.

These lines therefore have constant slope, equal to the absolute temperature corresponding to p; it follows that they are straight in the two-phase (L + V) region because the temperature is dependent on the pressure alone there, and the latter is a constant. The lines gradually steepen as they advance into the vapour (V, superheat) region.

Use of the $h{\sim}s$ diagram to determine the final state in an isentropic expansion. The Mollier diagram is particularly useful to steam-power engineers because it enables the calculation described in Example 13.1 on p. 259 to be replaced by a simple graphical construction, which is usually of sufficient accuracy. On Fig. 13.14 an isentropic expansion from state 1 is represented by a straight line dropped vertically downwards. Its intersection 2 with the given exhaust-pressure line p_2 is immediately found: this fixes state 2 (in particular, h_2) without reference to Steam Tables. The desired magnitude $W_{x,R}$ is equal to $(h_1 - h_2)$, according to eq. (13.1), i.e. to the vertical distance between 1 and 2, measured on the enthalpy scale.

Fig. 13.14 also shows, as a full line, a typical actual expansion between state 1 and pressure p_2. Its lower end-point 2′ is bound, because of the upward slope of the p_2 line, to be higher than 2. The actual shaft work W_x in this adiabatic process (with kinetic energy assumed to be negligible), which equals the vertical distance of 2′ below 1, is therefore less than $W_{x,R}$, as we should expect.

Understanding Entropy

The problem

It is one thing to follow the argument which shows how entropy is defined and determined, another to learn to use entropy in engineering calculations, and yet another to attain that grasp of the concept which we call "understanding". It is the last of these stages which many students find elusive; they are nagged by the question, "But what *is* entropy?" And, finding no satisfying answer, they lose confidence in their mastery of this part of thermodynamics.

Difficulty in comprehending entropy is so widely experienced that we devote the remainder of the chapter to examining, and trying to remove the causes of, this mental disquiet. It will be argued that the obstacles to understanding are not peculiar to entropy, but are of kinds that the student has encountered and surmounted *singly* on many occasions. What is unusual is the fact that the entropy concept happens to present several of these obstacles at the same time; the student is defeated by the effort of solving many difficulties simultaneously.

The nature of the question: "What *is* entropy?"

Similar questions. An inquirer after the nature of entropy could reasonably be directed to the early part of the present chapter for an answer to his question. If this answer does not satisfy him, it is probably because he unconsciously compares it with answers which he is accustomed to receiving to similar questions about other scientific concepts, for example:

Q. What *is* pressure?
A. The force exerted on unit area of a surface, measurable by means of a manometer or a Bourdon gauge.
Q. What *is* velocity?
A. The distance travelled in unit time, measurable by means of ruler and clock, by speedometer, or by Pitot tube.
Q. What is kinetic energy?
A. Something which increases with speed (as the square); it is useful because it enters equations.
Q. What *is* work?
A. The product of a force and the distance moved by its point of application; or, better, the number of standard weights that could be raised.
Q. What *is* energy?
A. The difference: $Q - W$; or the sum of the kinetic energy and the strain energy for all the molecules; or "the capacity for doing work".

Some remarks about the answers. 1. In each of these answers, there is an almost *direct connexion* between the words used and experiments which can be visualized.

2. *Numerous* answers can be supplied (some of them inexact, or even wrong); this richness gratifies the mind's proper desire to view the world from several angles.

3. Although the concepts are derived from experimental measurements by way of mathematical manipulations, these manipulations are those of *elementary arithmetic*; for example:

velocity equals distance *divided by* time;
kinetic energy is proportional to velocity *multiplied by* velocity;
work may be evaluated as force *multiplied by* distance;
energy increase is expressed as work *subtracted from* heat.

4. The *nearness to experience* of the quantities in question (except perhaps energy) makes it easy to accept that their values are worth knowing; one can scarcely doubt, for example, that steam at 6000 kN/m² has potentialities which are interestingly different from those of steam at 200 kN/m².

Answers to the question about entropy. 1. We may correctly say: "entropy is $\int[(dQ_R)/T]$ measured above some datum state". Immediately we note that, as in the above cases, there is clearly an experiment implied in the answer, followed by a mathematical manipulation. However:

(i) The experiment of measuring the heat transfer must be carried out under conditions of reversibility; this restriction is difficult to comply with.

(ii) The mathematical manipulation of the results of the experiment belongs to *advanced* mathematics, i.e. integral calculus, and not to elementary arithmetic.

2. The above is the only quantitative answer which is usually given. It is true that, in response to the need for a less abstract answer, it is often said that "entropy is a measure of disorder". Yet, although there is a sense in which the answer is valid, it usually fails to satisfy because:

(i) Disorder is not, for most people, a concept which is expressible in numbers; how, for example, could a person *measure* whether his desk is more or less untidy than his workshop?

(ii) It does not appear in the least obvious that saturated steam at 100 °C is less disordered than saturated steam at 40 °C; yet the Steam Tables show that it has the lesser entropy.

3. Another statement of the same kind is: "the entropy of a state is a measure of its probability". It is true that the theory of statistical mechanics does reveal a connexion between entropy and probability; however, the statement fails to augment understanding because:

(i) Although the student may have a quantitative idea of probability (connected with dice throwing, card games and the like), he will not, unless he actually studies statistical mechanics, manage to connect this idea with entropy.

(ii) One state can reasonably be said to be "more probable" than another only if they are both randomly occurring alternatives; yet there is no normally applicable sense in which saturated steam at 100 °C is this kind of alternative to saturated steam at 40 °C; chance does not enter at all.

4. In summary, the obstacles to understanding appear to reside in the facts that: entropy is related to measurements by a relationship involving integration; the defining relation carries a restriction (reversibility); no "physical picture" exists enabling the student to think, as he may do in respect of energy, that if he knew enough about the movements of the molecules he could calculate the entropy; it is not obvious that entropy is an important property, i.e. one connected to the interests of the engineer through force, "work-capacity", or other lively idea.

How to improve understanding. Mere enumeration of the obstacles to understanding may alleviate to some extent the distress which they cause. On the following pages we shall however seek ways of removing some and circumventing others. The discussion will incidentally provide a connexion with another formulation of the Second Law of Thermodynamics, namely that made by Caratheodory.

Entropy defined by an additive relationship

Analysis. Consider the situation illustrated in Fig. 13.15, in which the system with whose entropy change we are concerned is connected with a 'reservoir" at a standard temperature T_0 by way of a reversible engine E_R.

From the Second Law of Thermodynamics and the definition of absolute temperature, we have, for infinitesimal heat and work interactions:

$$\frac{dq}{T_0} = \frac{dQ_R}{T} \tag{13.53}$$

Now the definition of entropy, eq. (13.8), yields, for the entropy difference between two states, 1 and 2, of the system:

$$S_2 - S_1 = \int_1^2 \frac{dQ_R}{T} \tag{13.54}$$

because the heat transfer to the system is reversible.

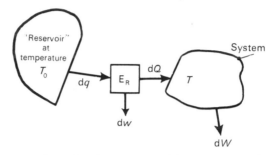

Fig. 13.15 Illustrating the apparatus which might be used for the measurement of entropy.

Combination of the two relationships yields:

$$S_2 - S_1 = \int_1^2 \frac{dq}{T_0} = \frac{1}{T_0} \int_1^2 dq$$

i.e.
$$S_2 - S_1 = \frac{q_{1\to2}}{T_0}, \text{ say.} \tag{13.55}$$

Here $q_{1\to2}$ denotes the total heat transfer to the reversible engine as the system changes from state 1 to state 2, while the engine executes a finite number of complete cycles. This relationship, which is entirely general, may be used, in place of eq. (13.54), as the definition of entropy. We shall see whether this shift of emphasis from the heat transferred to the system to the heat transferred to the reversible engine affects in any way the understandability of the concept.

Verbal re-definition of entropy. Let the standard temperature of the constant-temperature system ("reservoir"), T_0, be chosen as the temperature unit; e.g. let T_0 be 1 K. Then eq. (13.55) can be expressed in words as follows:

The increase in entropy in the system as it changes from state 1 to state 2 is equal to the heat transfer to a reversible engine (from a "reservoir" at unit

temperature) which, by interacting with the system, brings about the change in state.

Discussion. 1. Entropy defined in the above way is an *additive* quantity. Thus, if the change from state 1 to state 2 is succeeded by a change from state 2 to state 3, the total entropy increase is equal to the total heat transfer from the standard "reservoir"; for

$$S_2 - S_1 = q_{1\rightarrow2}/T_0 \tag{13.55}$$

$$S_3 - S_2 = q_{2\rightarrow3}/T_0 \tag{13.56}$$

and so,

$$S_3 - S_1 = (q_{1\rightarrow2} + q_{2\rightarrow3})/T_0$$

$$\equiv q_{1\rightarrow3}/T_0 \text{ , say.} \tag{13.57}$$

The additivity of the relationship brings the definition into the realm of arithmetic rather than that of calculus. One obstacle to understanding is thus removed.

2. We can now interpret quantitative information about entropy in concrete terms. For example:

From the Steam Tables, the entropy of 1 kg of steam increases from 7355·4 J/K to 8258·3 J/K, i.e. by 902·9 J/K when its state is changed from saturation at 100 °C to saturation at 40 °C.

This means that, in order to effect the change reversibly by means of an engine communicating with a "reservoir" at 1 K, it would be necessary for the heat transfer from the "reservoir" to be 902·9 J per kg of steam.

It is interesting to observe that this heat transfer is much smaller than, and of opposite sign to, the enthalpy or internal-energy increases that take place during the process; for example, the enthalpy *decreases* by 101·6 × 10³ J/kg.

3. An isentropic change is seen to be one which can be achieved without any heat transfer to, or of course from, the reversible engine; so an isentropic change may be both reversible and adiabatic, a fact that we already knew but now see from a slightly different angle.

4. The choice of T_0 as 1 K was made so that the entropy change would be numerically equal the heat transfer to the engine. If another temperature were chosen, the heat transfer would change in proportion; this is why the units of specific entropy are J/kg K. The K, it might be said, represents the temperature of the reservoir from which the J/kg are transferred.

5. If the First Law of Thermodynamics for a system is applied to a boundary enclosing both the system and the reversible engine in Fig. 13.15, we deduce:

$$w_{1\rightarrow2} + W_{1\rightarrow2} = q_{1\rightarrow2} - (E_2 - E_1)$$

$$= T_0(S_2 - S_1) - (E_2 - E_1) \tag{13.58}$$

where the subscript $1 \rightarrow 2$ on w and W means: "as the system changes from state 1 to state 2". From this relationship we can learn something about the "*worth*" of entropy, i.e. we can see whether it is desirable for S to be high or low.

Let us take, as state 2, one that can assuredly be attained; for example, let it be the state having the temperature and pressure of the atmosphere; then S_2 and E_2 are fixed.* From eq. (13.58) we deduce that the work which may be obtained by transforming the system from state 1 to state 2 by means of the reversible engine is equal to: $(T_0S_2 - E_2) + (E_1 - T_0S_1)$; further, this work quantity is the maximum possible, as may be shown by invoking the Clausius inequality and the fact that reversible engines are the most efficient of all. It follows that the most work is obtained from the process when the starting state of the system has the highest value of $(E_1 - T_0S_1)$.

This result at once clarifies the questions of the "worth" of energy and the "worth" of entropy: if maximum work production is the desideratum, a "good" system is one having high energy and low entropy; but knowledge of *both* these quantities is needed if one system is to be compared with another quantitatively.

The role of the restriction to reversibility

The remaining obstacle. The re-definition of entropy effected on page 277, eq. (13.55), may be felt to have eroded three of the obstacles to understanding:

(i) The integral relationship has been replaced by an additive one.

(ii) Consequently a conceptually simple measurement procedure can be envisaged, namely the measurement of the heat transfer from the "reservoir" to the reversible engine.

(iii) The "worth" of entropy has been revealed: the available work is greater, the larger is $E - T_0S$.

However, this progress has been achieved at the expense of making the restrictions on the measuring procedure even more stringent: previously it was necessary that the heat-transfer process should be reversible merely as far as the system is concerned; now it is necessary to use in addition a reversible engine. So this particular obstacle to understanding has actually been enlarged. It is therefore necessary to consider carefully the significance of this restriction and to enquire whether it is unprecedented.

It will be argued that *all* definitions of physical quantities involve measurements under restricted conditions; for the most part the restrictions are so obvious that they are not stated. The argument will proceed by way of examples.

---

* Unless it should happen that the temperature and pressure happen to correspond to two-phase equilibrium. In that case we must add that state 2 shall have the lowest energy of all those which the system can take up at the temperature and pressure in question.

The candle clock. Suppose that the duration of time is to be measured by observing the shortening of a burning candle which has a scale scribed along its length. It requires little knowledge of the laws of combustion to recognize that the "clock" will "tell the time" accurately only if the flame is shielded from draughts of air; for a light draught will accelerate the process of burning; a strong one will stop it altogether.

This restriction is too obvious to be stated; yet, if the relationship between time and the candle length were to be expressed formally, for example for some legal purpose, it would be well that the restriction should be mentioned.

The grain in a store. Suppose that we are concerned with measuring the quantity of grain in a store (Fig. 13.16). We might use the equation:

$$\text{Grain in store} = \Sigma \text{ supply} - \Sigma \text{ withdrawal} \qquad (13.59)$$

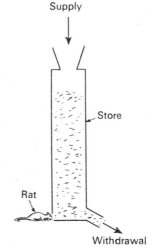

Fig. 13.16 Illustration of statement (13.60).

Supply

Store

Rat

Withdrawal

Suppose however that a rat has access to the store, but the precise amount that it eats cannot be observed; then eq. (13.59) ceases to be true. All that can be asserted is the inequality:

$$\text{Grain in store} \leqslant \Sigma \text{ supply} - \Sigma \text{ withdrawal} \qquad (13.60)$$

The equals sign is valid when the rat is fasting; otherwise, the $<$ sign is appropriate.

The inequality (13.60), though true, is too indefinite for measurement purposes. At any particular moment, whatever the rat has been doing, there is a definite amount of grain in the store; and we may wish to know how much this is. We can effect our purpose by devising a special process, designated "vermin-free", in which, either by the high speed of withdrawal, or by fumigation, the activities of the rat are made negligible. Then eq. (13.59) has a restored validity, written thus:

$$\text{Grain in store} = (\Sigma \text{ supply} - \Sigma \text{ withdrawal})_{\text{vermin-free}} \qquad (13.61)$$

The height of a flag. Fig. (13.17) illustrates the mechanism for raising a flag up a flag-pole; the main feature is a loop of cord passing around two pulley wheels, the lower one of which can be turned by hand; the flag is firmly sewn to the cord. For a reason shortly to be explained, the whole

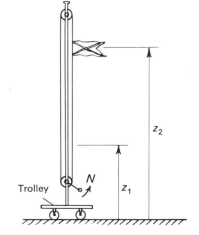

Fig. 13.17 Mechanism for raising a flag.

flag pole is mounted on a carriage, enabling it to be rolled along a horizontal plane.

If the cord is tight and inextensible, and if there is no slipping between the cord and the lower pulley, there is a simple relation between the increase in the height of the flag, $z_2 - z_1$, and the number of revolutions undergone by the lower pulley wheel during the change, N; it is:

$$z_2 - z_1 = \pi DN \qquad (13.62)$$

where D is the diameter of the lower pulley.

If slipping *does* occur, eq. (13.62) is not obeyed; all that can be asserted is that the increase of height is less than πDN, because the flag will slip down rather than up. In general we may write:

$$z_2 - z_1 \leqslant \pi DN \qquad (13.63)$$

Even when slipping has occurred, the flag still has a *definite* height; this can be measured by lowering the flag to some datum level, while taking precautions to prevent slipping, and counting the revolutions meanwhile. The relation used is the modified form of eq. (13.62):

$$z_2 - z_1 = (\pi DN)_{\text{no slip}} \qquad (13.64)$$

Conclusions. It is easy to invent further examples; indeed it is perhaps impossible to think of any measurement which is not subject to restrictions of some sort. Therefore, the fact that the restrictions are made quite explicit in the case of the definition of entropy should not occasion disquiet.

In real processes, the restrictions necessary for measurement are not always obeyed; but this does not in any way reduce the definiteness of the value of the property in question. Like the length of the candle, the amount of grain in the store, or the height of the flag, the entropy of a system has a definite value at any instant, no matter by what path the state has been reached; they are all *properties.*

Relation to Caratheodory's statement of the Second Law of Thermodynamics. Concerning the arrangement described in Fig. 13.17, it may be said:

(i) In the neighbourhood of each flag position there are other flag positions which are unattainable without turning the handle or allowing slip. (All higher and lower positions fall in this class; only those positions at the same level can be reached, by horizontal movement of the whole flag pole on its trolley.)

(ii) Some of the positions attainable in no-turn no-slip processes are attainable via no-turn processes *with* slip, but some are not. (The attainable positions are of course those at lower levels.)

Caratheodory's Statement of the Second Law. In 1909 Caratheodory developed a system of thermodynamics in which the Second Law was stated as follows:

(i) In the neighbourhood of each state of a system there exist other states which are unattainable by reversible adiabatic paths.

(ii) Some of the states which are unattainable via reversible adiabatic paths are attainable via irreversible adiabatic paths, but some are not.

Comparison. The parallel between the two pairs of statements is exact: the reversible adiabatic process corresponds to horizontal movement of the flag; the irreversible adiabatic process corresponds to downward slip, accompanied perhaps by horizontal movement. Indeed it can be said that Caratheodory's statement is a generalization of what we have learned from the flag example.

It thus appears that, in illuminating one aspect of the concept of entropy, we have stumbled across an alternative formulation of the laws of thermodynamics. The interested reader will be left to explore the parallel further with the aid of the bibliography.

BIBLIOGRAPHY

BUCHDAHL, H. A. On the Unrestricted Theorem of Caratheodory. *Amer. J. Phys.,* **17,** Nos. 1 and 4, 1949.

CARATHEODORY, C. Grundlagen der Thermodynamik. *Math. Annalen,* **67,** p. 355, 1909.

CARNOT, S. N. L. *Réflexions sur la Puissance Motrice du Feu et sur les Machines Propres à Développer cette Puissance.* Bachelier, 1824. (Translated by R. H. Thurston, A.S.M.E., 1943.)

CLAUSIUS, R. Ueber die bewegende Kraft der Wärme und die Gesetze, welche sich daraus für die Wärmelehre selbst ableiten lassen. *Poggendorfs Annalen*, **78**, pp. 368–500, 1850.

Enthalpy-Entropy Chart for Steam. Edward Arnold, London, 1962.

MOLLIER, R. *Neue Tabellen und Diagramme für Wasserdampf.* Springer, Berlin, 1906.

CHAPTER 13—PROBLEMS

13.1 (a) A fluid system at a temperature of 60 °C and a pressure of 90×10^3 N/m² undergoes a reversible process during which the temperature of the system remains constant. Given that the heat transfer to the fluid during the process is 120 kJ, evaluate the increase in entropy.

(b) The system in (a) has a mass of 2·31 kg. Evaluate the increase in the specific entropy of the system.

(c) A second fluid system, identical to that in (a), undergoes an irreversible isothermal process from the same initial state to the same final state as in (a). The heat transfer to the fluid in this irreversible process is 80 kJ. Evaluate the increase in the specific entropy of the fluid.

13.2 Using the Steam Tables given in Appendix B, obtain the magnitude and units for each of the following:

(a) Specific entropy of dry-saturated steam at a temperature of 150 °C.

(b) Specific entropy of saturated water at a pressure of 7×10^3 N/m².

(c) Specific entropy of steam at a pressure of 700×10^3 N/m², dryness 0·9.

(d) Specific entropy of steam at a pressure of 400×10^3 N/m² and a temperature of 600 °C.

13.3 (a) A system consists of a mixture of 0·1 kg of saturated water and 0·7 kg of saturated steam in equilibrium at 300×10^3 N/m². Evaluate the dryness fraction, the temperature and the specific entropy of the mixture.

(b) Steam at a temperature of 250 °C has a specific entropy of 6·97 kJ/kg K. Evaluate the pressure, the specific volume and the specific internal energy of the steam.

(c) Steam at a temperature of 250 °C has a specific entropy of 5·661 kJ/kg K. Evaluate the pressure, the specific volume and the specific internal energy of the steam.

13.4 Using the Steam Tables given in Appendix B, sketch approximately to scale the boundaries of the liquid- and vapour-phase regions of H_2O on charts with the following coordinates:

(a) $T \sim s$ (b) $h \sim s$.

On sketch (a) show a few lines of constant p and constant x, and on sketch (b) a few lines of constant p and constant T.

13.5 Reconsider problem 9.4 to obtain the entropy of the mixture.

13.6 One kilogram of saturated water at a pressure of 800×10^3 N/m² is contained in a cylinder fitted with a frictionless leakproof piston. The water undergoes a reversible (fully-resisted) constant-pressure process, as the piston moves slowly outwards, causing the contents of the cylinder to be converted to dry-saturated steam. The heat transfer during the process is 2046·5 kJ. The initial and final volume of the contents of the cylinder are 1115 cm³ and 0·240 26 m³.

Evaluate the increases in entropy and in internal energy of the contents of the cylinder and compare the values calculated with those given in the Steam Tables (Appendix B).

13.7 Steam at a temperature of 200 °C and a dryness 0·9 undergoes a reversible constant-temperature process to a pressure 200×10^3 N/m².

(a) Use the Steam Tables (Appendix B) to evaluate the increase in the entropy of the steam and hence, via the definition of entropy, the heat transfer per unit mass of steam during the process.

(b) Use the Steam Tables to evaluate the increase in internal energy and the increase in enthalpy of the steam.

(c) Using the results of (a) and (b), and assuming the process to take place in a piston-cylinder mechanism, determine the work done, per unit mass of steam, during the process Plot the path of the process to scale on a $p{\sim}v$ state diagram, evaluate $\int p \, dv$ and compare with the calculated value for the work done.

(d) Using the results of (a) and (b) and assuming the process to take place in steady flow, with negligible changes in the kinetic and potential energies, determine the external work done, per unit mass of steam, during the process, and compare it with $- \int v \, dp$.

13.8 A volume of 0·01 m³ of steam at a pressure of 1·0 × 10⁶ N/m² and a dryness of 0·95 is contained in a cylinder closed by a frictionless piston. The steam undergoes a fully-resisted expansion to a final pressure of 200 × 10³ N/m², according to $pV = $ constant, where p is the steam pressure and V the corresponding steam volume.

Evaluate: (a) the initial and final temperatures,
    (b) the work done,
    (c) the increase in entropy,
    (d) the heat transfer.

13.9 Steam is compressed reversibly and adiabatically from a pressure of 200 × 10³ N/m², dryness 0·9, to a pressure of 2 × 10⁶ N/m².

Determine: (a) the final temperature;
    (b) the increase in specific internal energy;
    (c) the increase in specific enthalpy.

Using these results, state:
    (d) the minimum work required to compress unit mass of steam adiabatically from a pressure of 200 × 10³ N/m², dryness 0·9, to a pressure of 2 × 10⁶ N/m²;
    (e) the minimum shaft work required if the compression were carried out adiabatically in steady flow, changes in the kinetic and potential energies being negligible.

13.10 A mass of 1 kg of steam at a pressure of 400 × 10³ N/m² and a temperature of 200 °C is contained in a cylinder. The steam undergoes an irreversible expansion process to a final pressure of 100 × 10³ N/m²; the initial and final entropies of the steam are equal.

(a) Given that the work done is 80 per cent of that for a reversible adiabatic process between the same end states, determine the magnitude and sign of the heat transfer during the process.

(b) In a second process between the same initial state and the same final pressure, the steam does an amount of work equal to that in the first irreversible process. The second process is adiabatic. Evaluate the increase in entropy in the second process.

13.11 (a) Steam at a pressure of 6 × 10⁶ N/m³ and a temperature of 400 °C flows steadily into a turbine and leaves at a pressure of 800 × 10³ N/m² and a temperature of 200 °C. The flow is adiabatic, and changes in kinetic energy and in elevation are negligible. Evaluate the external work done per unit mass of steam flowing.

(b) Steam at a pressure of 3 × 10⁶ N/m² and a temperature of 300 °C flows steadily into a turbine and leaves at a pressure of 800 × 10³ N/m². The flow is adiabatic and changes in kinetic energy and in elevation are negligible. Evaluate the maximum external work which could be done by the steam.

(c) Show the expansion processes (a) and (b) on a sketch of the enthalpy~entropy diagram and evaluate the isentropic efficiency of the turbine in (a).

13.12 Check that the entropy changes in the reversible adiabatic expansion and compression processes in problem **11.2** are zero.

13.13 A steady-flow direct heat engine using H_2O as the working fluid operates on the Carnot cycle between the pressure limits of 1·4 × 10⁶ N/m² and 100 × 10³ N/m². The working fluid changes from saturated liquid to dry-saturated steam during the isothermal heating process at the higher pressure.

(a) Calculate the heat transfer and the shaft work for each process.

(b) Compare the net shaft work done and the net heat transfer during the cycle.

(c) Evaluate the cycle efficiency (i.e. w/q_1) and compare it with the value obtained from the definition of the Absolute-Temperature Scale.

13.14 Recalculate problem **13.13** assuming the engine to be operating on the Rankine cycle, with the same pressure limits and with the H_2O in the saturated state at the beginning of both reversible adiabatic processes. Ignore the feed pump work.

13.15 (a) Making use of eq. (13.47), show that, for the steady, reversible flow of unit mass of an incompressible fluid when changes in kinetic energy and in elevation are negligible, the external work is given by

$$W_x = v(p_1 - p_2)$$

where v is the specific volume of the incompressible fluid and p_1, p_2 are the initial and final pressures of the fluid.

(b) Assuming water to be an incompressible fluid for which $v = 0.001\ 007\ m^3/kg$, evaluate the shaft work per unit mass required in a reversible feed pump taking in water from a condenser at a pressure of $7 \times 10^3\ N/m^2$ and delivering it to a boiler operating at $3 \times 10^6\ N/m^2$.

13.16 A steady flow of steam from a boiler enters an adiabatic turbine at a pressure of $3 \times 10^6\ N/m^2$ and a temperature of 350 °C and discharges into a condenser at a pressure of $7 \times 10^3\ N/m^2$. The condensed steam from the condenser enters the feed-pump as saturated liquid at a pressure of $7 \times 10^3\ N/m^2$ and is returned to the boiler at a pressure of $3 \times 10^6\ N/m^2$. The isentropic efficiencies of the turbine and of the pump are 70 per cent and 50 per cent respectively. The turbine and the pump may be assumed to be adiabatic and changes in kinetic energy and in elevation are negligible.

(a) Evaluate the dryness fraction of the steam leaving the turbine and the shaft work done by the steam in the turbine, per unit mass of steam flowing.

(b) Evaluate the work done on the water in the feed pump. Use the results from problem **13.15**.

(c) Evaluate the efficiency and the efficiency ratio with and without allowance for the feed-pump work.

13.17 A vapour-compression refrigerator (Fig. 10.5) uses methyl-chloride as the working fluid. The fluid flows steadily into the compressor at a pressure of $119 \times 10^3\ N/m^2$ and is delivered to the condenser as dry-saturated vapour at a pressure of $653 \times 10^3\ N/m^2$. The fluid leaves the condenser as saturated liquid at a pressure of $653 \times 10^3\ N/m^2$ and, after expansion in the throttle valve to a pressure of $119 \times 10^3\ N/m^2$, it flows through the evaporator and thence back into the compressor again. The compression process may be assumed to be reversible and adiabatic and the throttling process to be adiabatic. Changes in kinetic energy and in elevation are negligible.

(a) Evaluate the dryness fraction of the fluid entering the compressor and hence the shaft work done per unit mass of refrigerant.

(b) Evaluate the dryness fraction of the fluid after the throttling process.

(c) Evaluate the heat transfer per unit mass to the refrigerant in the evaporator.

Pressure N/m^2	Saturation temperature °C	Specific enthalpy kJ/kg		Specific entropy kJ/kg K	
		Saturated liquid	Saturated vapour	Saturated liquid	Saturated vapour
119×10^3	−20	30·1	455·2	0·124	1·803
653×10^3	30	108·6	478·7	0·406	1·627

(d) Evaluate the coefficient of performance of the refrigerator and compare it with the value for a reversed Carnot cycle operating between the given temperature limits.

Use the data for methyl-chloride given on p. 284.

13.18 (a) Steam at a pressure of $1 \cdot 0 \times 10^6$ N/m² and a temperature of 250 °C flows steadily into a horizontal nozzle with a velocity of 100 m/s. The steam leaves the nozzle at a pressure of 250×10^3 N/m². Given that the flow process is reversible and adiabatic, evaluate the dryness fraction and the velocity of the steam leaving the nozzle.

(b) A second nozzle receives steam in the same condition and expands it to the same final pressure as in (a). The flow process is adiabatic but in this case, because of friction, the increase in the kinetic energy of the steam is 90 per cent of that in the nozzle of (a) (i.e. the *nozzle efficiency* is 90 per cent). Evaluate the velocity and the dryness fraction of the steam leaving the nozzle. Evaluate also the increase in specific entropy of the steam.

(c) Compare the exit areas of the two nozzles.

13.19 A heat engine operates steadily on the following cycle. Saturated water at a temperature of 200 °C is pumped into a boiler and leaves as dry-saturated steam at a temperature of 200 °C. After adiabatic expansion through a turbine to a pressure of 100×10^3 N/m² the dryness fraction is 0·90. The exhaust steam from the turbine passes to a condenser and is partially condensed, leaving at a pressure of 100×10^3 N/m² with a dryness fraction of 0·15. The wet steam leaving the condenser is then compressed adiabatically in the feed pump before re-entering the boiler as saturated water at a temperature of 200 °C.

(a) Determine the specific entropy values around the cycle.

(b) State whether the turbine and pump processes are reversible or irreversible: give reasons.

(c) Evaluate the heat transfer per unit mass of steam in each component, and determine the efficiency of the heat engine.

(d) Determine the value of $\oint \dfrac{\mathrm{d}Q}{T}$ for unit mass of steam.

(e) What would have been the value of $\oint \dfrac{\mathrm{d}Q}{T}$ and of the efficiency if the turbine and pump had each been reversible with the same states at entry to and exit from the boiler and the same condenser pressure.

13.20 (a) Making use of eq. (13.44), show that for a substance for which h is a function of t and for which c_p is constant, the increase in entropy between initial and final states 1 and 2 is given by

$$s_2 - s_1 = c_p \ln \frac{T_2}{T_1} - \int_1^2 \frac{v}{T} \, \mathrm{d}p.$$

Hence show that, in a constant-pressure process, the increase in entropy is given by

$$s_2 - s_1 = c_p \ln \frac{T_2}{T_1}$$

(b) An open insulated vessel is divided into two parts by a vertical non-conducting partition. On one side of the partition are 5 kg of water at a temperature of 40 °C whilst on the other side are 10 kg at a temperature of 70 °C. When the partition is removed, the two masses of water mix; after a time conditions become uniform throughout the vessel. Assuming zero heat transfer to the atmosphere, and taking the specific heat of water at constant pressure to be $4 \cdot 19 \times 10^3$ J/kg K, evaluate the increase in entropy of the system comprising the 15 kg of water.

(c) An open insulated vessel contains 9 kg of water at a temperature of 20 °C. A mass of 1 kg of ice at a temperature of -4 °C is added to the water and after a time the temperature of the contents of the vessel becomes uniform. Assuming the heat transfer to the atmosphere to be zero, determine the increase in entropy of the system comprising the final contents of the vessel. Take the specific heat at constant pressure of ice to be 2·09 kJ/kg K and the enthalpy of fusion of ice at atmospheric pressure to be 333·5 kJ/kg.

(d) A steel tool of mass ·0·5 kg, at a temperature of 350 °C is plunged suddenly into an insulated vessel containing 10 kg of oil at a temperature of 20 °C. After a time the temperature of the contents of the vessel becomes uniform. Assuming the heat transfer to the atmosphere to be zero, and that none of the oil evaporates, evaluate the increase in entropy of the final contents of the vessel. Take the specific heats at constant pressure of the oil and of the steel to be 1·88 kJ/kg K and 0·48 kJ/kg K respectively.

13.21 By comparing the entropy increase with $\int \dfrac{\mathrm{d}Q}{T}$, show that the following processes are irreversible:

    (a) the expansion process in problem **13.11** (a);

    (b) the expansion process in problem **13.18** (b);

    (c) the mixing process in problem **13.20** (b);

    (d) the mixing process in problem **13.20** (c);

    (e) the quenching process in problem **13.20** (d).

14   Ideal Gases

Introduction

So far we have been concerned to apply the First and Second Laws of Thermodynamics to systems in general and, more particularly to pure substances. Steam has been chosen as the chief example of the latter, partly because of its engineering importance, and partly because it exhibits most of the peculiarities and irregularities which are likely to be encountered. When knowledge of the properties of the fluids has been required, it has been extracted from results of experimental research embodied in tables and charts.

Since many other fluids are important in engineering, it is fortunate that some of them, over restricted but practically interesting ranges of conditions, exhibit regularities in their properties which enable tabulation of these properties to be dispensed with: instead simple algebraic formulae express the relations between the properties with sufficient accuracy for many purposes. The use of the formulae facilitates analysis; algebra replaces arithmetic over a large part of the calculation, with corresponding increase in speed and generality.

Every fluid permits such algebraic approximation when its temperature sufficiently exceeds its critical temperature, or when its pressure is very much below its critical pressure; these are the conditions mentioned on p. 180, where it was noted that isotherms tend to become rectangular hyperbolae on a $p{\sim}v$ diagram and horizontal straight lines on diagrams with internal energy or enthalpy as ordinate. The important fluids are those for which these conditions are satisfied by moderate temperatures and pressures. Examples are the so-called "permanent" gases: oxygen, hydrogen, nitrogen, air, and carbon dioxide; even so, liquid air is a common substance in laboratories, while solid carbon dioxide ("dry ice") is carried by every ice-cream vendor.

For real gases the formulae are approximations; however we may imagine that there exist gases which obey the formulae exactly. Such *Ideal Gases* will be discussed in the present chapter; they are of two kinds, called: *Semi-Perfect Gases* and *Perfect Gases*. We discuss the features which a real gas must possess to approximate to an Ideal Gas, and demonstrate an important fact about the effect of temperature on such a gas: it is shown that thermometers using them would indicate the *absolute temperature*. Thereafter formulae will be derived which enable First- and Second-Law analyses to be made of processes executed by Ideal Gases.

Symbols

$\tilde{c}_v$	Specific heat at constant volume (mole basis).	p	Pressure.
c_v	Specific heat at constant volume (kilogram basis).	Q	Heat transfer.
		R	Gas constant.
$\bar{c}_v$	Mean specific heat at constant volume.	$\mathscr{R}$	Molar gas constant (universal gas constant)
$\tilde{c}_p$	Specific heat at constant pressure (mole basis).	s	Specific entropy.
		T	Absolute (thermodynamic) temperature.
c_p	Specific heat at constant pressure (kilogram basis).	t	Temperature.
		u	Specific internal energy.
$\bar{c}_p$	Mean specific heat at constant pressure.	V	System volume, Velocity.
g	Gravitational acceleration.	$\tilde{V}$	Molar volume.
g_c	Constant in Newton's Second Law.	v	Specific volume.
		W	Net work done by a system.
h	Specific enthalpy.	W_x	External work at a control surface.
M	Relative molecular mass (molecular mass)*	z	Height of a mercury column (example 14.3, p. 314).
m	Mass.	γ	Ratio of specific heats,
$\dot{m}$	Mass flow rate.		c_p/c_v, $\tilde{c}_p/\tilde{c}_v$
n	Number of moles. Constant in pv^n = a constant.		

Experimental Facts About "Permanent" Gases

Boyle's Law

Soon after the first means of pumping gases and preventing leakage became available, Boyle (in 1662) discovered that an approximately (he thought exactly) reciprocal relationship existed between the pressure and the volume of a fixed mass of "permanent" gas when expanded or compressed at constant temperature. We write this symbolically as

$$pv = \text{constant, at constant temperature}$$

$$= \text{f}(t), \text{ very closely,} \tag{14.1}$$

where f(t) indicates "some function of temperature". The lower case t is used, since we shall suppose for the present that we have no means of knowing temperature on the absolute scale. Boyle's Law will be shown below to assist in removing this ignorance.

Joule's Law

The equal-u experiment. In an attempt to establish the dependence of the internal energy of "permanent" gases on their density, Joule (in 1845) carried out an experiment similar to that discussed on p. 65 under the

* Formerly called *molecular weight*.

heading "Unresisted Expansion". Fig. 14.1 illustrates the apparatus, which consists of two vessels, A and B, immersed in a water bath; the vessels are inter-connected through a cock, which is closed initially. At first, container A holds gas but B is evacuated. The cock is suddenly opened and gas rushes into B; eventually the gas state becomes uniform throughout A and B. If the temperature of the gas changes in this process, a heat transfer will occur between the surrounding water bath and the gas; this will be reflected in a change in the water temperature.

Joule found no change in water temperature, signifying no detectable heat transfer. Since the experiment is one in which the initial and final internal

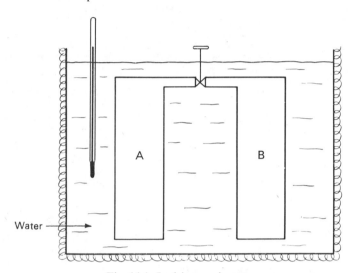

Fig. 14.1 Joule's experiment.

energies of the gas are equal ($Q = 0$, $W = 0$; $\therefore$ $\Delta u = Q - W = 0$), we may conclude:

$$u = f(t), \text{ within experimental accuracy} \qquad (14.2)$$

That is to say that u depends only on the temperature of the gas and not on its pressure or specific volume (both p and v change considerably in the above experiment). Here $f(t)$ means "some function of temperature", not, of course, the same function as that in eq. (14.1). Eq. (14.2) is sometimes known as *Joule's Law*.

The equal-h experiment. Not satisfied with the above negative result, Joule and Thomson (later Lord Kelvin) in 1852 devised a steady-flow experiment for the same purpose. Since pv was known from Boyle's Law to depend only on temperature, and since h equals $u + pv$, an equal-h experiment would serve their purpose as well as one at equal u; moreover it would be far more sensitive. Fig. 14.2 shows the apparatus. A porous plug of cotton wool is placed in a pipe. A gas such as air is forced steadily along the pipe and suffers a large decrease of pressure at the plug. The pipe is insulated thermally, and the steady temperatures of the gas upstream and downstream of the plug are measured; the kinetic energy of the gas is negligible both upstream and downstream. For these conditions, the enthalpy of the gas is the same on both sides of the plug, as may be seen by applying the Steady-Flow Energy Equation to the control surface C, Fig. 14.2.

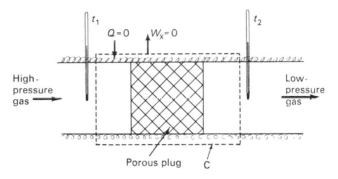

Fig. 14.2 The Joule-Thomson experiment.

In these experiments, small but definite temperature changes were detected, e.g. a *decrease* of about 0·3 K per atmosphere pressure difference for air, and an *increase* of about 0·03 K per atmosphere pressure difference for hydrogen, both gases being at about 0 °C.

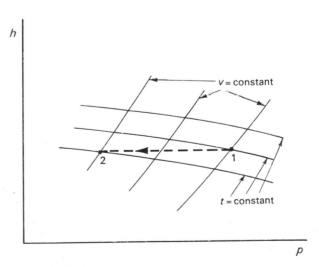

Fig. 14.3 Enthalpy~pressure diagram for a gas.

These results indicate a small dependence of enthalpy and internal energy on pressure (or specific volume) in addition to the major dependence on temperature. Nevertheless, for many purposes it suffices to ignore the pressure dependence and to quote the result as

$$h = \mathrm{f}(t), \text{ very closely,} \tag{14.3}$$

and

$$u = \mathrm{f}(t), \text{ very closely.} \tag{14.4}$$

Expressed graphically, these equations imply that on an $h{\sim}p$ or a $u{\sim}v$ diagram, the isotherms are very nearly horizontal lines. Fig. 14.3 illustrates the former case and shows how a slight temperature decrease can arise in the equal-enthalpy porous-plug experiment.

The Semi-perfect Gas

Definition of the semi-perfect gas

It is natural, in view of the good approximations with which eq. (14.1), (14.2), (14.3) and (14.4) describe the behaviour of real gases, to investigate the properties of an Ideal Gas, which obeys them exactly. The characteristics of the Ideal Gas can then be used to represent very closely those of real gases. We adopt the following definition, and then proceed to examine how the pressure and volume of such a gas depend on its temperature.

A semi-perfect gas is one which obeys exactly the equations

$$pv = \mathrm{f}(t) \tag{14.5}$$

$$u = \mathrm{f}(t) \tag{14.6}$$

The $p{\sim}v{\sim}T$ relation of such a gas will be established in two steps. In the first we show that the pressure of the gas is proportional to the *absolute* (thermodynamic) temperature, T, when the volume is kept constant. In the second we show that the expression pv/T is a constant under all conditions.

The $p{\sim}v{\sim}T$ relation of a semi-perfect gas

The $p{\sim}T$ relation in constant-volume heating. Eq. (14.5) and eq. (14.6) contain the properties p, v, u and t. The only general relation which we possess relating these is equation (13.42): $T\,ds = du + p\,dv$; this we put in the form

$$ds = \frac{1}{T}\,du + \frac{p}{T}\,dv \tag{14.7}$$

and then integrate, to give

$$s_2 - s_1 = \int_1^2 \frac{1}{T}\,du + \int_1^2 \frac{p}{T}\,dv \tag{14.8}$$

Now consider the $u \sim v$ diagram for the gas shown in Fig. 14.4. The isotherms on this diagram are horizontal lines, since u depends on temperature alone, and vice versa. Consider the states marked 1 and 2, and two paths connecting them, one via A and the other via B. The abscissae of 1 and 2 differ by the infinitesimal amount dv, but the difference of ordinate $(u_2 - u_1)$, is finite.

Considering the right-hand side of eq. (14.8), and invoking eq. (14.6), we note that the first integral is independent of the path of integration, i.e. has a definite value, since u depends only on t and so on T. But since entropy is a property, $s_2 - s_1$ is certainly also independent of the path of

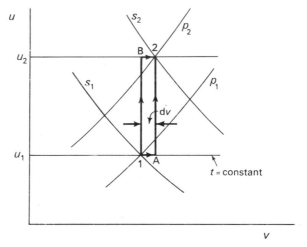

Fig. 14.4 Internal energy~volume diagram for a gas.

integration. It follows that the second integral on the right-hand side is independent of the path.

Examining this conclusion more closely, we see that it implies that the value of p/T along the line 1A is the same as that along the line B2. But the levels of 1 and 2 were arbitrarily chosen. It follows that p/T must be constant along a line of constant volume. We express this conclusion symbolically as

$$\frac{p}{T} = f(v) \qquad (14.9)$$

Comments. 1. It will have been noted that eq. (14.5) has not yet been invoked. Yet already we have the important result that a gas of which the internal energy depends upon temperature alone will, if used in a *constant volume* gas thermometer, register the *absolute temperature*. Thus if a 1 degree temperature rise produces a $P\%$ increase in gas pressure at constant volume, the temperature on the absolute scale with the chosen size of degree is $100/P$ degrees absolute.

2. To be satisfactory in this respect, the gas does not have to obey eq. (14.6) under all conditions, but merely under the conditions prevailing in the constant-volume gas thermometer.*

3. It is equally possible, though less useful, to prove that a gas obeying eq. (14.9) must obey eq. (14.6), i.e. to prove the converse of the above.

4. If an actual gas deviates from eq. (14.6) by a known amount, the magnitude of its deviation from eq. (14.9) can also be calculated.

Deduction of the "Ideal-Gas Rule". The second step of the argument involves the combination of eq. (14.5) and eq. (14.9) to give

$$\frac{pv}{T} = \text{constant} = R, \text{ say.} \tag{14.10}$$

This is directly obvious, since what other $p\sim v\sim T$ relation could satisfy both equations? For formality's sake, however, a proof is appended:

Substituting eq. (14.9) in eq. (14.5), we have

$$vT \cdot \text{f}(v) = \text{f}(T) \tag{14.11}$$

having recalled that f( ) merely means "some function of . . .", and that therefore t and T are interchangeable in f( ). Since v and T are independent properties, eq. (14.11) can be true only if

$$v \cdot \text{f}(v) = \text{constant} = \frac{1}{T} \cdot \text{f}(T)$$

Hence, from eq. (14.5)

$$\frac{pv}{T} = \text{constant} \qquad\qquad \text{Q.E.D.}$$

Eq. (14.10) is sometimes written in another form, namely as a relation between the mass m of a gas and the volume V which it occupies.

Since V equals mv, we have, from eq. (14.10)

$$pV = mRT \tag{14.12}$$

The Gas Constant R

Eq. (14.10) is known as the *Ideal-Gas Rule.* Its importance resides in the fact that the $p\sim v\sim T$ relations of real gases often obey it closely, just as they do equations (14.5) and (14.6), provided that the value of R is appropriately chosen. This quantity, known as the *Gas Constant,* must have a different value for each gas. For a number of technically important gases at low pressures, values of R are given in Table 14.1.

---

* We commonly say that for an incompressible fluid $u \approx \text{f}(t)$. The present proof shows that such a fluid in a constant-volume thermometer would register the absolute temperature T. However,

(a) the experimental difficulty is much greater with incompressible fluids than with gases, and

(b) the otherwise negligible dependence of u on p is not negligible in this application.

TABLE 14.1 *Gas Constant R in* J/kg K

Gas	Air	O_2	N_2 (atmospheric)	N_2	H_2	CO	CO_2	H_2O
R	287	260	295	297	4124	297	189	462

The units used in Table 14.1 are SI. *R* may also be expressed in other units; its value then changes in accordance with the usual rules relating the various unit systems. In the Imperial system the units of *R* are ft lbf/lb$_m$ °R.

Two gases included in Table 14.1 call for comment:

(i) *Atmospheric nitrogen* is the name given to the gaseous mixture which remains when all oxygen is removed from dry air. This mixture consists of pure nitrogen together with about 1·8 per cent by mass of argon and traces of carbon dioxide and other gases.

(ii) Steam has been included because *at low pressures* it obeys eq. (14.10) quite closely, even at temperatures well below that of its critical point.

Since real gases do not obey eq. (14.5) and eq. (14.6) exactly, they also exhibit departures from eq. (14.10). The order of magnitude of these departures is indicated by the values of $pv/(RT)$ for air, shown in Table 14.2; the *R* value is that given in Table 14.1.

TABLE 14.2 $\frac{pv}{RT}$ *for Air*

$\begin{matrix} & p \rightarrow \\ t \downarrow & \end{matrix}$	0	10	100 atm
0°C	1	0·9945	0·9699
200	1	1·0031	1·0364

It is evident that at low pressures the error involved in presuming pv/T to be a constant will often be negligible.

The Molar Gas Constant $\mathscr{R}$

Thermodynamics is not concerned with the microscopic structure of materials. However, we here introduce concepts which are explicable only in terms of the molecular nature of gases.

Molecular mass. The relative molecular mass, *M*, of a substance is defined by

$$M = \frac{\text{(mass of one molecule of the substance)}}{\frac{1}{12} \times \text{(mass of one atom of carbon-12)}} \tag{14.13}$$

For convenience, we shall from now on refer to this quantity as molecular mass; it was formerly called molecular weight.

The masses of the molecules may be determined or compared experimentally. The values of M for common gases given in Table 14.3 are the results of such experiments.

TABLE 14.3 *Relative Molecular Masses of Gases*

Gas	Air	O_2	N_2 (atmospheric)	N_2	H_2	CO	CO_2	H_2O
M	28·97	31·9988	28·17	28·0134	2·015 94	28·010 55	44·009 95	18·015 34

In Table 14.3 it is seen that, in addition to pure gases, values are given for gases such as air which are mixtures of different kinds of molecules. The significance of these is explained in Chapter 15, p. 329.

To explain why 1/12 times the mass of a carbon-12 atom is used in defining M, we first note that originally the mass of one hydrogen atom was taken as the reference. This made the molecular mass of hydrogen (M_{H_2}) equal to 2; it was thought that at that time the other molecular masses were also whole numbers; the molecular mass of carbon, for example, was thought to be 12. More careful experiments then showed that the molecular masses were not exactly whole numbers, for a reason connected with the composition of molecules and atoms from even smaller particles. Subsequently, by international agreement, it was decided to make molecular mass of carbon-12 equal to 12 exactly, so that M_{H_2} became 2·015 94 and the other substances took the values shown in Table 14.3. It will be seen, however, that they are still very nearly equal to whole numbers.

The mole. In this book we have used the kilogram as the unit of mass. Frequently, however, the analysis is simplified by the use of another unit of mass, the *mole*, defined as follows:

The mole is the amount of substance which contains as many molecules as there are carbon atoms in 12 grams of carbon-12.

This definition, combined with that for molecular mass (p. 294), therefore specifies that the mass, in grams, of a mole of any substance has the same numerical value as its molecular mass.

For example, one mole† (1 mol) of oxygen has a mass of 32 grams approximately.

The *molar mass* of a substance in a system containing n moles and mass m of the substance is given by m/n; its units are g/mol.

The introduction of a different mass unit for each material appears a retrograde step, similar to using both "troy weight" and "avoirdupois"*. The

† Formerly called gram-mole. A pound-mole (lb-mole) is **453·592 37** times as large as a mole.
* Troy weight is used for gold and has 12 ounces per lb_m, as compared with avoirdupois which is used for other substances and has 16 ounces per lb_m.

molar units however lead to great simplification in chemical calculations (see Chapter 16), because of their relation, just mentioned, to the molecular constitution of matter.

The Molar Gas Constant $\mathscr{R}$. The molecular mass has been introduced in the present chapter because of an experiment fact about gases which obey the Ideal-Gas Rule: it is found that the product of the Gas Constant and the molecular mass of such gases is independent of the nature of the gas. Symbolically

$$RM = \text{constant}$$

$$= \mathscr{R}, \text{ for } all \text{ ideal gases} \qquad (14.14$$

$\mathscr{R}$ is known as the *Molar Gas Constant* and is *defined* by eq. (14.14). I is sometimes called the Universal Gas Constant.

The numerical value of $\mathscr{R}$ depends on the units used. For example, ir the SI set of units

$$\mathscr{R} = 8314 \cdot 3 \text{ J/kmol K for } all \text{ Ideal Gases.}$$

It will be noted that the units of $\mathscr{R}$ contain kmol, corresponding to 100(moles; sometimes this latter amount of material is called a kilogram-mole For a particular Ideal Gas, eq. (14.14) gives R in J/kg K as:

$$R = \frac{8314 \cdot 3}{M}$$

Table 14.4 gives the values of $\mathscr{R}$ for various sets of units together with th corresponding units of R.

TABLE 14.4 *Values of the Molar (Universal) Gas Constant*

If R is in	$\mathscr{R}$ is
Nm/kg K	8314·3 J/kmol K
J/g K	8·3143 J/mol K
Btu/lb*m* °R	1·986 Btu/lb-mole °R

The Ideal-Gas Rule in molar units. The Ideal-Gas Rule has been stat as

$$pv = RT \qquad (14.1$$

or $$pV = mRT \qquad (14.$$

These equations may be written in terms of the mole as follows. Let $\tilde{V}$ the volume occupied by one mole of gas of molecular mass M. Th eq. (14.13) becomes

$$p\tilde{V} = MRT$$

Substituting from eq. (14.14) we have

$$p\tilde{V} = \mathscr{R}T \qquad (14.15)$$

or, if a mass of n moles occupies a volume V, then

$$pV = n\mathscr{R}T \qquad (14.16)$$

Remarks on $\mathscr{R}$ and M.

1. *Avogadro's Hypothesis.* The experimental fact $RM = \mathscr{R}$, stated in eq. (14.14), was first suspected by Avogadro in 1811, who expressed his hypothesis as: "Equal volumes of gases at equal pressures and temperatures contain equal numbers of molecules".

That this statement is consistent with eq. (14.14) may be shown by substituting from eq. (14.14) into eq. (14.12). Thus

$$m = \frac{pV}{RT} = M\frac{pV}{\mathscr{R}T}$$

so that

$$\frac{m}{M} = \frac{pV}{\mathscr{R}T} \qquad (14.17)$$

If now several gases occupy equal volumes V at equal pressures p and temperatures T, since $\mathscr{R}$ is a universal constant, from eq. (14.17) there results:

$$\frac{m}{M} = \text{constant, for all gases} \qquad (14.18)$$

This equation signifies that the masses of equal volumes of all gases at equal pressures and temperatures are proportional to the masses of their molecules. The numbers of molecules in each equal volume must therefore be equal.

In addition, eq. (14.18) also implies that equal volumes of gases at equal pressures and temperatures contain equal numbers of moles. This follows from eq. (14.16), (14.12) and (14.14) since

$$n = \frac{pV}{\mathscr{R}T} = \frac{mR}{\mathscr{R}} = \frac{mR}{MR} = \frac{m}{M}$$

The two interpretations of eq. (14.18) link together the mole concept and Avogadro's Hypothesis.

Avogadro's Hypothesis is obeyed by real gases to the same extent as is the Ideal-Gas Rule.

2. *Molar volume.* From eq. (14.15) and the information contained in Table 14.4, it follows that

$$1 \text{ mol of any Ideal Gas occupies } 0\cdot022\ 416 \text{ m}^3$$

when $p = 1$ atm and $t = 0\ °\text{C}$, or

$$1 \text{ lb-mole of any Ideal Gas occupies } 359 \text{ ft}^3$$

when $p = 14\cdot7$ lbf/in², and $t = 32\ °\text{F}$.

Charles' Law

A consequence of the Ideal-Gas Rule is that, at a fixed pressure, the specific volume of a gas obeying the Rule increases linearly with absolute temperature. This fact is often confused with Charles' Law, discovered in 1787, which may be expressed as: "The specific volume of a 'permanent' gas at constant pressure increases (approximately) linearly with the temperature measured on a uniformly divided mercury-in-glass thermometer."

Charles' Law was propounded well before the establishment of the First and Second Laws of Thermodynamics and the definition of the Absolute Temperature scale. Temperature therefore had to be defined in terms of a particular thermometric substance. Of course the gas itself could be used as the thermometric substance, in which case the "law" would reduce to a mere definition of a temperature scale, the only advantage of which would lie in the fact that *any* "permanent" gas could be used in the thermometer.

Charles' Law in itself tells us nothing about the absolute temperature. It is only the deductions from Boyle's and Joule's Laws, given above, that enable us to deduce that a temperature scale using an Ideal Gas as thermometric substance happens to measure the absolute temperature directly.

The specific heats of gases obeying the Ideal-Gas Rule

We have seen that many gases obey eq. (14.3) and eq. (14.4) very closely. It follows that property diagrams with internal energy or enthalpy as

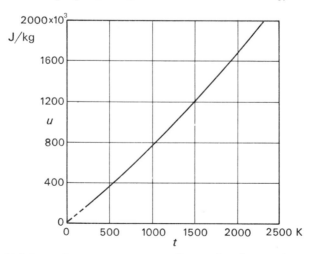

Fig. 14.5 Internal-energy~temperature diagram for air at moderate pressures.

ordinate, and temperature as abscissa, degenerate to single curves; this is illustrated by Figs. 14.5 and 14.6 which are drawn to scale for air.

Recalling the definition of c_v and c_p introduced in equations (7.6) and (7.8), we see that they may be re-written for a gas obeying eq. (14.10) without the conditions that v and p should respectively be kept constant in the differentiation; the partial derivatives are replaced by total derivatives because h and u are functions of t only. Thus, for a semi-perfect gas

$$c_v \equiv \frac{du}{dt}$$ (14.19)

$$c_p \equiv \frac{dh}{dt}$$ (14.20)

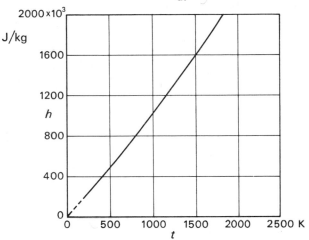

Fig. 14.6 Enthalpy~temperature diagram for air at low pressures.

c_v and c_p are the slopes of the curves illustrated by Figs. 14.5 and 14.6. Since these curves turn upwards with their slopes increasing, c_v and c_p are not constants but increase with temperature. This is the case for *all* gases.

Relation between c_v, c_p and R for Semi-Perfect Gases. Internal energy and enthalpy are connected by the definition, eq. (7.7)

$$h = u + pv$$

which, in view of eq. (14.10) may be written

$$h = u + RT$$ (14.21)

Differentiating with respect to temperature, and, remembering that R is a constant, we obtain

$$\frac{dh}{dt} = \frac{du}{dt} + R\frac{dT}{dt}$$

i.e. $$c_p = c_v + R$$ (14.22)

Herein it is supposed that the temperature is measured, as is always the case nowadays, on such a scale that t is equal to the absolute temperature minus a constant, so that dT/dt equals unity (see p. 239).

Eq. (14.22) is an important relation, which should be remembered. Of course it is necessary that consistent units should be used. Since c_p and c_v will normally be in J/kg K, R must be expressed in the same units.

Sometimes it is convenient to work with the mole as the unit of mass. Then *molar specific heats*, $\tilde{c}_v$ and $\tilde{c}_p$, are used, defined by

$$\tilde{c}_v \equiv Mc_v \tag{14.23}$$

$$\tilde{c}_p \equiv Mc_p \tag{14.24}$$

From equations (14.14) (14.22) (14.23) and (14.24), we deduce

$$\tilde{c}_p = \tilde{c}_v + \mathscr{R} \tag{14.25}$$

The specific-heat ratio, γ. It will later be found convenient to write eq. (14.22) and eq. (14.25) in terms of the ratio of the specific heats, γ, defined by

$$\gamma \equiv \frac{c_p}{c_v} = \frac{\tilde{c}_p}{\tilde{c}_v} \tag{14.26}$$

Substituting from eq. (14.26) into eq. (14.22), we have

$$\frac{R}{c_v} = \gamma - 1 \tag{14.27}$$

and

$$\frac{R}{c_p} = \frac{\gamma - 1}{\gamma} \tag{14.28}$$

Corresponding relations are obtained from eq. (14.25), namely

$$\frac{\mathscr{R}}{\tilde{c}_v} = \gamma - 1 \tag{14.29}$$

$$\frac{\mathscr{R}}{\tilde{c}_p} = \frac{\gamma - 1}{\gamma} \tag{14.30}$$

The specific-heat ratio γ is a property of the gas. For real gases, γ tends to decrease as the temperature rises, since both c_p and c_v increase, the difference between them remaining constant.

Since R and the specific heats are positive, it is clear that γ must be greater than unity. Values for common gases at moderate and high temperatures are given in Table 14.5.

A striking feature of Table 14.5 is that, at moderate temperatures, all the di-atomic molecules have identical values of γ. This is because γ is related to the number of degrees of freedom of the molecule. The subject is an advanced one however, particularly when the decrease of γ at high temperatures has to be explained; we shall not discuss it here.

Sometimes the symbol k is used in place of γ.

TABLE 14.5 *Values of γ for various gases*

Gas	Air	Atm N_2	N_2	O_2	H_2	CO	CO_2	A
γ at 15 °C	1·4	1·4	1·4	1·4	1·4	1·4	1·3	1·67
γ at 1100 °C	1·32	1·32	1·32	1·3	1·36	1·32	1·17	1·67
No. of atoms in molecule	≈ 2	≈ 2	2	2	2	2	3	1

Mean specific heats, $\bar{c}_v$ and $\bar{c}_p$. Formulae treating specific heats as *constants* are so convenient that we sometimes use them even though the $u{\sim}t$ and $h{\sim}t$ lines are curved, by defining *mean* specific heats, valid over specified temperature intervals. For any two states, 1 and 2, of the gas, the definitions are:

$$\bar{c}_{v_{12}} \equiv \frac{\int_1^2 c_v \, dt}{t_2 - t_1} = \frac{u_2 - u_1}{t_2 - t_1} \tag{14.31}$$

$$\bar{c}_{p_{12}} \equiv \frac{\int_1^2 c_p \, dt}{t_2 - t_1} = \frac{h_2 - h_1}{t_2 - t_1} \tag{14.32}$$

Often internal-energy and enthalpy data for gases will be found tabulated in this way. Then the lower temperature will have some specified standard value, t_0, such as 0 °C or 0 K, while the upper temperature of the range will be the argument t of the table. Any desired internal-energy difference, for example, is then calculated from

$$u_2 - u_1 - (u_2 - u_0) - (u_1 - u_0)$$
$$= \bar{c}_{v_{02}}(t_2 - t_0) - \bar{c}_{v_{01}}(t_1 - t_0) \tag{14.33}$$

Similarly the enthalpy difference is given by

$$h_2 - h_1 = \bar{c}_{p_{02}}(t_2 - t_0) - \bar{c}_{p_{01}}(t_1 - t_0) \tag{14.34}$$

The First Law of Thermodynamics for a Semi-Perfect Gas. We now illustrate the use of mean specific heats by writing the First Law, eq. (7.5), for a Semi-Perfect Gas. This, in differential form, becomes:

$$\frac{dQ}{m} = c_v \, dt + \frac{dW}{m} \tag{14.35}$$

and for a larger change

$$\frac{Q}{m} = \bar{c}_{v_{12}}(t_2 - t_1) + \frac{W}{m} \tag{14.36}$$

Likewise, the Steady-Flow Energy Equation becomes in differential form

$$\frac{dQ - dW_x}{m} = c_p \, dt + d \left(\frac{V^2}{2g_c} + \frac{gz}{g_c} \right) \tag{14.37}$$

and for a finite change

$$\frac{Q - W_x}{m} = \bar{c}_{p_{12}}(t_2 - t_1) + \Delta \left(\frac{V^2}{2g_c} + \frac{gz}{g_c} \right) \tag{14.38}$$

An *important point* to notice is that the specific heat at constant volume may be used in eq. (14.35) and eq. (14.36) even though, in general, the volume of the system will change; similarly, in eq. (14.37) and eq. (14.38) the specific heat at constant pressure is used, even though, in general, the pressure of the gas changes. This is permissible since neither the internal energy nor the enthalpy of a Semi-Perfect gas, which are expressed here in terms of c_v and c_p, are affected by changes in volume or pressure. There is, however, one useful formula which can be used only when the pressure remains constant: *for a non-flow, constant-pressure, reversible process of a Semi-Perfect Gas*, the First Law eq. (14.35) becomes

$$\frac{Q}{m} = \bar{c}_{v_{12}}(t_2 - t_1) + p(v_2 - v_1)$$

$$= \bar{c}_{v_{12}}(t_2 - t_1) + R(T_2 - T_1)$$

$$= \bar{c}_{p_{12}}(t_2 - t_1) \tag{14.39}$$

Tabulation of the properties of gases obeying the Ideal-Gas Rule. The variations of u and h for gases have to be established experimentally just as for less regular fluids. The results are contained in tables. However, since temperature is the only dependent variable, the information about a single substance can be tabulated in much less space. A widely used source of data

TABLE 14.6 c_p *for Air*

T (K)	200	250	300	400	500	1000	1500
c_p (J/kg K)	1003	1003	1005	1014	1030	1141	1211

is Keenan and Kaye's *Gas Tables*, from which the data given in Table 14.6 has been calculated.

c_v can be derived from these data, via eq. (14.22), by subtracting R in the appropriate units. For air this is 287 J/kg K.

An important deduction from Table 14.6 is that, at moderate temperatures, it is sufficiently accurate to assume that c_p is about 1005 J/kg K; correspondingly, c_v for air is fairly constant at $1005 - 287 = 718$ J/kg K.

This means that the slopes of the $h{\sim}t$ and $u{\sim}t$ curves are nearly constant. In this range it is usually sufficiently accurate to assume that the $h{\sim}t$ and $u{\sim}t$ curves are straight lines and to replace $\bar{c}_p$ by c_p and $\bar{c}_v$ by c_v. This greatly facilitates calculation.

The Perfect Gas

For a Semi-Perfect Gas, it has been shown that pv is equal to RT and that c_p and c_v depend upon its temperature only. The characteristics of real gases are represented adequately by these relationships over a *wide range* of conditions.

The near-constancy of the specific heats of real gases over practically important temperature ranges make it desirable to define a second Ideal Gas, the *Perfect Gas*, to include this characteristic also. The characteristics of a Perfect Gas will represent those of a real gas over a *restricted range* of conditions.

Definition of a Perfect Gas. A Perfect Gas obeys exactly the equations:

(1) $$pv = RT$$ (14.10)

(2) $$c_p = \text{constant}$$ (14.40)

Here part (1), the Ideal-Gas Rule, is a consequence of the definition of a Semi-Perfect Gas adopted earlier, (eq. (14.5) and eq. (14.6)), as has been proved. Specification of c_p in part (2) of the definition is arbitrary; we could equally well have stated $c_v = \text{constant}$, because of the relation (14.22).

The second part of the definition makes it possible to derive explicit algebraic expressions for the properties of a Perfect Gas, and for the heat and the work in various technically important processes. The remainder of the Chapter will be devoted to their derivation.

The properties of a Perfect Gas

The *internal energy and enthalpy* are obtained explicitly by integrating eq. (14.19) and eq. (14.20). Using the absolute temperatures, we obtain

$$u - u_0 = c_v(T - T_0)$$ (14.41)

$$h - h_0 = c_p(T - T_0)$$ (14.42)

where u_0 and h_0 are the values at some base temperature T_0. It should be noted that, if u_0 is arbitrarily made equal to zero, as is permissible, then h_0 will *not* be zero, but will equal RT_0 from the definition of h, eq. (14.21).

The *entropy of a Perfect Gas* is obtained by integrating one or other of the $T\,ds$ relations. Taking eq. (13.42) as our starting point, namely

$$T\,ds = du + p\,dv$$

we have

$$ds = c_v \frac{dT}{T} + \frac{p}{T} dv$$

$$= c_v \frac{dT}{T} + \frac{R \, dv}{v}$$

Since c_v is constant for the Perfect Gas, we have on integration*

$$s_2 - s_1 = c_v \ln \left(\frac{T_2}{T_1} \right) + R \ln \left(\frac{v_2}{v_1} \right) \tag{14.43}$$

By substitution from the Ideal-Gas Rule, eq. (14.43) may be written in two other ways, namely

$$s_2 - s_1 = c_p \ln \left(\frac{T_2}{T_1} \right) - R \ln \left(\frac{p_2}{p_1} \right) \tag{14.44}$$

and

$$s_2 - s_1 = c_p \ln \left(\frac{v_2}{v_1} \right) + c_v \ln \left(\frac{p_2}{p_1} \right) \tag{14.45}$$

Since these three relations are so readily derived, it is not recommended that they should be remembered. Which one should be used in practice depends on the problem. If temperatures and volumes are known, as in the following example, eq. (14.43) is preferable.

It may be remarked in passing that equations (14.43) to (14.45) are particular analytical instances of the fact that entropy is a property, i.e. that its value is specified, apart from a constant, when the values of two independent properties are fixed.

EXAMPLE 14.1

Problem. 1 kg of air at 600×10^3 N/m² and 300 K expands into an evacuated insulated container so that its volume is doubled (Fig. 14.7). What is the increase of entropy of the air?

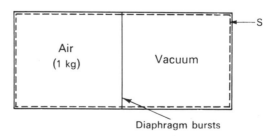

Fig. 14.7 Unresisted expansion of air.

* Note that to integrate the equation for a *Semi*-Perfect Gas we would need to know how c_v varies with temperature T.

Solution. A convenient system for the analysis of this problem is that enclosed by the boundary S, Fig. 14.7. For this system, the work and the heat are each zero and so, by the First Law, the internal energy has the same value before and after the process. Therefore, if the air is a Perfect Gas, no temperature change occurs. Substituting in eq. (14.43), we have

$$s_2 - s_1 = c_v \ln \left(\frac{300}{300} \right) + R \ln \left(\frac{2}{1} \right)$$

$$= 287 \ln 2$$

$$= 199 \text{ J/kg K} \qquad\qquad Answer$$

Comments. 1. The pressure and temperature (apart from the latter's constancy) were immaterial to the problem.

2. We note that the entropy *increases*, as was in fact to be expected of this isolated irreversible process (see p. 262).

Processes Executed by Perfect Gases

In the final section of this chapter, formulae are derived which replace, for Perfect Gases, the references to property tables which were necessary in applications of the First and Second Laws made in earlier Chapters. None of the formulae, except perhaps that involving pv^γ, is worth remembering; and blind application of the formulae should not be employed as a substitute for the analysis of a problem in terms of system and control-volume boundaries, the interactions across them, and their relations through the First and Second Laws. In all cases it is necessary to understand the restrictions on the validity of the formulae. All of them are subject to the restriction that the gas obeys *both* parts of the Perfect-Gas Definition.

The reversible adiabatic process (isentropic): $p{\sim}v{\sim}T$ relations

The importance of the reversible adiabatic process was stressed earlier (p. 257). For steam, calculation of the state at the end of an expansion starting from a given state was made via the Steam Tables. We now discuss the counterpart of this calculation for a Perfect Gas.

For a reversible adiabatic process, s does not change, so the right-hand side of, for example, eq. (14.43) can be put equal to zero:

$$c_v \ln \left(\frac{T_2}{T_1} \right) + R \ln \left(\frac{v_2}{v_1} \right) = 0$$

This may be re-written as

$$\frac{T_2}{T_1} \left(\frac{v_2}{v_1} \right)^{R/c_v} = 1 \qquad\qquad (14.46)$$

Use of eq. (14.44) or eq. (14.45) leads similarly to

$$\frac{T_2}{T_1} \left(\frac{p_2}{p_1} \right)^{-R/c_p} = 1 \qquad\qquad (14.47)$$

and

$$\frac{p_2}{p_1} \left(\frac{v_2}{v_1} \right)^{c_p/c_v} = 1 \qquad\qquad (14.48)$$

These equations may be written in terms of the specific-heat ratio, γ, by using the following $\gamma \sim R$ relations developed for a Semi-Perfect Gas (p. 300),

$$\gamma = \frac{c_p}{c_v} \tag{14.26}$$

$$\frac{R}{c_v} = \gamma - 1 \tag{14.27}$$

$$\frac{R}{c_p} = \frac{\gamma - 1}{\gamma} \tag{14.28}$$

Equations (14.26), (14.27) and (14.28) are valid for the Perfect Gas, which is merely a particular sort of Semi-Perfect Gas. Their special feature with regard to the Perfect Gas is that γ is constant and therefore independent of temperature. This result follows from part (2) of the definition of a Perfect Gas and contrasts with the Semi-Perfect Gas, for which γ depends upon temperature (p. 301).

Equations (14.46), (14.47) and (14.48) become respectively

$$\left(\frac{T_2}{T_1}\right) = \left(\frac{v_1}{v_2}\right)^{\gamma-1} \tag{14.49}$$

$$\left(\frac{T_2}{T_1}\right) = \left(\frac{p_2}{p_1}\right)^{(\gamma-1)/\gamma} \tag{14.50}$$

$$p_2 v_2{}^{\gamma} = p_1 v_1{}^{\gamma} \tag{14.51}$$

It should be noted that these three equations are *not applicable* to Semi-Perfect Gases; it is not permissible to use in them a mean value of γ defined as the ratio of the mean specific heats, viz. $\bar{\gamma}_{12} \neq \bar{c}_{p12}/\bar{c}_{v12}$. The reason is that γ has no significance for a change of state; γ has meaning at a *particular state* only.

Eq. (14.51) is the most often quoted of the three, perhaps because it gives the equation for the curve of the process on the $p \sim v$ diagram.

Since γ is greater than unity, reversible adiabatic processes are represented by curves on the $p \sim v$ diagram which are steeper than the isotherms ($pv =$ constant). Fig. 14.8 (p. 309) illustrates this. A consequence is that the gas temperature *falls* in a reversible adiabatic *expansion*, and *rises* in a reversible adiabatic *compression*.

Calculation of work done in adiabatic processes of a Perfect Gas

The First Law applied to a system comprising mass m of a pure substance in the absence of gravity, motion, etc., eq. (7.5), states

$$Q = m(u_2 - u_1) + W$$

If the pure substance is a Perfect Gas and the process is adiabatic ($Q = 0$), we have

$$\frac{W}{m} = u_1 - u_2$$

$$= c_v(T_1 - T_2)$$

$$= \frac{p_1v_1 - p_2v_2}{\gamma - 1} \tag{14.52}$$

Which of these three forms of the relation is the most useful depends on the variables in terms of which the problem happens to be expressed.

Here W represents *all* the work done by the system in changing from state 1 to state 2. In *steady-flow* problems this comprises displacement work associated with the flow, and external work. Usually we are interested in only the latter, W_x. For a Perfect Gas in adiabatic steady flow, the S.F.E.E. becomes

$$\frac{W_x}{m} + \Delta\left(\frac{V^2}{2g_c} + \frac{gz}{g_c}\right) = h_1 - h_2$$

$$= c_p(T_1 - T_2)$$

$$= \frac{\gamma}{\gamma - 1}(p_1v_1 - p_2v_2) \tag{14.53}$$

Eq. (14.52) and eq. (14.53) hold whether the processes are reversible or irreversible. To evaluate the formulae, two pieces of information must be specified for both final and initial states. If however we are told that the process is reversible, one piece of information may be omitted: for example, it suffices to know the initial state and the final pressure.

Reversible adiabatic processes. If reversibility is specified, the initial and final states are connected by the $p{\sim}v{\sim}T$ relations, equations (14.49), (14.50) and (14.51). The appropriate ones may be substituted into eq. (14.52) and eq. (14.53), giving, among other possible expressions

$$\frac{W}{m} = \frac{p_1v_1}{\gamma - 1}\left[1 - \left(\frac{p_2}{p_1}\right)^{(\gamma-1)/\gamma}\right]$$

$$= \frac{p_1v_1}{\gamma - 1}\left[1 - \left(\frac{v_1}{v_2}\right)^{\gamma-1}\right] \tag{14.54}$$

and

$$\frac{W_x}{m} + \Delta\left(\frac{V^2}{2g_c} + \frac{gz}{g_c}\right) = \frac{\gamma}{\gamma - 1}p_1v_1\left[1 - \left(\frac{p_2}{p_1}\right)^{(\gamma-1)/\gamma}\right]$$

$$= \frac{\gamma}{\gamma - 1}p_1v_1\left[1 - \left(\frac{v_1}{v_2}\right)^{\gamma-1}\right] \tag{14.55}$$

These relations can be derived in an *alternative way*, by evaluation of the $p\,dv$ or $v\,dp$ integrals. For a reversible process we have

$$\frac{W}{m} = \int_1^2 p\,dv \tag{14.56}$$

On insertion of p from eq. (14.51), this becomes

$$\frac{W}{m} = p_1 v_1{}^\gamma \int_1^2 \frac{dv}{v}$$

$$= \frac{p_1 v_1}{1-\gamma}\left[\frac{1}{v_2^{1-\gamma}} - \frac{1}{v_1^{1-\gamma}}\right]$$

$$= \frac{p_1 v_1}{\gamma-1}\left[1 - \left(\frac{v_1}{v_2}\right)^{\gamma-1}\right] \tag{14.54}$$

as before.

For reversible steady-flow processes in general, we have from eq. (13.47)

$$\frac{W_x}{m} + \Delta\left[\frac{V^2}{2g_c} + \frac{gz}{g_c}\right] = -\int_1^2 v\,dp \tag{14.57}$$

After substitution of v from eq. (14.51), this becomes

$$\frac{W_x}{m} + \Delta\left[\frac{V^2}{2g_c} + \frac{gz}{g_c}\right] = - v_1 p_1^{1/\gamma} \int_1^2 \frac{dp}{p^{1/\gamma}}$$

$$= -\frac{v_1 p_1^{1/\gamma}}{1-\gamma}\left[p_2^{1-(1/\gamma)} - p_1^{1-(1/\gamma)}\right]$$

$$= \frac{\gamma}{\gamma-1}p_1 v_1\left[1 - \left(\frac{p_2}{p_1}\right)^{(\gamma-1)/\gamma}\right] \tag{14.55}$$

as before.

"Polytropic" process of a Perfect Gas: $p{\sim}v{\sim}T$ relations

The formulae for adiabatic and reversible adiabatic processes are of such convenience in engineering calculations that we attempt to extend them to processes that are not adiabatic by considering the family of processes that have the pressure~volume relation.

$$pv^n = \text{constant} \tag{14.58}$$

when n is any number.

These processes are termed "*polytropic*". They are important because the pressure~volume relations of a number of technically important processes can be *approximately* represented by eq. (14.58). That is to say that it is possible to find a value of n which more or less fits the experimental results. It must be clearly understood however that whereas γ is a property of the gas, n is *not*: it depends upon the process. Fig. 14.8 illustrates the curves given by different values of n. Values close to unity are commonly encountered. Of course not all practically-encountered curves belong to

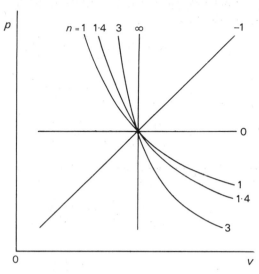

Fig. 14.8 Polytropic paths.

this family: examples of some that do not are given at the end of this section.

Relations connecting initial and final states in a polytropic process. To underline its similarity with eq. (14.51), eq. (14.58) will be written as

$$p_2 v_2{}^n = p_1 v_1{}^n \qquad (14.59)$$

By substitution from the Ideal-Gas Rule, two further relations can be derived, corresponding respectively to eq. (14.49) and eq. (14.50). They are

$$\frac{T_2}{T_1} = \left(\frac{v_1}{v_2}\right)^{n-1} \qquad (14.60)$$

$$\frac{T_2}{T_1} = \left(\frac{p_2}{p_1}\right)^{(n-1)/n} \qquad (14.61)$$

These relations may be used to relate the *initial* and *final* states of a polytropic process. If the process is *reversible*, eq. (14.58) can be used for evaluating work from $\int p\, dv$, or external work from $-\int v\, dp$, because in this case it is valid for the *intermediate* states also.

Entropy change in a polytropic process. In contrast to the reversible adiabatic ($pv^\gamma = $ constant) process, for which the entropy is constant, the entropy of a Perfect Gas usually changes in a polytropic process. We now calculate the entropy change, making use of the general relations for the entropy of a Perfect Gas, equations (14.43), (14.44), (14.45), and of the polytropic $p \sim v \sim T$ relations just derived. Substituting eq. (14.60) in eq. (14.43) for example, we obtain

$$s_2 - s_1 = c_v \ln \left(\frac{T_2}{T_1} \right) + \frac{R}{n-1} \ln \left(\frac{T_1}{T_2} \right)$$

$$= \left(\frac{1}{\gamma - 1} - \frac{1}{n-1} \right) R \ln \left(\frac{T_2}{T_1} \right)$$

$$= \frac{(n - \gamma)}{(\gamma - 1)(n-1)} R \ln \left(\frac{T_2}{T_1} \right) \tag{14.62}$$

Relations in terms of pressure and specific volume may also be derived, namely

$$s_2 - s_1 = \frac{(n - \gamma)}{n(\gamma - 1)} R \ln \left(\frac{p_2}{p_1} \right) \tag{14.63}$$

and

$$s_2 - s_1 = -\frac{(n - \gamma)}{\gamma - 1} R \ln \left(\frac{v_2}{v_1} \right) \tag{14.64}$$

Examination of each of these expressions shows that the entropy change is zero when n equals γ; this is in accordance with expectations.

It should be understood that, in a polytropic process, an entropy increase is not necessarily a sign that the process is irreversible, for the increase may have resulted from heat transfer from the surroundings. Entropy decreases are possible if the gas is cooled.

Heat and work in polytropic processes. Application of the First Law to a system containing mass m of a Perfect Gas in the absence of gravity, motion, etc., leads to

$$\frac{Q - W}{m} = u_2 - u_1$$

$$= c_v(T_2 - T_1)$$

$$= \frac{p_2 v_2 - p_1 v_1}{\gamma - 1} \tag{14.65}$$

If in addition the process is polytropic, we can derive further relations. For example, substituting from eq. (14.59) in eq. (14.65), we obtain

$$u_2 - u_1 = \frac{Q - W}{m} = \frac{p_1 v_1}{\gamma - 1} \left[\left(\frac{p_2}{p_1} \right)^{(n-1)/n} - 1 \right]$$

$$= \frac{p_1 v_1}{\gamma - 1} \left[\left(\frac{v_1}{v_2} \right)^{n-1} - 1 \right] \tag{14.66}$$

Application of the S.F.E.E. to the steady flow of a Perfect Gas leads to

$$\frac{Q - W_x}{m} - \Delta\left[\frac{V^2}{2g_c} + \frac{gz}{g_c}\right] = h_2 - h_1$$

$$= c_p(T_2 - T_1)$$

$$= \frac{\gamma}{\gamma - 1}(p_2 v_2 - p_1 v_1) \qquad (14.67)$$

If, in addition, the process is polytropic, further relations can be derived, namely

$$\frac{Q - W_x}{m} - \Delta\left[\frac{V^2}{2g_c} + \frac{gz}{g_c}\right] = \frac{\gamma}{\gamma - 1} p_1 v_1 \left[\left(\frac{p_2}{p_1}\right)^{(n-1)/n} - 1\right]$$

$$= \frac{\gamma}{\gamma - 1} p_1 v_1 \left[\left(\frac{v_1}{v_2}\right)^{n-1} - 1\right] \qquad (14.68)$$

These equations do not permit the separate determination of the heat and the work. For that to be possible, either one of them must be given, or a statement about the reversibility must be made.

Still without restriction to reversible processes, the integrals of $p \, dv$, $-v \, dp$ and $T \, ds$ will now be expressed in terms of properties pertaining to the beginning and the end of the process. The resulting relations are general for polytropic processes because they relate to properties.

Integral property relations in polytropic processes. By an integration procedure similar to that applied to eq. (14.56), we find that, in a process in which pv^n is constant,

$$\int_1^2 p \, dv = \frac{p_1 v_1}{n - 1}\left[1 - \left(\frac{v_1}{v_2}\right)^{n-1}\right]$$

$$= \frac{p_1 v_1}{n - 1}\left[1 - \left(\frac{p_2}{p_1}\right)^{(n-1)/n}\right] \qquad (14.69)$$

Likewise we derive

$$-\int_1^2 v \, dp = \frac{n}{n - 1} p_1 v_1 \left[1 - \left(\frac{v_1}{v_2}\right)^{n-1}\right]$$

$$= \frac{n}{n - 1} p_1 v_1 \left[1 - \left(\frac{p_2}{p_1}\right)^{(n-1)/n}\right] \qquad (14.70)$$

The integral of $T \, ds$ is obtained from eq. (13.42) namely

$$T \, ds = du + p \, dv$$

The integral of the first term of its right-hand side has already been evaluated in eq. (14.69); the integral of the second term is given in eq. (14.69). Combining these, we are led to

$$\int_1^2 T \, ds = \frac{(\gamma - n)}{(\gamma - 1)(n - 1)} p_1 v_1 \left[1 - \left(\frac{p_2}{p_1} \right)^{(n-1)/n} \right]$$

$$= \frac{(\gamma - n)}{(\gamma - 1)(n - 1)} p_1 v_1 \left[1 - \left(\frac{v_1}{v_2} \right)^{n-1} \right]$$

$$= \frac{(\gamma - n)}{(\gamma - 1)(n - 1)} R(T_1 - T_2) \tag{14.71}$$

The special case of n = 1. Each of the integral property relations becomes indeterminate when n equals unity; this value corresponds to an isothermal change of the Perfect Gas, either reversible or irreversible. The following relations are then valid:

$$\int_1^2 p \, dv = p_1 v_1 \int_1^2 \frac{dv}{v}$$

$$= p_1 v_1 \ln \left(\frac{v_2}{v_1} \right)$$

$$= p_1 v_1 \ln \left(\frac{p_1}{p_2} \right) \tag{14.72}$$

and

$$- \int_1^2 v \, dp = -p_1 v_1 \int_1^2 \frac{dp}{p}$$

$$= p_1 v_1 \ln \left(\frac{p_1}{p_2} \right)$$

$$= \int_1^2 p \, dv \tag{14.73}$$

and

$$\int_1^2 T \, ds = u_2 - u_1 + \int_1^2 p \, dv$$

$$= \int_1^2 p \, dv$$

Here there is no change in temperature or internal energy. The relations can therefore be summarized by

$$\int_1^2 T \, ds = - \int_1^2 v \, dp = \int_1^2 p \, dv$$

$$= p_1 v_1 \ln \left(\frac{v_2}{v_1} \right)$$

$$= p_1 v_1 \ln \left(\frac{p_1}{p_2} \right) \tag{14.74}$$

Reversible polytropic processes of a Perfect Gas. The integral property relations have been derived because of their importance when the process is *reversible.* For this case only we can write, *in general*

$$\int_1^2 p \, dv = \frac{W}{m}, \text{ in the absence of gravity, motion, etc.} \qquad (14.75)$$

$$-\int_1^2 v \, dp = \frac{W_x}{m} + \Delta \left(\frac{V^2}{2g_c} + \frac{gz}{g_c} \right) \text{ in steady flow} \qquad (14.76)$$

$$m \int_1^2 T \, ds = Q_R \qquad (14.77)$$

These equations, together with equation, (14.69) to (14.74), allow the heat and work quantities to be expressed in terms of the end-state properties of the process.

Examples

This Chapter will be concluded by two examples of calculations with Perfect Gases, which illustrate the pitfalls, firstly of unthinking application of formulae, and secondly of the lack of general application of the polytropic formulae.

EXAMPLE 14.2

Problem. Measurements of pressure and temperature at various stages in an *adiabatic* air turbine show that the air states lie on the line

$$pv^{1 \cdot 25} = \text{constant.}$$

Derive a formula for the shaft work per unit mass as a function of pressure, neglecting kinetic and gravitational potential energy. γ for air equals $1 \cdot 4$.

Wrong solution.

$$\frac{W_x}{m} = -\int_1^2 v \, dp \text{ and so from eq. (14.70)}$$

$$\frac{W_x}{m} = \frac{n}{n-1} p_1 v_1 \left[1 - \left(\frac{p_2}{p_1} \right)^{(n-1)/n} \right]$$

$$= 5 \cdot 0 p_1 v_1 \left[1 - \left(\frac{p_2}{p_1} \right)^{1/5} \right]$$

Correct solution.

$Q = 0$ and so from eq. (14.68)

$$\frac{W_x}{m} = \frac{\gamma}{\gamma-1} p_1 v_1 \left[1 - \left(\frac{p_2}{p_1} \right)^{(n-1)/n} \right]$$

$$= 3 \cdot 5 p_1 v_1 \left[1 - \left(\frac{p_2}{p_1} \right)^{1/5} \right] \qquad \qquad Answer$$

Comment. The discrepancy is a big one. The fault in the wrong solution lay in the first step. The process is not reversible, so $W_x < - \int_1^2 v \, dp$. Substitution in eq. (14.63) shows that there is an increase of entropy when p_2 is less than p_1, even though the process is adiabatic: this is an infallible sign of irreversibility in adiabatic processes.

EXAMPLE 14.3

Problem. A gas at 1 atm pressure and 450 K is enclosed in a 1·52-m long vertical tube. The upper end is closed and the lower end just rests in a tank of mercury (Fig. 14.9). As the gas cools, (*a*) what is the $p{\sim}v$ relation of the gas? (*b*) how high does the mercury rise if the gas cools to 300 K? (*c*) how much work per unit mass will the gas do if the mercury rises 0·38 m? The gas constant of the gas is 235 J/kg K.

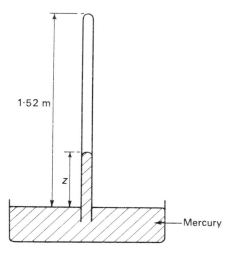

Fig. 14.9 Diagram used in the solution of Example 14.3.

Solution (a).

From hydrostatics:
$$p = 1 - \frac{z}{0.76}$$

where p is the gas pressure in atm, z is the height of the mercury in the tube in metres and 0·76 metres of mercury corresponds to 1 atm.

From geometry
$$\frac{v}{v_1} = \frac{1.52 - z}{1.52}$$

when v_1 is the initial specific volume of the gas.

By elimination of z,
$$p = \frac{2v}{v_1} - 1 \qquad\qquad \textit{Answer (a)}$$

Comment. This relation does not fit into the polytropic family. It is sketched in Fig. 14.10.

Solution (b).

From the Ideal-Gas Rule:
$$\frac{(pv)_{300}}{(pv)_{450}} = \frac{300}{450}$$

Hence
$$\left(1 - \frac{z_{300}}{0\cdot76}\right)\left(1 - \frac{z_{300}}{1\cdot52}\right) = \frac{300}{450}$$

The physically meaningful solution of this quadratic equation is

$$z_{300} = 0\cdot18 \text{ m} \qquad\qquad \text{\textit{Answer (b)}}$$

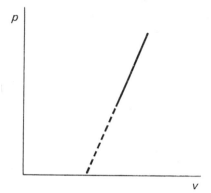

Fig. 14.10 Pressure~volume diagram
   for the process in Example 14.3.

Solution (c). Only displacement work is done, and the process is reversible, so

$$\frac{W}{m} = \int_1^2 p\,dv = \int_0^{0\cdot38} 101\cdot325 \times 10^3 \left(1 - \frac{z}{0\cdot76}\right)\left(-\frac{v_1}{1\cdot52}dz\right)$$

$$= -101\cdot325 \times 10^3 \times \frac{v_1}{1\cdot52}\left[z - \frac{z^2}{1\cdot52}\right]_0^{0\cdot38}$$

$$= -\frac{RT_1}{1\cdot52}\left[0\cdot38 - \frac{(0\cdot38)^2}{1\cdot52}\right]$$

$$= -\frac{235 \times 450}{1\cdot52} \times 0\cdot285$$

$$= -19\cdot85 \times 10^3 \text{ J} \qquad\qquad \text{\textit{Answer (c)}}$$

Comment. The polytropic relations were not useful here, and we had to return to fundamentals. Since this situation arises frequently in engineering problems, the fundamental relations are the only ones which are worth remembering.

BIBLIOGRAPHY

HILSENRATH, J. *et al.* *Tables of Thermodynamic and Transport Properties of Air, Argon, Carbon Dioxide, Carbon Monoxide, Hydrogen, Nitrogen, Oxygen and Steam.* Pergamon Press, Oxford, 1960. (Originally published as National Bureau of Standards Circular 564.)

KEENAN, J. H. and KAYE, J. *Gas Tables.* Wiley & Sons, New York, 1948.

WEAST, R. C. *et al.* *Handbook of Chemistry and Physics.* The Chemical Rubber Co., Cleveland, 1971.

CHAPTER 14—PROBLEMS

14.1 Examine the conditions under which (a) Boyle's Law and (b) the Ideal-Gas Rule may be applied to superheated steam by evaluating pv and pv/T from Table IV of the Steam Tables (Appendix B) for the following conditions:

(i) Temperature, 150 °C; pressures, $2 \cdot 0 \times 10^3$ N/m², 100×10^3 N/m² and 400×10^3 N/m².

(ii) Temperature, 400 °C; pressures, $2 \cdot 0 \times 10^3$ N/m², 100×10^3 N/m², 400×10^3 N/m², 5×10^6 N/m² and 20×10^6 N/m².

(iii) Temperature, 800 °C; pressures, $2 \cdot 0 \times 10^3$ N/m², 100×10^3 N/m², 400×10^3 N/m², 5×10^6 N/m² and 20×10^6 N/m².

14.2 Two rigid vessels, connected by a valve, are immersed in a constant-temperature bath. Initially one vessel is evacuated and the other contains steam. The valve is opened; after a time, conditions become uniform throughout the two vessels. Calculate the increase in the internal energy of the steam in each of the following cases:

(i) Far from the critical point.
    Initially: steam temperature = 800 °C, pressure = 100×10^3 N/m².
    Finally: steam pressure = 50×10^3 N/m².

(ii) Near the critical point.
    Initially: steam temperature = 350 °C, pressure = 8×10^6 N/m².
    Finally: steam pressure = 4×10^6 N/m².

14.3 Steam is expanded in adiabatic steady flow through a porous plug, the final pressure being one-half of the initial pressure. On the assumptions that changes in kinetic energy and in elevation are negligible, estimate the changes in the temperature of the steam for the following initial conditions.

(i) Far from the critical point: temperature = 800 °C, pressure = 100×10^3 N/m².

(ii) Near the critical point: temperature = 350 °C, pressure = 8×10^6 N/m².

14.4 Taking the value of the Molar Gas Constant $\mathscr{R}$ as 8314 J/kmol K and using the values of molecular mass given in Table 14.3 (p. 295), confirm that the values of the Gas Constant R for each gas given in Table 14.1 are correct.

14.5 Each of the cylinders of a marine diesel engine has a bore of 400 mm and a clearance volume (see page 56) of $4 \cdot 5$ dm³. The piston stroke is 530 mm.

At the end of the induction stroke (i.e. with the piston in its outermost position) the pressure and temperature of the gas in the cylinder are 98×10^3 N/m² and 40 °C respectively. On the assumption that the gas obeys the Ideal-Gas Rule and that for this gas $R = 286$ J/kg K, calculate the mass and the number of moles of the gas contained in the cylinder at the end of the induction stroke.

14.6 A boiler drum of volume $1 \cdot 8$ m³ contains steam at a temperature of 200 °C and a pressure of 200×10^3 N/m². Calculate the mass of steam contained in the drum,

(i) assuming steam to obey the Ideal-Gas Rule. Use the value of R given in Table 14.1;

(ii) using the Steam Tables (Appendix B).

Which value is correct? and why?

14.7 (a) A mass of $1 \cdot 5$ kmol of an Ideal Gas of molecular mass 25 is contained in a rigid vessel of volume 4 m³ at a temperature of 100 °C. Evaluate the mass, the pressure, the specific volume and the Gas Constant of the gas.

(b) For this condition the gas has a value of $\gamma = 1 \cdot 38$. Evaluate the corresponding values of c_p and c_v.

(c) Subsequently the gas cools to atmospheric temperature, 15 °C. Evaluate the final pressure of the gas.

(d) Assuming the gas to be perfect, evaluate the increase in internal energy, the increase in enthalpy and the magnitude and sign of the heat transfer.

14.8 Using the definition of a Perfect Gas (page 303), sketch the following property charts for such a gas, showing a few lines of constant volume in (i), and a few lines of constant temperature in (ii), (iii) and (iv).

(i) p, t;   (ii) h, p;   (iii) p, v;   (iv) u, v.

14.9 One kilogram of nitrogen at a temperature of 150 °C occupies a volume of 0·2 m³ The nitrogen undergoes a fully-resisted constant-pressure expansion, without friction, to a final volume of 0·36 m³. Evaluate the final pressure, the final temperature, the work done and the heat transfer.

Assume nitrogen to be a Perfect Gas for which $c_v = 743$ J/kg K and $R = 297$ J/kg K.

14.10 In a chemical plant, carbon monoxide flows steadily into a turbine at a pressure of 800×10^3 N/m² and a temperature of 240 °C; there it expands reversibly and adiabatically to a final pressure of 100×10^3 N/m². Changes in velocity and elevation are negligible.

(i) Calculate the final specific volume, the final temperature and the increase in entropy.

(ii) Given that the mass flow rate is 10 kg/s, evaluate the heat-transfer rate from the gas and the power delivered by the turbine.

Assume carbon monoxide to be a Perfect Gas for which $M = 28·01$, $c_v = 20·82$ kJ/kmol K. Take $\mathscr{R} = 8314$ J/kmol K.

14.11 A volume of 0·36 m³ of oxygen initially at a temperature of 220 °C and a pressure of 400×10^3 N/m² is compressed reversibly and isothermally to a final volume of 0·06 m³. Calculate the mass, the final pressure, the increase in the internal energy, the work done, the heat transfer and the change in entropy.

Assume oxygen to be a Perfect Gas for which $R = 260$ J/kg K.

14.12 (a) One kilogram of air at a pressure of $1·0 \times 10^6$ N/m² and a temperature of 100 °C undergoes a reversible polytropic process which may be represented by $pv^{1·1}$ = constant. The final pressure is 200×10^3 N/m².

(i) Evaluate the final specific volume, the final temperature and the increase in entropy.

(ii) Evaluate the work done and the heat transfer.

Assume air to be a Perfect Gas for which $R = 287$ J/kg K and $\gamma = 1·4$.

(b) Repeat (a) assuming the process to be irreversible and adiabatic.

14.13 The indicator diagram of a reciprocating air-compressor shows that during the compression stroke the volume of air in the cylinder is 0·8 dm³ when the pressure is 140×10^3 gauge. Correspondingly, when the pressure is 520×10^3 N/m² gauge the volume is 0·36 dm³. The mass of air contained in the cylinder at both points is $2·2 \times 10^{-3}$ kg. The barometric pressure is 773 mm Hg.

(i) Calculate the temperature of the air at each point.

(ii) On the assumption that the compression process may be represented by pV^n = constant, where n is a constant, evaluate n.

(iii) Assuming the process to be reversible and polytropic, evaluate the magnitude and sign of the work and of the heat transfer.

Assuming air to be a Perfect Gas for which $R = 287$ J/kg K and $c_p = 1000$ J/kg K.

14.14 (a) Air flows steadily into a compressor at a temperature of 16 °C and a pressure of 100×10^3 N/m² and leaves at a temperature of 246 °C and a pressure of 600×10^3 N/m². There is no heat transfer to or from the air as it flows through the compressor; changes in velocity and elevation are negligible. Evaluate the external work done per unit mass of air, assuming air to be a Perfect Gas for which $R = 287$ J/kg K and $\gamma = 1·4$.

(b) Evaluate the minimum external work required to compress the air adiabatically from the same initial state to the same final pressure.

(c) Evaluate the isentropic efficiency of the compressor (see page 260).

14.15 Helium at a pressure of 90×10^3 N/m² and a temperature of 60 °C flows steadily into a rotary compressor, and leaves at a pressure of 210×10^3 N/m². Changes in velocity and elevation are negligible; the compression process may be assumed to be adiabatic.

Given that the isentropic efficiency of the compressor is 73 per cent, calculate:

(a) the temperature of the helium leaving the compressor;

(b) the power required to compress 6 kg of helium per minute.

Assume helium to be a Perfect Gas for which $c_p = 5·19$ kJ/kg K and $M = 4$. Take $\mathscr{R}$ = 8314 J/kg K.

14.16 (*a*) A gas turbine develops a power of 7000 kW when the gas mass flow rate is 25 kg/s. The gas flows steadily into the nozzles of the turbine at a pressure of 400 × 10^3 N/m² and a velocity of 150 m/s. The gas leaves the turbine at a pressure of 105 × 10^3 N/m², a temperature of 500 °C and a velocity of 280 m/s. The expansion process may be assumed to be adiabatic.

Assume the gas to be Perfect and take c_p = 1150 J/kg K and R = 287 J/kg K.

Calculate: (i) the temperature of the gas entering the turbine;

               (ii) the flow area at entry to the turbine nozzles.

(*b*) The gas leaving the turbine flows into a cooler; at exit from the cooler the gas pressure is 100 × 10^3 N/m², its temperature is 65 °C and its velocity is 60 m/s. Determine the heat transfer from the gas in the cooler.

14.17 (*a*) A cylinder, fitted with a piston, contains 20 dm³ of methane gas (CH_4) at a pressure of 500 × 10^3 N/m² and a temperature of 115 °C. The methane undergoes a process to a pressure of 100 × 10^3 N/m² during which the work done by the gas is 10 000 N m and the heat transfer to it is 1 kJ. Determine the final temperature of the methane and the increase in entropy.

(*b*) If the gas had executed a reversible adiabatic process between the same initial state and the same final pressure, what then would have been the final temperature and the work done?

(*c*) If the gas had executed a process between the same initial state and the same final pressure such that the final and initial entropies were equal, and if during this process the heat transfer was −1·5 kJ, what then would have been the final temperature and the work done?

Assume methane to be a Perfect Gas for which c_p = 2234 J/kg K. Take $\mathscr{R}$ = 8314 J/kmol K.

14.18 A rigid vessel of volume 0·3 m³ contains air at a pressure of 35 × 10^3 N/m² and a temperature of 15 °C. Atmospheric air at a pressure of 100 × 10^3 N/m² and a temperature of 15 °C leaks slowly into the vessel. When the pressure of the air in the vessel has risen to 70 × 10^3 N/m² the leak is stopped; it is then found that the temperature of the air in the vessel is 50 °C.

(i) Calculate the mass of air which has leaked in.

(ii) Calculate the magnitude and sign of the heat transfer between the contents and the vessel walls.

(iii) Subsequently the contents cool to the atmospheric temperature. Calculate the corresponding pressure of the contents.

Assume air to be Perfect Gas for which γ = 1·4 and R = 287 J/kg K.

14.19 A well-insulated, rigid vessel contains 0·3 kg of a Perfect Gas (M = 20, c_p = 1·80 × 10^3 J/kg K) at a pressure of 100 × 10^3 N/m² and a temperature of 20 °C. A coiled tube within the vessel is connected to a reservoir containing dry and saturated steam at a pressure of 100 × 10^3 N/m². The steam condenses in the coil and collects therein as saturated water at a pressure of 100 × 10^3 N/m².

(i) Calculate the maximum pressure reached by the gas.

(ii) Calculate the quantity of steam condensed in the coil to achieve this maximum pressure.

14.20 The compression ratio r of a reciprocating internal-combustion engine is defined by the relation $r = (V_c + V_{sw})/V_c$ where V_c is the cylinder clearance volume and V_{sw} the cylinder swept volume ("the cylinder capacity"). One limitation to the value of r which may be used in a gasoline engine (to avoid "pre-ignition") is the temperature at which a gasoline-air mixture is supposed to ignite spontaneously. For a typical motor fuel this temperature may be taken as 450 °C.

In a particular engine, the pressure and temperature of the mixture at the beginning of the compression stroke (outer dead-centre) are 96 × 10^3 N/m² and 100 °C respectively; further, the compression process may be represented by $pV^{1·34}$ = constant.

(i) Determine the maximum value of r which may be used if pre-ignition is to be avoided.

(ii) Given that the bore and stroke of the engine are 66 mm and 94 mm respectively and assuming the process to be fully resisted, evaluate the work done and the heat transfer during the compression process when the maximum r is used. Assume the gasoline-air mixture to be a Perfect Gas with $c_p = 1 \cdot 0 \times 10^3$ J/kg K and $R = 262$ J/kg K.

14.21 A nozzle is supplied with a steady stream of a Perfect Gas ($\gamma = 1 \cdot 4$) at a pressure of 700×10^3 N/m², a temperature of 45 °C, a density of 4·5 kg/m³ and a velocity of 120 m/s.

(i) Determine the Gas Constant, R, and the specific heat at constant pressure, c_p, in J/kg K.

(ii) Assuming the flow to be reversible and adiabatic, determine the temperature and velocity of the gas at the nozzle exit where the pressure is 400×10^3 N/m².

(iii) If the mass flow rate is 1 kg/s, determine the cross-sectional areas at entry to and exit from the nozzle.

14.22 (a) A Perfect Gas flows steadily into a convergent-divergent nozzle (Fig. 8.8) from a reservoir where the absolute pressure is p_1, and the absolute temperature is T_1. As the gas expands reversibly and adiabatically in the nozzle, its pressure falls continuously; at any downstream section, where the area is A, the pressure is p. Derive expressions for the velocity V and the area A in terms of the mass flow rate $\dot{m}$ and of p_1, T_1, p, γ and R.

(b) For fixed inlet conditions and mass flow rate, plot A against p/p_1. Note that A first decreases and then increases.

(c) By differentiating the expression for A with respect to p, determine the values of p/p_1 and V at the minimum value of A.

(d) A *convergent* nozzle of fixed shape and size is connected to a supply at a fixed absolute pressure p_1 and absolute temperature T_1 and a reservoir of variable pressure p_2. A Perfect Gas flows reversibly and adiabatically through the nozzle; the entry velocity is negligible. Show that the mass flow rate $\dot{m}$ is a maximum when

$$\frac{p_2}{p_1} = \left(\frac{2}{\gamma + 1}\right)^{\frac{\gamma}{\gamma - 1}}$$

(N.B. The expression for $\dot{m}$ suggests that, as p_2/p_1 is reduced below this "critical" value, the mass flow decreases. This does not really happen. The reason is that for values of $p_2/p_1 <$ critical value, the formula applies only if the nozzle becomes convergent-divergent (see above), i.e. *changes* in shape.)

14.23 A hot-air engine operates on the Carnot cycle between the temperature limits of 250 °C and 50 °C; all the processes take place in a piston-cylinder mechanism. The pressure and volume at the start of isothermal expansion are 600×10^3 N/m² and 0·1 m³ respectively. The pressure after isothermal expansion is 300×10^3 N/m². Assume air to be a Perfect Gas with $\gamma = 1 \cdot 4$ and $R = 287$ J/kg K.

(i) Show the cycle on sketches of the $p \sim v$ and $T \sim s$ diagrams. Mark on each sketch the numerical values of the co-ordinates at each "corner" in the cycle diagram.

(ii) Evaluate the efficiency of the cycle and the work done per cycle.

(iii) Evaluate the power developed by the engine when it performs 96 cycles/min.

14.24 A cyclic process, known as the Air-Standard Diesel cycle, is often used as a standard of comparison for reciprocating oil engines. The cycle is executed by unit mass of air contained in a cylinder closed by a frictionless leakproof piston and consists of four processes in sequence, exemplified in the following.

(a) Initially, with the piston in its outermost position (outer dead-centre), the cylinder contains air at pressure p_1, temperature T_1 and specific volume v_1 (state 1).

(b) As the piston is pushed slowly inwards to its innermost position (inner dead-centre), the air is compressed reversibly and adiabatically to state 2 such that $v_1 = rv_2$ where r is the "compression ratio" (see problem **14.20**).

(c) The piston now commences its outward stroke. During the first part of this stroke the pressure remains constant and heat transfer occurs until the air is at state 3, where $v_3 = \lambda v_2$; λ is the "cut-off ratio" ($\lambda < r$). Subsequently, as the piston continues its outward stroke to the outer dead-centre, the air expands reversibly and adiabatically to state 4; $v_4 = v_1$.

(d) Finally, with the piston at the outer dead-centre (volume $= v_4$), heat transfer from the air at constant volume restores the air to its original state. The sequence may then be repeated.

Assuming air to be a Perfect Gas:

    (i) Sketch the $p{\sim}v$ and $T{\sim}s$ state diagrams.

    (ii) Derive an expression for the efficiency η of the cycle in terms of r, λ and the ratio of the specific heats of the working fluid, γ.

    (iii) Evaluate η when $r = 12$, $\lambda = 1.5$ and $\gamma = 1.4$.

    (iv) Evaluate the temperature at each state point given that $t_1 = 25°C$.

    (v) Evaluate the work done per cycle assuming $R = 287$ J/kg K.

14.25 A cyclic process, known as the Joule (constant-pressure) cycle, is often used as a standard of comparison for gas-turbine engines. The cycle consists of four processes which take place in sequence in steady flow, exemplified in the following. (See Fig. 10.3.)

(a) Initially, air at pressure p_1, temperature T_1 (state 1) is compressed reversibly and adiabatically in a rotary compressor to state 2; p_2 equals rp_1 where r is the "pressure ratio."

(b) Heat transfer at constant pressure to the air in the heater raises the air temperature to T_3 (state 3).

(c) The air then undergoes a reversible adiabatic expansion in a turbine to state 4 such that p_4 equals p_1.

(d) Finally, heat transfer from the air, at constant pressure, restores the air to its initial state before it re-enters the compressor to repeat the process.

Assuming air to be a Perfect Gas and neglecting changes in velocity and in elevation:

    (i) Sketch the $p{\sim}v$ and $T{\sim}s$ state diagrams.

    (ii) Derive an expression for the efficiency, η, of the cycle in terms of r and the ratio of the specific heats of the working fluid, γ.

    (iii) Evaluate η when $r = 5$ and $\gamma = 1.4$.

    (iv) Derive an expression for the net shaft work per unit mass of air in terms of r, γ, T_1 and T_3.

    (v) Evaluate the shaft work per unit mass of air when $t_1 = 40\ °C$ and $t_2 = 760\ °C$. For air take $R = 287$ J/kg K.

    (vi) In what respects would your answers be different if the air underwent the same changes of state in a piston-cylinder apparatus?

14.26 A closed-cycle gas-turbine engine (Fig. 10.3) employs air as the working fluid. The pressure and temperature at entry to the rotary compressor are 140×10^3 N/m² and 40 °C respectively. The pressure ratio is 5 and the temperature at entry to the turbine is 760 °C. The isentropic efficiency of the compressor is 83 per cent and that of the turbine is 87 per cent. There are no pressure losses in the heater or in the cooler and all heat transfers may be assumed to be negligible except those in the heater and cooler.

Assuming air to be a Perfect Gas with $\gamma = 1.4$ and $R = 287$ J/kg K:

    (i) Sketch the $T{\sim}s$ state diagram.

    (ii) Evaluate the temperatures at the four points in the cycle.

    (iii) Evaluate the efficiency of the engine.

    (iv) Evaluate the shaft work done per unit mass of air.

14.27 (a) A mass of 1.5 kg of a Semi-Perfect Gas at a pressure of 100×10^3 N/m² and a temperature of 100 °C occupies a volume of 3.3 m³.

    (i) Evaluate the Gas Constant R for the gas.

    (ii) Given that the specific heat at constant pressure of the gas at this state is 1920 J/kg K, evaluate γ for this condition.

(iii) The mean specific neat at constant pressure of the gas between the temperatures of 0 °C and 100 °C is 1880 J/kg K. Evaluate the specific enthalpy and the specific internal energy of the gas for the given condition.

(b) The gas undergoes a process to a final pressure of 400 × 10³ N/m² and a final temperature of 310 °C. The specific heat at constant pressure corresponding to the final state is 2930 J/kg K; the mean specific heat at constant pressure between temperatures 0 °C and 310 °C is 2340 J/kg K.

(i) Evaluate γ corresponding to the final state of the gas.

(ii) Evaluate the final specific volume, the final specific enthalpy and the final specific internal energy of the gas.

(iii) Evaluate the mean values of the specific heat at constant pressure and of the specific heat at constant volume between the temperatures 100 °C and 310 °C.

(c) If the process in (b) may be represented by pV^n = constant, find the value of n.

(d) Assuming the process to take place reversibly and polytropically in a piston-cylinder mechanism, evaluate the work done and the heat transfer during the process.

(e) Assuming the process to take place adiabatically and polytropically in steady flow with negligible changes in velocity and elevation, evaluate the external work done per unit mass of gas.

14.28 Re-calculate problem **14.27** assuming the gas to be Perfect. Take the specific heat at constant pressure as 1840 J/kg K.

14.29 A steady flow of air enters a heater at a pressure of 400 × 10³ N/m², a temperature of 40 °C and a velocity of 60 m/s. The air leaves the heater at a pressure of 370 × 10³ N/m² and a temperature of 550 °C. The flow area at exit is 22 × 10⁻³ m³ and the air mass flow rate is 5 kg/s. Given that R for air equals 287 J/kg K, evaluate the exit velocity and the heat-transfer rate to the air by the use of each of the following sets of data.

(i) Take air to be a Perfect Gas with specific heat at constant pressure = 1000 J/kg K.

(ii) Take air to be a Semi-Perfect Gas having the following characteristics:

Temperature (°C)	0	100	200	300	400	500	600
Enthalpy (kJ/kg K)	272·9	373·8	475·3	578·8	684·3	791·9	901·7

In this case also evaluate the mean specific heat at constant pressure for the given temperature range.

14.30 Air flows steadily at the rate of 18 kg/s throughout a rotary air compressor, entering with a velocity of 100 m/s, a pressure of 96 × 10³ N/m² and a density of 1·19 kg/m³, and leaving with a velocity of 160 m/s, a pressure of 340 × 10³ N/m² and a density of 2·68 kg/m³. The compression process may be assumed to be adiabatic. Given that R for air is 287 J/kg K evaluate the ratio of the inlet area to the outlet area and the power input by the use of each of the sets of data given in (i) and (ii) of problem **14.29**.

15   Gaseous Mixtures

Introduction

In the last chapter it was stated that air, which is a mixture of oxygen, nitrogen and other gases, can be treated for many purposes as though it were a single chemical substance. We even went so far as to speak of the "molecular mass" of air, despite the fact that there is no such thing as an air molecule. This procedure will be justified in the present chapter, which is devoted mainly to the thermodynamics of gas-phase mixtures.

The discussion will explain a number of commonly observed facts that, at first sight, do not seem to accord with statements made earlier. For example, water is known to "disappear into thin air", i.e. to vaporize, at temperatures much below the boiling-point for the prevailing pressure. Ice, too, can turn directly into vapour, even at atmospheric pressure, though this is over a hundred times as great as that of the triple point.

The main reason for discussing gas-phase mixtures here, however, is that they occur in many engineering processes. Thus, in steam-power-plant practice, the cooling water for the condenser is often cooled in its turn by coming into direct contact with air in a so-called cooling tower; the vaporization of a small proportion of the water into the air assists materially in lowering the temperature of the remainder. In the combustion chambers of furnaces and engines, it is essential that fuel, air and combustion products should be mixed in order that the chemical reaction should proceed; to design combustion equipment we need to calculate the properties of the mixtures. Many mechanical engineers are concerned with the design of processing plant: textile dryers, air-conditioning plant, paper-making machinery and the like; in all these devices, the behaviour of mixtures of air and water vapour has to be predicted quantitatively. Such problems assume perhaps their greatest importance in chemical engineering, where the mixing, and still more the separating, of different substances are central tasks.

The main thermodynamic idea introduced in this chapter is *Dalton's Law* with the refinements made to it by Gibbs. This law enables the techniques introduced in earlier chapters to be brought to bear on mixtures, by showing how the properties of a mixture are related to the properties of its components. We shall exemplify Dalton's Law by discussing, first, mixtures of Ideal Gases and, second, mixtures of Ideal Gases with a condensable vapour. The way will then be clear for the application of the First and Second Laws of Thermodynamics to systems comprising mixtures.

Representation of the thermodynamic properties of mixtures is conveniently made with the help of diagrams. We discuss three types of these. Although water and air are used as examples, the methods are general.

Symbols

c_v	Specific heat at constant volume.	V	Volume of a pure substance.
c_p	Specific heat at constant pressure.	$\tilde{V}$	Molar volume.
H	Enthalpy of a pure substance.	v	Specific volume of a pure substance.
h	Specific enthalpy of a pure substance.	y	Mass fraction of air in an air-H_2O mixture.
M	Molecular mass.	ϕ	Relative humidity.
m	Mass.	γ	Ratio of specific heats, c_p/c_v.
n	Number of moles.	ρ	Density.
p	Pressure.	ω	Specific humidity.
R	Gas Constant.	*Subscripts*	
$\mathscr{R}$	Universal Gas Constant.	a, b, … n	Components a, b, … n of a mixture. Symbols without alphabetic subscripts represent mixture properties.
S	Entropy of a pure substance.		
s	Specific entropy of a pure substance.		
T	Absolute temperature.	w	Liquid water.
t	Temperature.	G	Air-steam mixture.
U	Internal energy of a pure substance.	v	Water vapour.
		g	Saturated steam.
u	Specific internal energy of a pure substance.	1,2	Initial, final state of a fluid.

The Gibbs-Dalton Law

Dalton's Law

Consider the apparatus shown in Fig. 15.1. A rigid vessel, fitted with a thermometer and a pressure gauge, is connected by means of pipes fitted with stop-cocks to storage bottles containing various pure chemical substances.

Suppose that the vessel is at first evacuated, but that thereafter each of the stop-cocks is opened for a short period in turn. As a result the vessel finally holds some oxygen, some nitrogen, some steam, and so on.

We consider the case in which the final contents of the vessel, which we will call the system, are entirely gaseous, are uniform in pressure and temperature, and are uniform and invariant in composition and chemical aggregation. This means that the gases are completely mixed, and have no tendency to react chemically with each other. The system is therefore a *pure substance* as defined in Chapter 7, page 117, even though it is not chemically pure.

The thermometer and the pressure gauge are in contact with all the components of the mixture simultaneously; their readings give respectively the temperature and pressure of the *mixture*. No thermometer can measure the "temperature" of a single component of the mixture, say the nitrogen, so this term has no meaning. In the same way, the "pressure" of an individual component cannot be measured, although the "partial pressure" of a component, defined below, is often thought of as being the same as the pressure exerted by that component.

In preceding chapters, we have been concerned with how the easily observable properties, pressure, temperature and volume, are related to the properties which feature in the First and Second Laws of Thermodynamics. It has been shown that these relations must be determined experimentally;

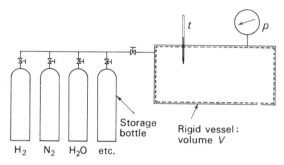

Fig. 15.1 Illustrating the mixing of pure substances.

the results of the experiments are embodied in tables, charts and, in the case of the Ideal Gases, algebraic formulae. If the same procedure is necessary for mixtures, the number of experiments which must be carried out becomes enormous, because of the great variety of mixtures which are of practical interest. It is therefore natural to look for a short cut, preferably one which enables us to calculate the thermodynamic properties of a mixture knowing only the properties of the components and the proportions in which they are mixed together.

Dalton's Law. Fortunately such a short cut is available. It rests on a regularity in the behaviour of gaseous mixtures first perceived and formulated by Dalton (1802), who expressed it in the phrase now known as Dalton's Law, namely:

Any gas is as a vacuum to any other gas mixed with it.

This compact statement means that each component of a gaseous mixture acts, and has the same properties, as if it alone filled the vessel. The statement is based on experiments which show that, although departures from Dalton's Law do exist, they are usually small enough to be ignored in engineering calculations. Certainly Dalton's Law is much more closely obeyed by most substances than is the Ideal-Gas Rule.

It should be remarked that Dalton's Law should be applied to only the *thermodynamic* properties of a mixture. The so-called *transport properties*, viscosity, thermal conductivity and diffusion coefficient, do not obey it.

Dalton's Law, together with Dalton's implicit assumption that each component acts as if it were at the temperature of the mixture, has been expanded into a form suitable for engineering use by Gibbs (1875), whose re-formulation of the principle is known as the *Gibbs-Dalton Law*.

The Gibbs-Dalton Law

Gibbs' formulation is in two parts. The first is concerned with the intensive property, pressure; the second with the extensive properties, internal energy, enthalpy and entropy:

1. The pressure of a gaseous mixture is the sum of the pressures which each component would exert if it alone occupied the volume of the mixture at the temperature of the mixture.

The pressure which a component would exert in these circumstances is known as the *partial pressure* of the component. It has to be measured in a separate experiment in which the mass of the component present in the mixture actually does occupy the volume by itself at the mixture temperature.

2. The internal energy, the enthalpy and the entropy of a gaseous mixture are respectively equal to the sums of the internal energies, the enthalpies and the entropies which each component of the mixture would have, if each alone occupied the volume of the mixture at the temperature of the mixture.

These component properties also must be measured, in the ways indicated in earlier chapters, by separate experiments on the appropriate masses of the isolated components.

Algebraic representation. The Gibbs-Dalton Law is expressed symbolically by the following equations, in which the subscripts a, b, ... n each refer to a particular component of the mixture. The symbol for a mixture property is left without suffix.

The assumption about temperature implicit in Dalton's Law is written as

$$t = t_a = t_b = \ldots = t_n \tag{15.1}$$

Since by Dalton's Law each component is thought of as occupying the whole volume V, we write

$$V = V_a = V_b = \ldots = V_n \tag{15.2}$$

or, in terms of the specific volumes, v,

$$mv = m_a v_a = m_b v_b = \ldots = m_n v_n \tag{15.3}$$

where the m's are the masses.

Since the mass of the mixture is equal to the sum of the masses of the components, i.e.

$$m = m_a + m_b + \ldots + m_n \tag{15.4}$$

the relation between the specific volumes can be written, from eq. (15.3) and eq. (15.4), as

$$\frac{1}{v} = \frac{1}{v_a} + \frac{1}{v_b} + \dots + \frac{1}{v_n} \qquad (15.5)$$

The reciprocal of the specific volume is the density, so eq. (15.5) can also be written as

$$\rho = \rho_a + \rho_b + \dots + \rho_n \qquad (15.6)$$

which means that the density of the mixture is equal to the sum of the densities of the components.

The *first part of the Gibbs-Dalton Law* is expressed symbolically as

$$p = p_a + p_b + \dots + p_n \qquad (15.7)$$

where p with a suffix signifies the *partial* pressure of a component.

The *second part of the Gibbs-Dalton Law* is written

i.e. $\qquad\qquad U = U_a + U_b + \dots + U_n$

$$H = H_a + H_b + \dots + H_n$$

and $\qquad\qquad S = S_a + S_b + \dots + S_n$

or $\qquad\qquad mu = m_a u_a + m_b u_b + \dots + m_n u_n \qquad (15.8)$

$$mh = m_a h_a + m_b h_b + \dots + m_n h_n \qquad (15.9)$$

and $\qquad\qquad ms = m_a s_a + m_b s_b + \dots + m_n s_n \qquad (15.10)$

Use of the Gibbs-Dalton Law. By means of the equations (15.1) to (15.10), we can predict the thermodynamic properties of a gaseous mixture from the properties of the components, provided that we know the composition of the mixture, and two other properties, say its volume and temperature. The following example illustrates the method.

EXAMPLE 15.1

Problem. A mixture consisting of 2 kg of air and 4 kg of nitrogen at a temperature of 25 °C occupies a volume of 1 m³. Evaluate the specific volume, the pressure, the specific enthalpy, the specific internal energy and the specific entropy of the mixture, assuming air and nitrogen to be Ideal Gases for which R is 287 and 297 J/kg K respectively.

Solution. Denoting air and nitrogen by subscripts "a" and "b" respectively, we have:

For 2 kg air: $\qquad V = 1 \text{ m}^3; \qquad\qquad t_a = 25 \text{ °C}$

$$v_a = \frac{1}{2} = 0.5 \text{ m}^3/\text{kg}$$

∴ $\qquad$ From the Ideal-Gas Rule $\qquad p_a = \dfrac{287 \times 298}{0.5} = 171.1 \times 10^3 \text{ N/m}^3$

Then from tables of properties of air (Keenan and Kaye) we obtain

$$h_a = 298 \cdot 52 \text{ kJ/kg}$$
$$u_a = 212 \cdot 90 \text{ kJ/kg}$$
$$s_a = 2 \cdot 3596 \text{ kJ/kg K}$$

For 4 kg nitrogen: $V = 1 \text{ m}^3; \qquad t_b = 25 \,°C$

$$v_b = \frac{1}{4} = 0 \cdot 25 \text{ m}^3/\text{kg}$$

correspondingly $p_b = \dfrac{297 \times 298}{0 \cdot 25} = 354 \cdot 0 \times 10^3 \text{ N/m}^2$

Hence from tables of properties of nitrogen (Keenan and Kaye) we have

$$h_b = 309 \cdot 64 \text{ kJ/kg}$$
$$u_b = 221 \cdot 11 \text{ kJ/kg}$$
$$s_b = 6 \cdot 4644 \text{ kJ/kg K}$$

Then *for the 6 kg of mixture:* $V = 1 \text{ m}^3 \qquad t = 25 \,°C$

$$v = \frac{1}{6} = 0 \cdot 1667 \text{ m}^3/\text{kg} \qquad\qquad\qquad\qquad \textit{Answer}$$

$$p = p_a + p_b \text{ from eq. (15.7)}$$
$$= (171 \cdot 1 + 354 \cdot 0) \times 10^3 = 525 \cdot 1 \times 10^3 \text{ N/m}^3 \quad \textit{Answer}$$

From eq. (15.9)
$$h = \frac{(2 \times 298 \cdot 52 + 4 \times 309 \cdot 64)}{(2 + 4)}$$
$$= 305 \cdot 93 \text{ J/kg} \qquad\qquad\qquad\qquad \textit{Answer}$$

From eq. (15.8)
$$u = \frac{(2 \times 212 \cdot 90 + 4 \times 221 \cdot 11)}{(2 + 4)}$$
$$= 218 \cdot 37 \text{ kJ/kg} \qquad\qquad\qquad\qquad \textit{Answer}$$

From eq. (15.10)
$$s = \frac{(2 \times 2 \cdot 3596 + 4 \times 6 \cdot 4644)}{(2 + 4)}$$
$$= 5 \cdot 0961 \text{ kJ/kg K} \qquad\qquad\qquad\qquad \textit{Answer}$$

Remarks. 1. It should be noted that the numerical values of h, u and s depend on the state chosen as datum. In this case, for both components, the zero of h in the tables used is 0 K and, since $h = u + pv$, this fixes the datum for u also; the zero of s is at a temperature of 0 K and a pressure of one atmosphere. Of course, in the evaluation of h, u and s, the same datum must be used for each of the components.

2. If the given mixture properties are pressure and, say, temperature, the equations (15.1) to (15.10) have to be solved simultaneously instead of one by one. This often involves trial and error.

The specific heats of a gaseous mixture. The specific heats at constant volume and pressure are respectively the differentials with respect to temperature of u and h (Chapter 7, pp. 125, 127). Symbolically we have

$$c_v = \left(\frac{du}{dt}\right)_v ; \qquad c_p = \left(\frac{dh}{dt}\right)_p$$

Now by differentiating eq. (15.8) with respect to temperature we obtain

$$m\left(\frac{du}{dt}\right)_v = m_a\left(\frac{du_a}{dt}\right)_v + m_b\left(\frac{du_b}{dt}\right)_v + \ldots + m_n\left(\frac{du_n}{dt}\right)_v$$

It follows therefore that

$$mc_v = m_a c_{v_a} + m_b c_{v_b} + \ldots + m_n c_{v_n} \qquad (15.11)$$

Similarly from eq. (15.9) we deduce

$$mc_p = m_a c_{p_a} + m_b c_{p_b} + \ldots + m_n c_{p_n} \qquad (15.12)$$

where c_{v_a}, c_{p_a}, etc. are the specific heats of the individual components when occupying the whole volume of the mixture at the temperature of the mixture.

Application of the Gibbs-Dalton Law to mixtures of Ideal Gases

The algebraic rules for the properties of gaseous mixtures become particularly simple when the components are Ideal Gases; for the properties of such gases themselves obey algebraic formulae, as seen in Chapter 14. We therefore exemplify the Gibbs-Dalton Law first by reference to Ideal-Gas mixtures.

Proof that a mixture of Ideal Gases is itself an Ideal Gas. Suppose that masses m_a, m_b, ... m_n of different Ideal Gases form a homogeneous mixture at pressure p and absolute temperature T. We will investigate whether this mixture is itself an Ideal Gas, by deriving its $p \sim v \sim T$ relation.

For the individual components, we have

$$p_a v_a = R_a T$$

$$p_b v_b = R_b T \quad \text{etc.} \qquad (15.13)$$

where R_a, R_b, etc., are the respective Gas Constants.

The combination of eq. (15.13) with eq. (15.7) then gives the pressure p of the mixture:

$$p = \frac{R_a T}{v_a} + \frac{R_b T}{v_b} + \ldots + \frac{R_n T}{v_n} \qquad (15.14)$$

The v's can be eliminated from this equation by means of eq. (15.3), which expresses the fact that each component occupies the mixture volume at mixture temperature T. We obtain

$$p = \frac{m_a}{mv} \cdot R_a T + \frac{m_b}{mv} \cdot R_b T + \ldots + \frac{m_n}{mv} \cdot R_n T$$

$$= \frac{T}{mv}(m_a R_a + m_b R_b + \ldots + m_n R_n) \qquad (15.15)$$

where m is the total mass of the mixture. If we now introduce a symbol R, defined by

$$R \equiv \frac{1}{m}(m_a R_a + m_b R_b + \ldots + m_n R_n) \tag{15.16}$$

eq. (15.15) becomes

$$p = \frac{RT}{v}$$

or

$$pv = RT \tag{15.17}$$

Now R is a constant for a mixture of given composition; for m_a/m, m_b/m, etc. are constants, and R_a, R_b, etc. are in fact the Gas Constants of the components. Eq. (15.17) therefore asserts that *the mixture of Ideal Gases is itself an Ideal Gas, and that R is its Gas Constant.* It will be noted that eq. (15.16) shows that R is an average, formed by "weighting" the individual R's with respect to the mass proportions of the components.

This completes the required proof. There is no need to prove separately that the internal energy of the mixture, for example, is dependent on temperature alone: this follows from the fact that the mixture obeys eq. (15.17), by reason of the arguments in Chapter 14.

The molecular mass of a mixture. The Gibbs-Dalton relations take a similar form to the above when the quantities of the components are expressed in molar units. For example, since $n_a = m_a/M_a$, $\tilde{V}_a = M_a v_a$, etc., equations (15.3) and (15.4) become,

$$n\tilde{V} = n_a \tilde{V}_a = n_b \tilde{V}_b = \ldots = n_n \tilde{V}_n \tag{15.18}$$

and

$$n = n_a + n_b + \ldots + n_n \tag{15.19}$$

Here n_a is the number of moles of substance "a" and $\tilde{V}_a$ is its molar volume; and so on for the other pure components. n and $\tilde{V}$ for the mixture, on the other hand, must be taken as being *defined* by equations (15.18) and (15.19); there is no such thing as a molecule of the mixture.

It is useful to talk about the number of moles of mixture, n, and the molar volume of a mixture $\tilde{V}$, only if these quantities satisfy the same relations as do n_a and $\tilde{V}_a$, etc.; we must therefore satisfy ourselves that

$$p\tilde{V} = \mathscr{R}T \tag{15.20}$$

and

$$V = n\tilde{V} \tag{15.21}$$

The second of these relations is easily established by inspection of eq. (15.18); for the mixture volume, V, is equal to the volume of each component at the mixture temperature, T, i.e. to $n_a \tilde{V}_a$, etc. To demonstrate eq.

(15.20) however, we must first re-write eq. (15.15) in terms of molar quantities as

$$p = \frac{T}{mv}(n_a \mathscr{R} + n_b \mathscr{R} + \ldots n_n \mathscr{R})$$

$$= \frac{\mathscr{R}T}{mv}(n_a + n_b + \ldots n_b) \tag{15.22}$$

$$= \frac{n}{mv}\mathscr{R}T, \text{ from eq. (15.19)}$$

$$= \frac{\mathscr{R}T}{\tilde{V}} \tag{15.23}$$

This is the relation, equivalent to eq. (15.20), which we wished to prove.

For the pure component "a", we have

$$p_a v_a = \frac{\mathscr{R}T}{M_a} = R_a T \tag{15.24}$$

where M_a is its molecular mass. By analogy, we *define* the molecular mass of the mixture M so that

$$pv = \frac{\mathscr{R}T}{M} \tag{15.25}$$

Comparison of eq. (15.22) and eq. (15.25) then shows that

$$M = \frac{m}{n_a + n_b + \ldots + n_n} = \frac{m}{n} \tag{15.26}$$

i.e. that the molecular mass of the mixture equals the mass of the mixture divided by the number of moles, as indeed it should.

In order to relate M to the molecular masses of the individual components directly, we note that $n_a = m_a/M_a$, etc. Substituting in eq. (15.26), we obtain

$$M = \frac{m}{\dfrac{m_a}{M_a} + \dfrac{m_b}{M_b} + \ldots + \dfrac{m_n}{M_n}}$$

or more symmetrically

$$\frac{m}{M} = \frac{m_a}{M_a} + \frac{m_b}{M_b} + \ldots + \frac{m_n}{M_n} \tag{15.27}$$

To complete the parallelism between a mixture of Ideal Gases and a chemically-pure Ideal Gas, the relation between M, R and $\mathscr{R}$ will be derived. Since $M = \mathscr{R}/R_a$, etc., substitution in eq. (15.27) gives

$$\frac{m}{M} = \frac{m_a R_a}{\mathscr{R}} + \frac{m_b R_b}{\mathscr{R}} + \ldots + \frac{m_n R_n}{\mathscr{R}} \tag{15.28}$$

Re-arrangement and substitution from eq. (15.16) yields

$$\frac{1}{M} = \frac{1}{\mathcal{R}m}(m_a R_a + m_b R_b + \ldots + m_n R_n)$$

$$= \frac{R}{\mathcal{R}} \tag{15.29}$$

Hence $$M = \frac{\mathcal{R}}{R} \tag{15.30}$$

in complete conformity with a chemically-pure Ideal Gas

Adiabatic mixing of Ideal Gases

So far in this chapter we have been concerned with the relationships between the *properties* of a mixture and its components. We now illustrate some special features of Ideal-Gas mixtures by considering a mixing *process*. Fig. 15.2 shows the system under consideration; it comprises an

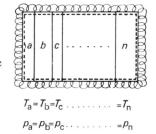

Fig. 15.2 Adiabatic, constant-volume
mixing of Ideal Gases.

$$T_a = T_b = T_c \ldots \ldots \ldots = T_n$$

$$p_a = p_b = p_c \ldots \ldots \ldots = p_n$$

adiabatic rigid container separated into compartments of various size by removable partitions. Each compartment is filled with a separate Ideal Gas, designated a, b, . . . n; the pressures and temperatures prevailing in each compartment are the same, p and T respectively.

The process consists of the removal of the partitions, which is followed by the mixing of the gases as a result of the molecular motions. At the end of the process, all the gases are uniformly mixed. We will examine whether the pressure and temperature of the system change in this adiabatic constant-volume process.

During the process the work and the heat transfer are zero. From the First Law therefore, the internal energy remains constant. But the internal energy of each component is a function of temperature alone because the gases are Ideal. Since the internal energy of the mixture is equal to the sum of the internal energies of the components in the mixed state, and moreover each component is to be regarded as at the mixture temperature, the only possibility is that *the final temperature is equal to the initial temperature.*

We now evaluate the final mixture pressure, p'. Let the volumes of the various compartments be V_a, V_b, . . . V_n, the masses of the components be

m_a, m_b, ... m_n, and the final partial pressures be p_a, p_b, ... p_n. Then from the Gibbs-Dalton Law

$$p' = p_a + p_b + \ldots + p_n \qquad (15.31)$$

and from geometrical considerations the mixture volume V is given by

$$V = V_a + V_b + \ldots + V_n \qquad (15.32)$$

Now the gases obey the Ideal-Gas Rule. Since there is no change of temperature, and the initial pressure of each component is p, we have

$$\left. \begin{array}{l} pV_a = p_aV \\ pV_b = p_bV \\ \text{etc.} \end{array} \right\} \qquad (15.33)$$

for each gas finally occupies the whole volume V. Combining eq. (15.31) and eq. (15.33), we obtain

$$p' = \frac{pV_a}{V} + \frac{pV_b}{V} + \ldots + \frac{pV_n}{V}$$

$$= \frac{p}{V}(V_a + V_b + \ldots + V_n)$$

$$= p, \qquad (15.34)$$

from eq. (15.32).

This proves that the final mixture pressure is equal to the initial pressure of the components: *the pressure in the system does not change.* A corollary is that, if we had postulated constancy of pressure during the process, we would have found that the volume does not change.

The specification of the composition of an Ideal-Gas mixture. Because of the results of the last section, it is permissible and useful to speak of the *volume analysis* of a gas mixture. For example, air is said to consist of 21 per cent of oxygen and 79 per cent of (atmospheric) nitrogen, *by volume,* even though both oxygen and nitrogen fill the whole volume. This means that, if 0·21 m³ of oxygen and 0·79 m³ of (atmospheric) nitrogen at equal temperatures and pressures are mixed in the manner indicated above, 1 m³ of air at the same temperature and pressure will result.

It will now be demonstrated that *the volume analysis is identical with the molar analysis.* To fix ideas however we first introduce the following definitions:

(a) The *mass fraction* of the mixture which has been considered is:

$$\frac{m_a}{m} \text{ for gas a,}$$

$$\frac{m_b}{m} \text{ for gas b,}$$

etc.

(b) The *volume fraction* of the mixture is

$$\frac{V_a}{V} \text{ for gas a,}$$

$$\frac{V_b}{V} \text{ for gas b,}$$

etc.

$\frac{V_a}{V}, \frac{V_b}{V}$, etc. are sometimes referred to as the *partial volumes* of the components of the mixtures.

(c) The *mole fraction* of the mixture is

$$\frac{n_a}{n} \text{ for gas a,}$$

$$\frac{n_b}{n} \text{ for gas b,}$$

etc.

Proof. For "a", the Ideal-Gas Rule in molar units gives

$$p_a V = n_a \mathscr{R} T \tag{15.35}$$

Eliminating p_a by means of eq. (15.33), we have

$$p V_a = n_a \mathscr{R} T \tag{15.36}$$

i.e.

$$n_a = V_a \cdot \frac{p}{\mathscr{R}T} \tag{15.37}$$

Thus the number of moles of gas "a" is proportional to the volume of that gas present before adiabatic mixing at constant pressure and temperature took place; for $p/\mathscr{R}T$ is a constant, equal to n/V, according to eq. (15.23). Thus

$$\frac{n_a}{n} = \frac{V_a}{V}$$

$$\frac{n_b}{n} = \frac{V_b}{V} \quad \text{and so on.} \tag{15.38}$$

The volume and molar analyses are therefore identical. It should be mentioned that, for non-Ideal Gases, the volume analysis ceases to have any useful meaning; for mixing two such gases at constant pressure brings about a change in volume in general.

Entropy change when Perfect* Gases are mixed adiabatically. In Chapter 14, p. 304, a formula, eq. (14.43), was derived for the dependence of the

* N.B. Not Ideal Gases in general.

entropy of a Perfect Gas on the temperature and specific volume. Applying this to the specific entropy increase, Δs_a, of gas "a", in the above constant-temperature mixing process, we obtain

$$\Delta s_a = R_a \ln \left(\frac{V}{V_a} \right) \tag{15.39}$$

Since V is greater than V_a, we note that the entropy of this component, and so of all components, has increased. *Mixing is therefore an irreversible process.* This is easier to understand if it is realized that it is possible, in principle and sometimes in practice also, to devise a piston permeable to all except one gas, and so to obtain work from the interpenetration of the gases; in the actual mixing process above we fail to obtain this work.

EXAMPLE 15.2

Problem. What is the entropy increase when 0·232 kg of oxygen and 0·768 kg of (atmospheric) nitrogen, initially at equal temperates and pressures, mix adiabatically a constant pressure to form 1 kg of air?

Solution. If the oxygen and nitrogen are given suffixes a and b respectively, we have from Table 14.1

$$R_a = 260 \text{ J/kg K}$$

$$R_b = 295 \text{ J/kg K}$$

Now for constant-pressure, adiabatic mixing, the final mixture temperature is equal to the initial temperature of the components. Denoting this temperature by T, we have

$$V_a = m_a \times R_a \, T/p$$

$$V_b = m_b \times R_b \, T/p.$$

If the final mixture volume is V, which is equal to $V_a + V_b$, then

$$\frac{V}{V_a} = \frac{0 \cdot 232 \times 260 + 0 \cdot 768 \times 295}{0 \cdot 232 \times 260} = 4 \cdot 756$$

$$\frac{V}{V_b} = \frac{0 \cdot 232 \times 260 + 0 \cdot 768 \times 295}{0 \cdot 768 \times 295} = 1 \cdot 266$$

The entropy increase is

$$\Delta S = 0 \cdot 232 \Delta s_a + 0 \cdot 768 \Delta s_b$$

$$= 0 \cdot 232 \times 260 \ln 4 \cdot 756 + 0 \cdot 768 \times 295 \ln 1 \cdot 266$$

$$= 147 \cdot 5 \text{ J/kg K} \qquad\qquad \textit{Answer}$$

Comment. An interesting question arises from the fact that mixing of two different gases causes an entropy increase. Suppose that the two gases are identical. Then, when the partition is removed, molecular motions ensue and cause mixing, so that eventually molecules which initially were in the separate compartments can be found at all parts of the vessel. Yet this time there is no entropy increase; for adding, say, 1 kg of oxygen to 1 kg of oxygen at the same pressure and temperature merely produces 2 kg of oxygen. Why is there an entropy increase when the gases differ in chemical properties and not when they do not?

This is known as *Gibbs' Paradox*. The answer is that when the gases are identical it is no longer possible, even in principle, to devise means to extract work from the inter-penetration of the gases; for a semi-permeable piston cannot distinguish between two molecules of the same sort.*

Mixtures of an Ideal Gas with a Condensable Vapour

Air-H$_2$O mixtures: gaseous phase

The most important mechanical engineering example of mixtures of Ideal Gases with a condensable vapour is the air-H$_2$O system. This will be dealt with exclusively from now on; but it should be understood that the methods and results are applicable to any other combinations of appropriate substances.

Single-phase mixtures. When only the gaseous phase is present, i.e. the mixture is composed of air and steam, the Gibbs-Dalton law is applicable without modification. A single example should suffice as illustration.

EXAMPLE 15.3

Problem. 0·01 kg of steam and 0·99 kg of air from a gaseous mixture at a pressure of 100×10^3 N/m^2 and a temperature of 20 °C. Determine (*a*) the partial pressures of the steam and air, (*b*) the specific volume of the mixture, and (*c*) the enthalpy of the mixture.

Solution (a) and (b). These have to be solved simultaneously. We have

$$p = 100 \times 10^3 \text{ N/m}^2 = p_{\text{steam}} + p_{\text{air}}$$

For air:
$$p_{\text{air}} = \frac{287 \times (273 + 20)}{v_{\text{air}}} \text{ N m/kg from the Ideal-Gas Rule.}$$

For steam: p_{steam} is related to v_{steam} by the Superheated Steam Table.

Further,
$$1 \times v = 0\cdot01 \, v_{\text{steam}} = 0\cdot99 \, v_{\text{air}}, \text{ from eq. (15.3).}$$

These five relations permit us to determine the five unknowns. However, examination of the Superheated-Steam Table (Appendix B) shows that it does not extend into the low-pressure region which will certainly be required. We therefore *assume* that at low pressures steam can be treated as an Ideal Gas, so replacing the Steam Table by

$$p_{\text{steam}} = \frac{462 \times (273 + 20)}{v_{\text{steam}}} \text{ N m/kg from the Ideal-Gas Rule.}$$

where 462 J/kg K is the Gas Constant for H$_2$O (see Table 14.1).

* What about mixing two isotopes of the same gas? These have the same chemical properties but differing nuclear ones. These nuclear differences could be used, in principle, to effect a reversible interpenetration, so mixing causes an entropy increase. However, if the practical possibility of reversible interpenetration is excluded, no difficulties arise from neglecting the entropy increase, provided that this is done consistently. Entropy is a man-made concept and we can trim its definition to suit our convenience and capabilities.

Then
$$\frac{p_{steam}}{p_{air}} = \frac{462}{287} \times \frac{v_{air}}{v_{steam}}$$

$$= \frac{462}{287} \times \frac{0\cdot01}{0\cdot99} = 0\cdot016\ 26$$

Now from eq. (15.7) $p_{steam} + p_{air} = 100 \times 10^3$ N/m².

∴
$$p_{steam} + \frac{p_{steam}}{0\cdot016\ 26} = 100 \times 10^3$$

or $$p_{steam} = \frac{0\cdot016\ 26}{1 + 0\cdot016\ 26} \times 100 \times 10^3 = 1\cdot60 \times 10^3 \text{ N/m}^2$$ *Answer (a)*

Hence $$p_{air} = 100 \times 10^3 - 1\cdot60 \times 10^3 = 96\cdot40 \times 10^3 \text{ N/m}^2$$ *Answer (a)*

and so $$v_{steam} = \frac{462 \times 293}{1\cdot60 \times 10^3} = 84\cdot6 \text{ m}^3/\text{kg}$$

therefore $$v = 0\cdot01 \times 0\cdot846 \text{ m}^3/\text{kg air.}$$ *Answer (b)*

Note. Solution by means of the Steam Tables would be a trial-and-error process. The steps are as follows:
   (i) Select a value of p_{steam}.
   (ii) From the Steam Tables, at $t = 20$ °C, look up the corresponding value of v_{steam}.
   (iii) Evaluate p_{air} from eq. (15.7).
   (iv) Evaluate v_{air} from the Ideal-Gas Rule.
   (v) The values of v_{steam} and v_{air} have to satisfy eq. (15.3) viz. $0\cdot01\ v_{steam} = 0\cdot99\ v_{air}$. Select values of p_{steam} until agreement is reached.

Solution (c). The enthalpy of the mixture is given by

$$h = 0\cdot01\ h_{steam} + 0\cdot99\ h_{air}$$

Since the Superheated-Steam Table (Appendix B) does not extend to the value of p_{steam}, we take h_{steam} equal to h_g at 20 °C, in conformity with our Ideal-Gas assumption that $h = f(t)$. For air, we take $c_p = 1\cdot0$ kJ/kg K and $R = 0\cdot287$ kJ/kg K. The datum states we select are zero values of *internal energy* for saturated water at the triple point and for air at 0·01 °C.

Then $$h = [0\cdot01 \times 2537\cdot5] + [0\cdot99 \times 1\cdot0(20 - 0\cdot01)] + 0\cdot287(0\cdot01 + 273\cdot15)]$$

$$= 25\cdot38 + 19\cdot80 + 78\cdot27 = 123\cdot45 \text{ kJ/kg mixture}$$ *Answer (c)*

Note. The selected datum state for air requires the inclusion of the third term in the expression for h; it corresponds to the value of the enthalpy of air at the datum state.

Steam-water-air mixtures

In the last example, the partial pressure of the steam (1600 N/m²) was less than the saturated-steam pressure (2336·6 N/m²) which Steam Tables show to prevail at 20 °C. If the calculation had shown a partial pressure in excess of 2336·6 N/m², as would have happened if the mass of H_2O per unit mass of mixture had been appreciably greater, the assumption of the calculation, viz. that only one phase was present, would have been wrong; for the saturation pressure corresponding to the mixture temperature is the highest pressure at which the vapour can exist in the mixture at that temperature.

To explain this, the transition between phases will be discussed in a similar way to that employed for pure substances in Chapter 9. We consider once more a series of constant-pressure experiments; but this time we suppose heat to be transferred *from* the system.

Constant-pressure cooling experiments. Fig. 15.3 illustrates the apparatus. A cylinder fitted with a frictionless leak-proof piston surmounted by a weight contains a mixture of air and steam. We suppose that initially the temperature of the system (the air and the steam) exceeds the boiling-point of water

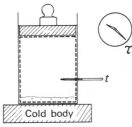

Fig. 15.3   Constant-pressure cooling of an air-H_2O mixture.

at the pressure of the system. The base of the cylinder is placed in thermal contact with a very cold body and the variation of the system temperature is observed and plotted against time.

Fig. 15.4 shows typical curves resulting from such experiments. The curve b corresponds to a mixture containing only a small amount of

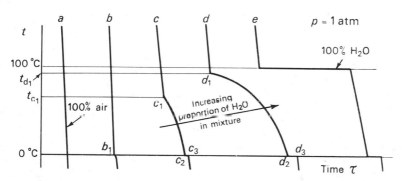

Fig. 15.4 Temperature–time curves for air-H_2O mixtures cooling at atmospheric pressure (to scale).

steam. To fix ideas, we suppose that the pressure within the cylinder is atmospheric. The temperature-time curve consists of two approximately straight sections; the change of slope occurs abruptly at 0 °C.* In contrast to the curves obtained for a pure substance (Fig. 9.2), the slope of the curve is finite (and negative) everywhere; there are no horizontal steps and the temperature falls continuously.

If the interior of the cylinder is observed during this first experiment, it is

* Actually the temperature would be slightly below 0 °C and would vary with the mixture composition. The difference may be neglected for engineering purposes.

seen that during the initial period of steep slope (i.e. $t > 0\ °C$) only the gaseous phase is present. At the temperature corresponding to the kink in the curve, point b_1, tiny crystals of ice begin to form on the base of the cylinder; these continue to grow in size, though at a decreasing rate, as the temperature falls further. The experiment has reproduced the conditions for the formation of *hoar frost*.

The next experiment is supposed to be carried out at the same pressure as before, but with a greater mass ratio of steam to air. This time, curve c is obtained; in this, the kink (c_1) occurs earlier, i.e. at a higher temperature than $0\ °C$. However, when $0\ °C^*$ is reached, the curve exhibits a second kink c_2 and becomes horizontal for a period, $c_2 c_3$; thereafter the curve bends downward once more. Observation of the interior of the cylinder reveals that, at the first kink c_1, small droplets of water appear on the base; these grow in size until the second kink c_2 is reached. At $0\ °C$, the droplets freeze, the temperature remaining unchanged during this process. The third kink c_3 occurs when all the condensed water has turned to ice; thereafter the quantity of ice increases as a result of further deposition from the gas phase and the temperature falls continuously. This experiment illustrates the formation of *dew*, which subsequently freezes.

If successive experiments are carried out, with the same pressure of 1 atmosphere and increasing proportions of steam to air, the temperature at which the first kink occurs rises continuously, and the horizontal extension of the curve increases. Eventually, when the proportion of air has been reduced to zero, the curve degenerates to the form already encountered for pure H_2O; a broad step occurs at $100\ °C$ and a narrower one occurs at $0\ °C$. At the other extreme, when there is no steam present, curve a is obtained. Fig. 15.4 illustrates these trends.

Further series of experiments can be carried out at other pressures. The general tendency of the temperature~time curves is the same, but the temperature of the upper kink increases as the pressure rises. When the pressure is below that of the triple point of H_2O ($611\cdot2\ N/m^2$), the upper kink does not occur, and all the changes of slope are confined to the region below $0\ °C$; this means that no water is formed, whatever the initial steam concentration, as will be understood by recalling the behaviour of pure steam at low pressures.

In studying the properties of a pure substance, in Chapter 9, a single series of experiments sufficed, pressure being the only variable. The necessity to carry out several series for air-H_2O mixtures is a consequence of the fact that such a mixture is not a pure substance; *three* properties are needed to fix its state, e.g. temperature, specific volume *and composition*.

The enthalpy~composition diagram. Pursuing the parallel with the procedure of Chapter 9 still further, we note that measurements of heat

* See footnote on the previous page.

transfer can be made during the constant-pressure cooling experiments; these yield the enthalpy changes directly. The results for a given pressure can then be plotted on a diagram with specific enthalpy of the mixture as ordinate and composition as abscissa. The composition will be characterized by a quantity y, defined as

$$y = \frac{\text{mass of air}}{\text{mass of mixture}} \qquad (15.40)$$

Thus $y = 1$ denotes pure air; $y = 0$ denotes pure H_2O.

Fig. 15.5 shows such a diagram, approximately to scale, for a pressure of one atmosphere, with the enthalpy base chosen as 0.01 °C for saturated water and for air. Inspection of the diagram shows that there are four regions: the upper one corresponds to states of purely gaseous phase (air and steam); the next lower one corresponds to mixtures in which the gaseous and liquid phases are present (air + steam + water); next comes a region comprising states involving the gaseous, liquid and solid phases simultaneously (air + steam + water + ice); the lowest region covers states in which only the gaseous and solid phases are present (air + steam + ice). Pure-air states are found on the right-hand vertical border, a, where $y = 1$; pure-steam, steam + water, pure-water, water + ice, and pure-ice states occupy successively lower stretches of the left-hand vertical border, e, where $y = 0$.

Lines of constant temperature are drawn on the $h{\sim}y$ diagram of Fig. 15.5. They consist of nearly straight lines because, for the conditions under consideration, the enthalpies of air, steam, water and ice depend on the mixture temperature alone. The lines exhibit kinks at the phase boundaries. The pattern made by the isotherms is best understood by direct inspection of the diagram and by thought about its relation to the constant-pressure-cooling experiments described above. Since the composition of the system undergoing such a process remains fixed, the state point moves downward on the diagram along a vertical line. The lines a, b, c, d, e, of Figs. 15.4 and 15.5 correspond; the kinks in these lines in Fig. 15.4 occur when those in Fig. 15.5 cross the phase boundaries. Fig. 15.5 should be compared with Fig. 15.7 which shows, to scale, the $h{\sim}y$ diagram for air-H_2O at a pressure of one atmosphere with the enthalpy base chosen as 0.01 °C; it will be noted that, for clarity, the scale of the lower part of Fig. 15.5 has been distorted.

Because composition is a third independent variable, a single $h{\sim}y$ diagram cannot cover all possible states of the system; a family of such diagrams is needed, one for every pressure. The higher the pressure, the higher rises the lower boundary of the gas-phase region. At lower pressures the size of the regions representing mixtures containing liquid water decreases; these regions are altogether absent at pressures below that of the triple point.

The saturation line. For most engineering purposes, only the air-steam-water regions are important; interest is restricted to the regions adjoining

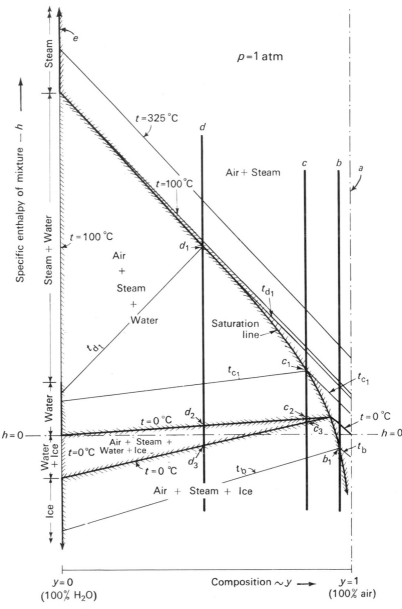

Fig. 15.5 Enthalpy~composition diagram for the air-H₂O system at atmospheric pressure, approximately to scale. For clarity, the scale in the regions where ice is present has been distorted. This diagram should be compared with Fig. 15.7, which is drawn to scale. The datum states are zero values of enthalpy for saturated water at the triple point and for air at 0·01 °C.

the upper phase boundary. This boundary is known as the *saturation line*. States lying on it comprise air and saturated steam; so the mixtures exhibit no tendency to change when brought into contact with saturated water at the same temperature and pressure.

Mixtures of air and saturated steam are known as *saturated air*. The term is to some extent misleading, for it suggests that the air *soaks up* water vapour. In reality, according to Dalton's Law, the steam ignores the presence of air altogether: it is the *space* that is saturated.

The composition of saturated air is calculated from the condition that the steam is saturated: its partial pressure is therefore that appearing in the Saturated-Steam Tables opposite the prevailing temperature.

EXAMPLE 15.4

Problem. Calculate y and h for saturated air at 20 °C and 1 atm.
Solution. From the Gibbs-Dalton Law

$$p_{steam} + p_{air} = 101\cdot33 \times 10^3 \text{ N/m}^2 \qquad \text{[eq. (15.7)]}$$

$$y \cdot v_{air} = (1 - y)v_{steam} \qquad \text{[eq. (15.3)]}$$

$$h = y \cdot h_{air} + (1 - y)h_{steam} \qquad \text{[eq. (15.9)]}$$

From the Ideal-Gas Rule, for air at 20 °C

$$p_{air} = \frac{287 \times (273 + 20)}{v_{air}} \text{ N m/kg} \qquad \text{[eq. (14.10)]}$$

From Table II, Appendix B, for saturated steam at 20 °C

$$p_{steam} = 2\cdot3366 \times 10^3 \text{ N/m}^2$$

$$v_{steam} = 57\cdot838 \text{ m}^3/\text{kg}$$

$$h_{steam} = 2538\cdot2 \text{ kJ/kg}$$

We obtain first the partial pressure of air, as

$$p_{air} = (101\cdot33 - 2\cdot34) \times 10^3$$
$$= 98\cdot99 \times 10^3 \text{ N/m}^2$$

This leads to the specific volume of air, as

$$v_{air} = \frac{287 \times 293}{98\cdot99 \times 10^3}$$
$$= 0\cdot850 \text{ m}^3/\text{kg}$$

and so to y, as

$$\frac{y}{1 - y} = \frac{57\cdot84}{0\cdot85} = 68\cdot05$$
$$y = 0\cdot9855 \text{ kg air/kg mixture} \qquad \textit{Answer}$$

Finally, taking c_p for air as $1\cdot0$ kJ/kg K, and selecting the datum states as zero values of *enthalpy* for saturated water at the triple point, and for air at 0·01 °C, we have

$$h = 0\cdot9855 \times 1\cdot0 (20 - 0\cdot01) + 0\cdot0145 (2537\cdot5 - 0\cdot000611\,3)$$
$$= 19\cdot71 + 36\cdot78 = 56\cdot49 \text{ kJ/kg mixture} \qquad \textit{Answer}$$

Note. The tabulated value of h_{steam} has to be reduced by 0·000 611 3 kJ/kg because the Steam Tables in Appendix B are based on a datum of zero internal energy for saturated water at the triple point; the reduction is h_t at the triple point, given in Table I. The adjustment is of course negligible for present purposes.

Constructing an h~y diagram. Example 15.4 shows how an enthalpy~ composition diagram can be constructed using Steam Tables, thermo-dynamic data for air, and the Gibbs-Dalton Law. When the saturation line and the temperature intercepts on the pure-H_2O and pure-air boundaries have been drawn, isotherms can be drawn as straight lines between points of equal temperature without appreciable error, because of the pressure-independence of enthalpy noted above.

It is also possible to draw on the $h{\sim}y$ diagram lines of constant specific volume, constant entropy, or any other desired property.

Other graphical representations of air-steam-water properties. Because of the technical importance of the air-H_2O system, several other ways have been developed for representing the system properties graphically. Two which are frequently encountered will be mentioned.

One, introduced by Mollier, also has enthalpy and mass composition as co-ordinates. It differs from the $h{\sim}y$ diagram however in basing both quan-tities on unit mass of *air* instead of on unit mass of *mixture*; the mass quantity is specially emphasized by the name "dry air". The enthalpy of the Mollier chart therefore exceeds h by an amount which increases with the proportion of H_2O present. The composition is characterized by the ratio: (mass of H_2O)/(mass of dry air). The Mollier chart therefore represents a *projection* of the $h{\sim}y$ diagram carried out in such a way that the left-hand boundary goes to infinity; the left-hand part of the diagram is necessarily left undrawn.

The second graphical representation is the so-called *psychrometric chart*, Fig. 15.6. This has temperature as abscissa and composition as ordinate, usually in the form of (mass of H_2O)/(mass of dry air). This represents a *distortion* of the $h{\sim}y$ diagram which is useful for some purposes. The relation of the psychrometric chart of Fig. 15.6 to the $h{\sim}y$ chart of Fig. 15.5 may be seen if the latter is imagined as being rotated clockwise through 90° and as then having all its isotherms "straightened" and made vertical.

Care should always be taken to establish whether enthalpy and composi-tion are being expressed in the units of h and y or in other units.

Special terms in use in psychrometry

Psychrometry is the study of steam-air mixtures and is particularly concerned with measurement of their properties. It is of special importance to heating-and-ventilating engineers; for the moisture (steam) content, i.e. the humidity, of the atmosphere greatly influences human comfort.

Specific humidity, ω, is the measure of moisture content used in the Mollier and psychrometric charts. Its definition is therefore

$$\omega \equiv \frac{\text{mass of } H_2O}{\text{mass of (dry) air}}$$

$$= \frac{1-y}{y} \qquad (15.41)$$

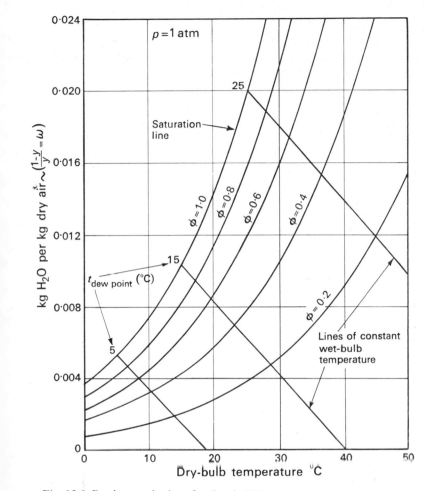

Fig. 15.6 Psychrometric chart for the air-H_2O system at 1 atm (to scale).

At the temperatures and pressures encountered in heating-and-ventilating practice, steam can be regarded as an Ideal Gas. It is therefore permissible to write, for gas-phase mixtures,

$$\omega = \frac{m_{H_2O}}{m_{air}}$$

$$= \frac{v_{air}}{v_{H_2O}} \qquad \text{from eq. (15.3)}$$

$$\approx \frac{\mathscr{R}T}{M_{air}p_{air}} \cdot \frac{M_{steam}p_{steam}}{\mathscr{R}T} \qquad \text{from the Ideal-Gas Rule, eq. (14.10)}$$

$$\approx \frac{p_{steam}}{p_{air}} \times \frac{18}{29} \qquad (15.42)$$

the numerical factor representing the ratio of the molecular weights (rounded values) of steam and air.

If the specific humidity is small, the partial pressure of the air p_{air} is only slightly less than p, the pressure of the mixture. In that case eq. (15.42) becomes

$$\omega \approx 0.622 \frac{p_{steam}}{p} \qquad (15.43)$$

Relative humidity. It is convenient to compare the humidity of the air with the humidity of saturated air at the same pressure and temperature. The *relative humidity*, ϕ, is therefore defined as the ratio of the specific volume of saturated steam at the mixture temperature (v_g) to the specific volume of the steam in the mixture (v_{steam}).

$$\phi = \frac{v_g}{v_{steam}} \qquad (15.44)$$

As a consequence of the definition, ϕ equals unity for saturated air.

Since from eq. (15.3)

$$v_{steam} = \frac{v_{air}}{\omega} \qquad (15.45)$$

eq. (15.44) can be written

$$\phi = \frac{\omega v_g}{v_{air}} \qquad (15.46)$$

$$= \omega \frac{p_{air} v_g}{R_{air} T} \qquad (15.47)$$

When steam is treated as an Ideal Gas, this reduces to

$$\phi \approx \omega \frac{R_{steam}}{R_{air}} \cdot \frac{p_{air}}{p_g}$$

$$\approx \omega \frac{29}{18} \frac{p_{air}}{p_g} \qquad (15.48)$$

where p_g is the saturation pressure of the steam at the prevailing temperature.

Other definitions of ϕ.

(1) ϕ may be defined as the ratio of the mass of water vapour in a given *volume* of mixture to the mass of water vapour in an *equal volume* of *saturated* mixture, at the mixture temperature. This means that ϕ is equal to the ratio of the density of the steam in the mixture to the density of saturated steam at the mixture temperature; this definition is therefore equivalent to that expressed by eq. (15.44).

(2) ϕ may be defined as the ratio of the partial pressure of the water vapour in the mixture to the saturation pressure of water vapour at the mixture temperature. If water vapour is assumed to be an Ideal Gas, this definition is equivalent to (1) above and so to that expressed by eq. (15.44).

(3) ϕ is sometimes defined as the ratio of the specific humidity of a given *volume* of mixture to the specific humidity of an *equal volume* of *saturated* mixture at the mixture temperature and pressure. This definition gives ϕ a slightly different magnitude from that of eq. (15.44); if the Ideal-Gas assumption is made, its value will be $\dfrac{(p_{air})_{sat.\ mixture}}{(p_{air})_{mixture}}$ times as big as the value given by eq. (15.44).

Dew-point. The moisture content of a mixture is sometimes measured by just such a constant-pressure cooling process as has been described above (p. 337): the mixture is slowly cooled and the temperature at which condensation of water first occurs is noted. This temperature, at which the mixture has become saturated, is known as the dew-point; it corresponds to the temperature of the point at which a vertical through the mixture state-point on an $h{\sim}y$ diagram crosses the saturation line. In Fig. 15.7, the line $1 \rightarrow 2$ illustrates such a process during which, it is to be noted, both y and ω remain constant.

Evaluation of y or ω from a dew-point measurement can be effected by way of eq. (15.41) and eq. (15.3); for at the dew-point we have:

$$\omega = \frac{1-y}{y} = \frac{v_{air}}{v_g}$$

$$= \frac{R_{air}T_{dew\text{-}point}}{v_g(p - p_g)} \tag{15.49}$$

where v_g and p_g are respectively the specific volume and pressure of saturated steam at the dew-point temperature, $T_{dew\text{-}point}$; p is the mixture pressure.

Dry-bulb temperature. The dry-bulb temperature of a steam-air mixture is simply the temperature of the mixture. The special term arises from the practice of measuring the moisture content of a mixture by means of a *wet-and-dry-bulb hygrometer*. This consists of two thermometers exposed to a stream of the mixture (Fig. 15.8). The bulb of one of the thermometers is kept wet by a wick dipping into water; the other is dry, and gives the (dry-bulb) temperature of the mixture.

Wet-bulb temperature. The temperature registered by the thermometer wetted by the wick is known as the *wet-bulb temperature*. This is normally lower than the dry-bulb temperature; the difference between the two is a measure of the humidity of the air.

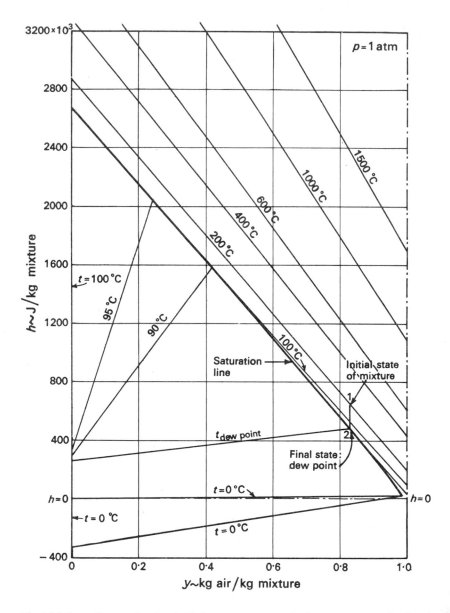

Fig. 15.7 $h \sim y$ diagram for the air-H_2O system at atmospheric pressure, to scale, showing the constant-pressure cooling of a given air-H_2O mixture to obtain the dew point (process 1 → 2).

The datum states are zero values of enthalpy for saturated water at the triple point and for air at 0·01 °C.

The relation between the two temperatures is fixed by processes which lie outside thermodynamics: the so-called *rate processes*, of heat transfer from the mixture stream to the wick surface, and of mass transfer (diffusion of steam) from the surface into the stream. Heat transfer must take place at a rate sufficient to provide the enthalpy of vaporization of steam leaving the wick. However it is a good approximation to take the wet-bulb temperature as equal to the *adiabatic saturation temperature*; this is dealt with below (p. 352) as one of the applications of the First Law.

The ordinate and abscissa of the psychrometric chart are, as may be seen from Fig. 15.6, respectively the specific humidity ω and the dry-bulb

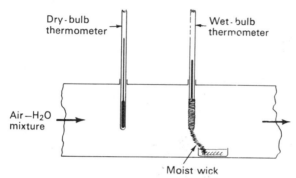

Fig. 15.8 Wet-and-dry-bulb hygrometer.

(mixture) temperature t; lines of constant ϕ and constant wet-bulb temperature are plotted on the chart. The chart may therefore be used to obtain values of ω and ϕ directly from the readings of the wet- and dry-bulb thermometers.

The Orsat Gas-Analysis Apparatus. We have deferred until this point discussion of a method of determining experimentally the composition of a given Ideal-Gas mixture, because the experiment is always carried out with gas which is first saturated with water vapour. There are several variants of the apparatus, of which the Orsat type, Fig. 15.9, is most familiar. It consists of a graduated cylinder containing the gas, a series of absorption vessels which remove one component after another from the mixture, and a levelling bottle for restoring the pressure of the mixture to atmospheric when measuring its volume.

In effect, the Orsat carries out the mixing process of p. 323 in reverse. Let the volume of the mixture be initially V, and let the volume decreases resulting from absorption, in turn, of the components be V_a, V_b, etc., measured at the initial pressure and temperature of the mixture; then the volumes measured in the graduated cylinder are successively V, $V-V_a$, $V-V_a-V_b$, etc.; from these measurements the volume analysis of the gas is easily calculated.

The only subtlety lies in the fact that, although the mixture is saturated with steam, only the permanent gases feature in the analysis. This is often called the "dry-gas analysis" for emphasis. The explanation is as follows. The pressure of the mixture does not feature in the expressions for the volume analysis provided that it is constant. The partial pressures of the Ideal Gases add up to a total that is less than the mixture pressure, the deficit being the steam partial pressure. Provided that the gas remains saturated however, and the temperature is constant, the partial pressure of steam

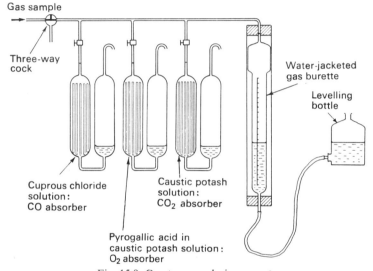

Fig. 15.9 Orsat gas-analysis apparatus.

remains unchanged. The sum of the partial pressures of the Ideal Gases therefore is also constant. The presence of the water vapour consequently has no influence on the results, because, as the volume decreases, a proportionate amount of water condenses out.

Applications of the First Law to Mixtures

Because, as we have seen, the internal energy and other properties of mixtures are related to the properties of the components, the computation of the heat transfer and work in processes undergone by mixtures is a simple matter of applying the First Law of Thermodynamics. A few examples should be sufficient illustration.

An Ideal-Gas example

EXAMPLE 15.5

Problem. A rigid insulated vessel of 1 m³ capacity contains pure hydrogen at a pressure of 104×10^3 N/m² and a temperature of 16 °C. It is connected to a compressed-air main in which the air is at a gauge pressure of 280×10^3 N/m^2 (atmospheric pressure 102×10^3 N/m^2) and at a temperature of 25 °C. Air enters the vessel until the pressures in the main and in the vessel are equal. What is the final temperature of the mixture in the vessel?

N.B. This problem is important because, if a mixture of hydrogen and air becomes hot enough, it will explode.

Solution. The data needed are:

$$\text{For air:} \qquad R = 287 \text{ J/kg K}$$
$$c_v = 717 \text{ J/kg K}$$
$$\text{For hydrogen:} \qquad R = 4160 \text{ J/kg K}$$
$$c_v = 10\,080 \text{ J/kg K}$$

The only difficulty in this problem is to organize the calculation in such a way as to retain clarity. This involves choosing a convenient notation, and working in symbols until the last possible moment. We use suffix "a" to represent air, and suffix "b" to

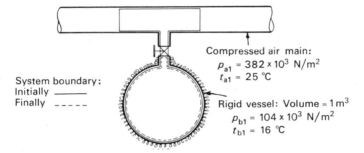

System boundary:
Initially ———
Finally - - - - -

Compressed air main:
$p_{a1} = 382 \times 10^3$ N/m^2
$t_{a1} = 25$ °C

Rigid vessel: Volume $= 1$ m^3
$p_{b1} = 104 \times 10^3$ N/m^2
$t_{b1} = 16$ °C

Fig. 15.10 Diagram used in the solution of Example 15.5.

represent hydrogen; the suffixes 1 and 2 represent the initial and final states respectively; the symbols for the properties of the mixture carry no alphabetic suffix; V represents the vessel volume. The system chosen is the mass of hydrogen, m_b, plus the mass of air, m_a, finally in the vessel (Fig. 15.10).

From the Ideal-Gas Rule,

$$m_b = \frac{p_{b_1} V}{R_b T_{b_1}} \tag{15.50}$$

$$= \frac{p_{b_2} V}{R_b T_2} \tag{15.51}$$

$$m_a = \frac{p_{a_2} V}{R_a T_2} \tag{15.52}$$

$$p_{a_1} v_{a_1} = R_a T_{a_1} \tag{15.53}$$

From the Gibbs-Dalton Law,

$$p_2 = p_{a_2} + p_{b_2} = p \tag{15.54}$$

From the First Law

$$Q - W = \Delta U \tag{15.55}$$

In this case

$$Q = 0$$

W, the work done by the moving boundary, $= -p_{a_1}(m_a \cdot v_{a_1})$

and $\Delta U = m_b c_{v_b}(T_2 - T_{b_1}) + m_a c_{v_a}(T_2 - T_{a_1})$ from the Gibbs-Dalton Law.

Hence eq. (15.55) becomes

$$m_a p_{a_1} v_{a_1} = m_b c_{v_b}(T_2 - T_{a_1}) + m_a c_{v_a}(T_2 - T_{a_1})$$

and so $$m_a[p_{a_1} v_{a_1} - c_{v_a}(T_2 - T_{a_1})] = m_b c_{v_b}(T_2 - T_{b_1}) \qquad (15.56)$$

Known are: $p_{a_1}, p_{b_1}, T_{a_1}, T_{b_1}, V, c_{v_a}, c_{v_b}, R_a, R_b$

Unknown are: $m_a, m_b, v_{a_1}, T_2, p_{a_2}, p_{b_2}$

The six equations suffice to yield the six unknowns. We are required to determine T_2 so we combine the six equations to eliminate the other five unknowns.

Substituting in eq. (15.54) from eq. (15.51) and eq. (15.52), we have

$$(m_a R_a + m_b R_b)\frac{T_2}{V} = p_{a_1}$$

and hence $$m_a = \frac{1}{R_a}\left(\frac{p_{a_1} V}{T_2} - m_b R_b\right) \qquad (15.57)$$

Then substitution for m_a in eq. (15.56), by reference to eq. (15.57), gives

$$\frac{1}{R_a}\left(\frac{p_{a_1} V}{T_2} - m_b R_b\right)\left[p_{a_1} v_{a_1} - c_{v_a}(T_2 - T_{a_1})\right] = m_b c_{v_b}(T_2 - T_{b_1}) \qquad (15.58)$$

Finally substitutions in eq. (15.58) for m_b from equation (15.50) and for $p_{a_1} v_{a_1}$ from eq. (15.53) produce

$$\frac{1}{R_a}\left(\frac{p_{a_1} V}{T_2} - \frac{p_{b_1} V}{T_{b_1}}\right)\left[R_a T_{a_1} - c_{v_a}(T_2 - T_{a_1})\right] = \frac{p_{b_1} V}{R_b T_{b_1}} c_{v_b}(T_2 - T_{b_1})$$

which reduces to

$$\left(\frac{p_{a_1}}{T_2} - \frac{p_{b_1}}{T_{b_1}}\right)\left[T_{a_1} - \frac{c_{v_a}}{R_a}(T_2 - T_{a_1})\right] = \frac{p_{b_1}}{T_{b_1}}\cdot\frac{c_{v_b}}{R_b}(T_2 - T_{b_1}) \qquad (15.59)$$

T_2 is the only unknown in eq. (15.59). Inserting the given data in eq. (15.59) we have

$$\left(\frac{382 \times 10^3}{T_2} - \frac{104 \times 10^3}{289}\right)\left[298 - \frac{717}{287}(T_2 - 298)\right]$$

$$= \frac{104 \times 10^3}{289} \times \frac{10\,080}{4160}(T_2 - 289)$$

Hence $$T_2 = 371\text{ K} = 98\text{ °C} \qquad \textit{Answer}$$

Remarks. 1. A quadratic in T_2 is obtained. The second value of T_2 which satisfies the equation is inadmissible as a solution because it corresponds to a negative value for the partial pressure of the air finally in the vessel.

2. The final mixture temperature T_2 is independent of the vessel volume V, since V does not appear in eq. (15.59). This means that for given initial conditions the gases will mix in the same mass *ratio* to produce the same final conditions, irrespective of the magnitude of V; we would expect this in view of the Gibbs-Dalton relations. V will, of course, affect the magnitudes of both m_a and m_b, but not their ratio.

Adiabatic saturation of air

As our first example of the application of the First Law to a mixture of an Ideal Gas and a condensable vapour, we consider the addition of water to air in a constant-pressure adiabatic process resulting in a saturated mixture. The process, which is called *adiabatic saturation*, is most easily followed on the $h{\sim}y$ diagram for the prevailing pressure. First, however, we prove a general result relating to mixing processes represented on such diagrams.

The mixing rule. If X and Y are the state-points of two air-H_2O systems of an $h{\sim}y$ diagram (Fig. 15.11), and these two systems are mixed together

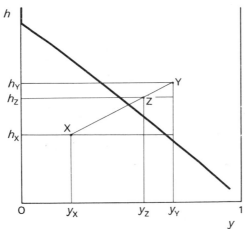

Fig. 15.11 Illustrating the mixing rule.

adiabatically at constant pressure in the mass proportions m_X to m_Y, the state-point Z for the resultant system lies on the line XY and divides it so that

$$\frac{ZY}{XZ} = \frac{m_X}{m_Y} \qquad (15.60)$$

Proof. From the First Law applied to the constant-pressure adiabatic process, Δh equals zero; therefore

$$m_X h_X + m_Y h_Y = (m_X + m_Y)h_Z$$

$$\frac{m_X}{m_Y} = \frac{h_Y - h_Z}{h_Z - h_X} \qquad (15.61)$$

The principle of conservation of mass applied to air in the process yields

$$m_X y_X + m_Y y_Y = (m_X + m_Y)y_Z$$

i.e.

$$\frac{m_X}{m_Y} = \frac{y_Y - y_Z}{y_Z - y_X} \qquad (15.62)$$

Equations (15.61) and (15.62), interpreted geometrically on the $h{\sim}y$ diagram, provide the required result.

Adiabatic saturation on the h~y diagram. Fig. 15.12 shows an $h{\sim}y$ diagram. G represents the initial state of the air-steam mixture while W represents the state of the water which is to be mixed with it. How much water must be added, if the resultant mixture is to comprise saturated air?

The above mixing rule shows that the state-point for the resultant mixture must lie on the line WG. But it must also lie on the saturation line. The required state-point is therefore S, where WG cuts the saturation line.

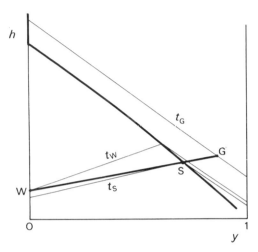

Fig. 15.12 Enthalpy~composition diagram for the air-H_2O system
showing an adiabatic saturation process.

The required mass ratio m_W/m_G is given by the mixing rule as the ratio of two lengths

$$\frac{m_W}{m_G} = \frac{SG}{WS} \tag{15.63}$$

Wet-bulb temperature. It has already been stated, on p. 347 above, that the wet-bulb temperature is approximately that corresponding to adiabatic saturation. But clearly the latter temperature, for a given gas condition, G, depends on the temperature of the water which is supplied. The experimental conditions in wet-and-dry-bulb hygrometers are usually such that the water supplied to the wick is already at the wet-bulb temperature. S is therefore fixed by the condition that W and S must lie on the same isotherm.

In hygrometry the usual problem is to evaluate the moisture content of the air from the measured values t_S and t_G. Fig. 15.13 shows that these values fix the line WS (which now is the isotherm for t_S in the mixed-phase region), and also the gas-phase isotherm, t_G, on which G lies. Hence G is determined uniquely as the intersection of the prolongation of WS and the isotherm t_G; and so y_G is obtained.

The same result can of course be obtained by means of tables of properties of steam and air alone. This however usually involves a considerable amount of trial-and-error, which is avoided by use of the $h \sim y$ diagram.

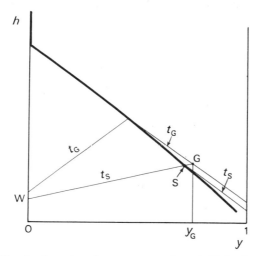

Fig. 15.13 Showing how the mixture content of an air-H_2O mixture may be evaluated from wet-and-dry-bulb-hygrometer readings.

The Steady-Flow Energy Equation applied to air-water mixtures

As our final illustration of the First Law applied to mixtures, we consider a problem arising in the design of a cooling tower. Fig. 15.14 illustrates a natural-draught cooling tower used for cooling the water circulating through the condenser of a steam power plant. The warm water from the condenser enters the tower and is sprayed on to a packing constructed of wooden slats; the water splashes from one row of slats to the next below it, and at last drips from the bottom of the packing, having been cooled by contact with air; it is collected in the pond at the base of the tower and returned to the condenser. The air enters at the bottom of the tower, and rises up through the packing, increasing meanwhile in temperature and moisture content, and leaves through the top of the tower. The circulation of air is caused by the buoyancy of the warmed air; the difference in hydrostatic pressure between the top and bottom of the packing exactly balances the flow resistance caused by the packing.

EXAMPLE 15.6

Problem. Water enters a cooling tower at 20 °C and leaves at 10 °C. The air enters at 6 °C and 50% relative humidity, and leaves, saturated, at 16 °C. Determine the ratio of mass flow rates of entering water and air, and the percentage of entering water which leaves with the air as steam. The pressure may be taken as $101 \cdot 3 \times 10^3$ N/m² throughout.

Solution. This problem is most easily solved analytically if we adopt as our basis 1 kg of *dry air* entering the cooling tower; for the mass of dry air leaving the tower is the same as that entering. This means that it is simplest to express steam concentrations in terms of specific humidity. We first calculate these humidities.

At air entry, section 2, ω_2 is given by equation (15.48)

$$\omega_2 = 0 \cdot 5 \times \frac{18}{29} \times \frac{p}{101 \cdot 3 \times 10^3 - p}$$

$$= 0 \cdot 5 \times \frac{18}{29} \times \frac{0 \cdot 935}{101 \cdot 3 - 0 \cdot 935} \qquad \text{from Steam Tables}$$

$$= 0 \cdot 002 \ 89 \text{ kg } H_2O/\text{kg dry air}$$

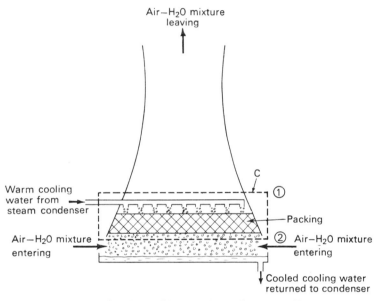

Fig. 15.14 Illustrating the problem on the cooling tower, Example 15.6.

At air exit, section 1, ω_1 is given by

$$\omega_1 = 1 \times \frac{18}{29} \times \frac{p}{101 \cdot 3 \times 10^3 - p}$$

$$= \frac{18}{29} \times \frac{1 \cdot 817}{101 \cdot 3 -- 1 \cdot 817} \qquad \text{from Steam Tables}$$

$$= 0 \cdot 011 \ 34 \text{ kg } H_2O/\text{kg dry air}$$

The mass of water vaporizing, m_V, is therefore

$$m_V = \omega_1 - \omega_2$$

$$= 0 \cdot 011 \ 34 - 0 \cdot 002 \ 89$$

$$= 0 \cdot 008 \ 45 \text{ kg/kg dry air}$$

To calculate the mass of water entering, m_W, we apply the Steady-Flow Energy Equation to a control volume enclosing the cooling tower. The Gibbs-Dalton Law is used implicitly by treating the enthalpies of the components separately. There is no heat transfer or external work at the control surface, and the gravitational and kinetic energies will be neglected. The enthalpy of the entering steam (mass ω_2) will be taken as the saturation enthalpy for the entering air temperature h_{g_2}, since for superheated steam under

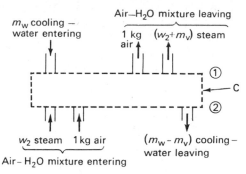

Fig. 15.15 Diagram showing the streams entering and leaving the control surface C of Fig. 15.14.

these conditions the enthalpy depends on temperature alone (See Example 15.3). Fig. 15.15 shows the various streams. The S.F.E.E. gives

$$0 = 1 \times 1 \times 10^3 (t_{a_1} - 0) + (\omega_2 + m_v)h_{g_1} + (m_w - m_v)h_{w_2}$$
$$- 1 \times 1 \times 10^3 (t_{a_2} - 0) - \omega_2 h_{g_2} - m_w h_{w_1}$$

$$\therefore \quad m_w = \frac{1 \times 10^3 (t_{a_1} - t_{a_2}) + \omega_2(h_{g_1} - h_{g_2}) + m_v(h_{g_1} - h_{w_2})}{h_{w_1} - h_{w_2}}$$

$$= \frac{1 \times 10^3 (16 - 6) + 0.002\,89\,(2530.9 - 2512.6) \times 10^3 + 0.008\,45\,(2530.9 - 42.0) \times 10^3}{(83.9 - 42.0) \times 10^3}$$

$$= \frac{10.0 + 0.0529 + 21.03}{41.9}$$

$$= \frac{31.08}{41.9}$$

$$= 0.742 \text{ kg water/kg dry air.} \qquad\qquad \textit{Answer}$$

N.B. The contribution of the vaporized steam to the energy equation, $m_v(h_{g_1} - h_{w_2})$, is an important one (21.03 in 31.08). This is one reason why direct contact of water and air is preferred to a construction in which vaporization is prevented. (Another reason is cheapness of construction.)

The required answers are now quickly obtained. The mass-flow ratio is 0.742 as just shown. The percentage of water vaporized is

$$100 \times \frac{m_v}{m_w} = \frac{100 \times 0.00\,845}{0.742}$$

$$= 1.14\% \qquad\qquad \textit{Answer}$$

N.B. This vaporized water, together with any that is carried away by the air in droplet form, has to be made up from some external supply. However the quantity to be supplied is obviously very much less than if no cooling tower were used and all the water had to be replaced from, say, a river.

The Second Law Applied to Gaseous Mixtures

In Chapter 14, when discussing reversible adiabatic processes undergone by air, we have already implicitly applied the Second Law of Thermodynamics to systems comprising gaseous mixtures. We conclude the present chapter by justifying the procedure of Chapter 14 by reference to the Gibbs-Dalton Law. The discussion is restricted to isentropic processes. Firstly Perfect-Gas mixtures are discussed, and secondly mixtures including a condensable vapour.

Reversible adiabatic expansion of a Perfect-Gas* mixture

Pure substances satisfying the Perfect-Gas definition have simple algebraic relations between their properties when undergoing isentropic processes, as shown in Chapter 14. For example, for such a substance and process,

$$pv^\gamma = \text{constant.} \tag{15.64}$$

It will now be shown that the same is true of *mixtures* of Perfect Gases.

$p{\sim}v{\sim}T$ *relations in an isentropic process.* For a single component of a gas, designated by suffix a, we have

$$s_{a_2} - s_{a_1} = c_{v_a} \ln \left(\frac{T_2}{T_1}\right) + R_a \ln \left(\frac{V_2}{V_1}\right) \tag{15.65}$$

Therefore for a mixture of masses m_a, m_b, m_c ... of gases a, b, c ... , the entropy change in any process $(s_2 - s_1)$ is given by the Gibbs-Dalton Law as

$$(m_a + m_b + m_c + \ldots)(s_2 - s_1) = (m_a c_{v_a} + m_b c_{v_b} + \ldots) \ln \left(\frac{T_2}{T_1}\right)$$

$$+ (m_a R_a + m_b R_b + \ldots) \ln \left(\frac{V_2}{V_1}\right) \tag{15.66}$$

Here T and V do not have letter subscripts because they are the same for each component.

Now if we define c_v and R for the mixture by the weighted averaging equations (see pp. 328, 329)

$$c_v \equiv \frac{m_a c_{v_a} + m_b c_{v_b} + \ldots}{m_a + m_b + \ldots} \tag{15.67}$$

and

$$R \equiv \frac{m_a R_a + m_b R_b + \ldots}{m_a + m_b + \ldots} \tag{15.68}$$

* This section is not valid for Ideal Gases in general.

eq. (15.66) becomes

$$s_2 - s_1 = c_r \ln \left(\frac{T_2}{T_1}\right) + R \ln \left(\frac{V_2}{V_1}\right)$$ (15.69)

This is of course the same equation as holds for a chemically pure substance.

Since for any component we have

$$c_{p_\mathrm{a}} - c_{r_\mathrm{a}} = R_\mathrm{a}$$ (15.70)

it is easy to show that a mixture specific heat at constant pressure, defined by

$$c_p \equiv \frac{m_\mathrm{a} c_{p_\mathrm{a}} + m_\mathrm{b} c_{p_\mathrm{b}} + \ldots}{m_\mathrm{a} + m_\mathrm{b} + \ldots}$$ (15.71)

obeys the equation

$$c_p = c_v + R$$ (15.72)

This completes the parallelism; it remains only to define a mixture specific-heat ratio γ, by

$$\gamma = \frac{c_p}{c_v} = \frac{m_\mathrm{a} c_{p_\mathrm{a}} + m_\mathrm{b} c_{p_\mathrm{b}} + \ldots}{m_\mathrm{a} c_{v_\mathrm{a}} + m_\mathrm{b} c_{v_\mathrm{b}}}$$ (15.73)

and to carry out the steps already shown in Chapter 14 (pp. 305, 306), in order to demonstrate that the mixture obeys the relations

$$pv^\gamma = \text{constant etc.,}$$ (15.74)

when carrying out a process of constant entropy.

Warning. It should be noted that, γ is *not* formed from the γ's of the components by a direct weighting procedure; thus:

$$\gamma \neq \frac{m_\mathrm{a} \gamma_\mathrm{a} + m_\mathrm{b} \gamma_\mathrm{b} + \ldots}{m_\mathrm{a} + m_\mathrm{b} + \ldots}$$

Energy interchange between components. Suppose two Perfect Gases having different γ's are mixed together; the mixture γ will have an intermediate value. If not mixed, the gas with the higher γ would become cooler, on expanding isentropically through a fixed pressure ratio from a given initial temperature, than the gas with the lower γ. If they are mixed before the expansion takes place, both components must have equal temperatures at all times. We infer that there is an energy interchange* between the components to bring this temperature equality about. A consequence is that although the entropy of the mixture is constant, the entropy of the gas with the lower γ decreases, while that of the gas with the higher γ increases during the expansion.

* Since this interchange takes place on the molecular scale it cannot be classified as heat or work.

Reversible adiabatic expansion of steam and air

In conclusion, a method will be indicated for calculating the $p{\sim}v{\sim}T$ relations for an isentropically expanding mixture of steam and air. The fundamental principles are: (i) the entropy *of the mixture* remains constant, not however the entropies of the components; (ii) the volume occupied by the air is the same as that occupied by the steam; (iii) the temperatures of the two components are equal at all times; (iv) the pressure of the mixture is equal to the sum of the partial pressures.

Typical problem. Given a mixture of steam and air at an initial pressure and temperature, determine the $p{\sim}v{\sim}T$ relations during isentropic expansion.

Method of solution. It is most convenient to derive the volume-temperature relation first. A suitable procedure is as follows. For any given specific volume of the mixture, guess the corresponding temperature. This yields the air entropy from the algebraic formula eq. (14.43), with air assumed to be a Perfect Gas, and the steam entropy obtained from interpolation in Steam Tables. The latter is somewhat complicated because the steam pressure is not known; an $h{\sim}s$ chart with constant-volume and constant-temperature lines drawn on it would enable s for steam to be found more easily.

The mixture entropy is thus calculated and compared with that of the initial mixture. If the two quantities are equal, the guessed temperature is correct; if not, a new guess is made. In this way the temperature corresponding to the chosen specific volume is found by trial-and-error.

The partial pressures of steam and air are then obtained respectively from Steam Tables (or charts) and the Ideal-Gas Rule. The mixture pressure is the sum of the partial pressures.

By such means the complete $p{\sim}v{\sim}T$ relation for the expansion can be worked out. The procedure is tedious however and, if condensation does not occur, is usually replaced by the approximate assumption that steam is a Perfect Gas.

BIBLIOGRAPHY

GIBBS, J. W., *Collected Works*, Longmans, London, 1931.
INSTITUTION OF HEATING AND VENTILATING ENGINEERS, *Psychrometric Chart*.
MOLLIER, R., *Z.V.D.I.*, **67** (1923) pp. 869–872, **73** (1929) pp. 1 009–1 013.
SPALDING, D. B., *Convective Mass Transfer*, Edward Arnold, London, 1963.

CHAPTER 15—PROBLEMS

15.1 A rigid vessel contains a mixture of 1 kg of carbon monoxide (CO) and 1 kg of hydrogen (H_2) at a pressure of 200×10^3 N/m² and a temperature of 18 °C.

Assuming CO and H_2 to be Ideal Gases, evaluate:
  (i) the partial pressures of the components;

  (ii) the volume and specific volume of the mixture;

  (iii) the volume analysis;

  (iv) the molar analysis;

  (v) the Gas Constant, the specific heats and γ of the mixture.

Use the following data:

Atomic masses: $C = 12 \cdot 01$, $H = 1 \cdot 008$, $O = 16$. Specific heat at constant pressure: for CO, $c_p = 1042$ J/kg K; for H_2, $c_p = 14\ 310$ J/kg K. The Molar Gas Constant $\mathscr{R}$ $= 8314$ J/kmol K.

15.2 Heat transfer to the mixture of the problem **15.1** at constant volume raises its pressure to 400×10^3 N/m². Assuming the mixture to be a Perfect Gas, evaluate:

  (i) the final temperature of the mixture;

  (ii) the increases in the specific enthalpy, the specific internal energy and the specific entropy;

  (iii) the magnitude of the heat transfer.

15.3 Re-consider problem **15.2**, given that the change of state was brought about by stirring work in the absence of heat transfer. In this case evaluate the magnitude of the stirring work.

15.4 (*a*) A cylinder, fitted with a piston, contains $0 \cdot 01$ kg of a Perfect Gas ($M = 15$, $c_p = 800$ J/kg K) at a pressure of 100×10^3 N/m² and a temperature of 15 °C. Calculate the volume of the gas.

  (*b*) With the piston held stationary, $0 \cdot 015$ kg of a second Perfect Gas ($M = 40$, $c_p = 1040$ J/kg K) is introduced into the cylinder. It is found that the temperature of the contents of the cylinder has increased to 40 °C. For the mixture, evaluate:

  (i) the final pressure;

  (ii) the specific heats and γ.

  (*c*) The mixture now undergoes a fully-resisted expansion to a final specific volume of 2 m³/kg as the piston moves slowly outwards; the process may be represented by $pV^{1\cdot4}$ $=$ constant, where p is the mixture pressure and V the corresponding mixture volume. Evaluate the final pressure, the work done by the mixture and the magnitude and sign of the heat transfer. Evaluate also the increase in specific entropy during the process.

  (*d*) What would have been the final pressure and the work done, if the expansion in (*c*) had been reversible and adiabatic. In this case evaluate the increase in the entropy of each component.

The Molar Gas Constant $\mathscr{R}$ equals $8 \cdot 314 \times 10^3$ J/kmol K.

15.5 Methane gas and air are mixed adiabatically in steady flow in the mass ratio of 1 : 20. The air stream flows steadily into the mixing chamber at a pressure of 105×10^3 N/m²; the methane is at a pressure of 250×10^3 N/m² and the mixed-gas steam flows steadily out of the mixing chamber at a pressure of 100×10^3 N/m². The temperature of both the methane and of the air at entry to the mixing chamber is 15 °C. Flow velocities and changes in elevation are negligible. Evaluate:

  (i) the temperature of the mixture stream;

  (ii) the partial pressure of the gases in the mixture stream;

  (iii) the increase in the entropy of the gases due to mixing, per unit mass of mixture.

Assume methane and air to be Perfect Gases and use the following data:

Molecular mass: CH_4, $16 \cdot 04$: Air, $28 \cdot 97$.

Specific heat at constant pressure: CH_4, $c_p = 2230$ J/kg K; air, $c_p = 1000$ J/kg K.

The Molar Gas Constant $\mathscr{R}$ equals $8 \cdot 314$ kJ/kmol K.

15.6 (*a*) A rigid vessel containing $0 \cdot 1$ m³ of carbon monoxide at a temperature of 15 °C and a pressure of 160×10^3 N/m², is connected via a valve and a short pipe, to a second rigid vessel containing $0 \cdot 15$ m³ hydrogen at a temperature of 15 °C and a pressure of 160×10^3 N/m². The vessels, the valve and the pipeline are well insulated. The valve, initially shut, is opened to allow the two gases to mix; after a time, conditions become uniform throughout the vessels. Assuming carbon monoxide and hydrogen to be Perfect Gases and using the data given in problem **15.1**, evaluate:

(i) the temperature of the mixture;

(ii) the partial pressures of the components;

(iii) the specific volume of the mixture;

(iv) the mass analysis;

(v) the molar analysis;

(vi) the Gas Constant, the specific heats and γ of the mixture;

(vii) the increase in the entropy of the system comprising the two gases.

(b) The two vessels in (a) are connected, through a valve, to a third insulated vessel of volume 0·25 m³. By the opening of the valve, the third vessel, which initially is evacuated, is put into communication with the two inter-connected vessels containing the gas mixture. What are the temperature and pressure when conditions throughout the three vessels have become uniform?

(c) What would have been the final temperature and pressure if the three vessels had been put into communication simultaneously?

15.7 A rigid insulated vessel contains 0·1 kg of hydrogen at a pressure of 100×10^3 N/m² and a temperature of 15 °C. The vessel is connected to a pipeline containing carbon monoxide; the pressure and temperature in the pipeline are maintained at 400×10^3 N/m² and 20 °C respectively. Carbon monoxide is allowed to flow into the vessel until the pressure of the mixture in the vessel has risen to 350×10^3 N/m². Evaluate the temperature and the mass of the mixture in the vessel at this time, assuming conditions throughout the vessel to be uniform.

Assume carbon monoxide and hydrogen to be Perfect Gases and use the data given in Problem **15.1**.

15.8 (a) 0·02 kg of steam and 0·18 kg of air form a gaseous mixture at a temperature of 150 °C and a pressure of 200×10^3 N/m². Determine (i) the partial pressures of the steam and air, (ii) the specific volume, and (iii) the specific internal energy of the mixture. In the latter case specify the datum states used.

Assume air and Superheated steam to be Perfect Gases and use the following data: for air $R = 287$ J/kg K and $c_p = 1000$ J/kg K; for steam $R = 462$ J/kg K and $c_p = 2000$ J/kg K.

(b) Recalculate (a) using the Superheated-Steam Table (Appendix B). In this case again specify the datum used.

15.9 (a) A vessel of 5 m³ capacity initially contains a mixture of air and saturated water vapour at a temperature of 40 °C and a pressure of 15×10^3 N/m². Evaluate the mass of air and vapour in the vessel.

(b) Subsequently 1·2 kg of air leak into the vessel and 0·16 kg of water vapour condenses. Determine the temperature and pressure of the mixture in the vessel.

Assume air to be an Ideal Gas for which $R = 287$ J/kg K.

15.10 A closed rigid vessel of capacity 1·5 m³ contains an air-H_2O mixture at a pressure of 600×10^3 N/m² and a temperature of 120 °C. The H_2O, which is present in both the liquid and vapour phases, has a total mass of 6 kg.

Heat transfer from the contents of the vessel reduces the temperature to 10 °C. Evaluate

(i) the final pressure;

(ii) the mass of water which condenses;

(iii) the magnitude of the heat transfer;

(iv) the increase in the entropy of the contents of the vessel.

Assume air to be a Perfect Gas with $R = 287$ J/kg K and $c_p = 1000$ J/kg K.

15.11 Evaluate h and y for a saturated steam-air mixture at a temperature of 90 °C and a pressure of 1 atm. Check your result against Fig. 15.7.

15.12 A vessel contains a mixture of air and steam at a pressure of 1 atm and a temperature of 25 °C. The mass ratio of air to steam is 99 : 1.

Evaluate:

(i) the partial pressures of the air and the steam;

(ii) the specific volume of the mixture;

(iii) the specific enthalpy of the mixture;

(iv) the specific internal energy of the mixture;

(v) the specific entropy of the mixture.

Assume air and superheated steam to be perfect gases and use the following data: R for steam $= 462$ J/kg K; R for air $= 2871$ J/kg K; c_p for steam $= 2010$, for air $= 1000$ J/kg K. Values of h, u and s, for steam should be obtained from Table II. The datum states to be used are zero values of the internal energy for saturated water at the triple point, and for air at 0.01 °C.

15.13 For the mixture of problem **15.12**, evaluate:

(i) the composition, y;

(ii) the specific humidity, ω;

(iii) the relative humidity, ϕ;

(iv) the dew point;

(v) the temperature to which the air must be cooled at constant volume to reach saturation. Check the answers to (ii), (iii) and (iv) against the Psychrometric Chart, Fig. 15.6.

15.14 For the mixture of problem **15.12**, compare the values of relative humidity corresponding to the various definitions given on pp. 344, 345.

15.15 One cubic decimetre of the mixture of problem **15.12** is contained in a well-insulated cylinder fitted with a non-conducting piston. Liquid water is added to the mixture as the piston moves slowly to keep the pressure constant at 1 atm. Evaluate the adiabatic saturation temperature and the mass of water added, when the water temperature is (a) 40 °C, and (b) 10 °C. Compare the values of saturation temperature with the wet-bulb temperature obtained from the Psychrometric Chart, Fig. 15.6.

15.16 An air-H_2O mixture at a pressure of 1 atm and a temperature of 25 °C has a relative humidity of 20 per cent. The mixture flows steadily into a cooler at the rate of 0.4 m³/s and emerges at a temperature of 4 °C. The pressure drop in the cooler is negligible. Evaluate:

(a) the specific humidity of the mixture entering and leaving the cooler;

(b) the relative humidity of the cooled mixture;

(c) the heat-transfer rate from the mixture.

15.17 Recalculate problem **15.16** given that the relative humidity of the mixture entering the cooler is 80 per cent. In this case determine also the quantity of liquid water leaving the cooler.

15.18 An air-H_2O mixture at a pressure of 1 atm, a temperature of 0 °C and a relative humidity of 40 per cent, is to be "conditioned" to a temperature of 25 °C and a relative humidity of 60 per cent, by successively passing it, in steady flow, through a heater and an adiabatic spray chamber. Water at a temperature of 4 °C is sprayed into the heated air in the proportion 0.015 kg/kg dry air, any excess being drained from the spray chamber at a temperature of 4 °C. The air duct is well insulated and the pressure may be assumed to be 1 atm throughout.

Evaluate:

(i) the rate at which water is drained from the spray chamber;

(ii) the temperature of the mixture leaving the heater;

(iii) the heat transfer to the mixture in joules per kilogram of mixture in the heater.

15.19 One kilogram of an air-H_2O mixture at a pressure of 1 atm, a temperature of 200 °C and a composition y of 0.8 is mixed with 0.5 kg of an air-H_2O mixture at a pressure of 1 atm, a temperature of 90 °C and a composition of 0.2. Evaluate the composition and enthalpy of the resultant mixture and the quantity of liquid water contained in it. Check the calculation against the $h{\sim}y$ chart, Fig. 15.7.

15.20 Plot to scale the saturation line and a few isotherms of an $h{\sim}y$ chart for the air-H_2O system at a pressure of 1 atm. The chart should be about 25 cm square.

15.21 Steam escapes from a safety valve at a pressure of 1 atm with 30 K superheat into dry air at a temperature of 25 °C. Will the jet be visible? (Solve with the aid of the chart of problem **15.20**, assuming that the various states in the jet are found by adiabatic mixture of the steam and air that their state points lie on the line joining the initial steam and air state points. If that line crosses the saturation line, condensation occurs and the droplets enable the jet to be seen.)

15.22 Air flowing in from an ocean is at a temperature of 10 °C and has a relative humidity of 80 per cent. It passes over a mountain range. Assuming that the air-steam mixture changes isentropically (with $\gamma = 1\cdot4$), determine the height of the mountain range which will just cause clouds to form at the peaks. The pressure-altitude relation may be taken as $p = 762 - 7\cdot5 \times 10^{-2} z$ where p is in millimetres of Hg when z is the height in metres above sea level.

16 Fuels and Combustion

Introduction

The engineering importance of combustion

We have seen that one way of producing mechanical power is to construct a heat engine which interacts with two systems: one at a high temperature and the other at a low temperature. Now, except in rare circumstances, systems of widely differing temperature are not found in nature close together; they must therefore be constructed by man. The most common procedure is to use the atmosphere or a river as the low-temperature system, and to provide a high-temperature system by burning fuel with air. It is in this way that a considerable proportion of the world's power is produced, usually in conjunction with steam power plant.

However this is not the only means of producing power from the combustion of fuel. Internal-combustion engines operate on the different principle of compressing the fuel and air, burning it at a pressure above that of the atmosphere, and then expanding the products of combustion into the atmosphere. Cars, locomotives and aircraft are propelled in this way.

It is therefore important for the engineer to know the properties of available fuels and of their combustion products, and to be able to calculate the temperature which can be reached in the combustion chamber and the maximum work that can be obtained from a given quantity of fuel. To this end, we now make an excursion into chemical thermodynamics.

This is a subject which is vital to other branches of engineering also. For example, the operations of the metallurgical industry can be understood only through knowledge of the interactions between the fuel used in smelting the metal-bearing ores, and the various metals which make up an alloy. Chemical thermodynamics is even more important to the chemical engineer, who can plan his operations only with the aid of quantitative knowledge of what reactions are possible, of how they depend on pressure and temperature, and of what changes in composition and energy accompany them.

The science of chemical change

Broadly speaking, the scientific knowledge needed in designing plant for carrying out combustion or other chemical reactions consists of two parts. First it is necessary, as just implied, to calculate the direction which a reaction can be expected to take, together with the associated composition and energy changes. Secondly, we have to predict *how fast* the reaction will be and so *how big* a piece of equipment must be provided for a given output.

It is the first body of knowledge which comprises *chemical thermodynamics*, and it is only this aspect of chemical change which will be discussed in the present book. Our aim, moreover, is to introduce merely the fundamental ideas and their application to mechanical engineering.

The second body of knowledge is known as the science of *rate processes*; this may be regarded as thermodynamics with a time scale. The rate of chemical reaction is studied in the part of this science known as *chemical kinetics*; it is beyond the scope of this book.

Other parts of the science of rate processes have already been noted. Thus, although heat has been extensively discussed above, the calculations have been about its magnitude in any process, not the time taken for the process to occur; the latter theme belongs to the subject known as *heat transfer*. In Chapter 15, the thermodynamics of mixtures was discussed; how long a given mixing process will take, and how this time depends on the area of contact between the mixing substances, can be discovered only by studying the subject known as *mass transfer*.

The nature of chemical thermodynamics

Our discussion of combustion will reveal that chemical thermodynamics has three main sections, corresponding to the three natural laws: the Law of Conservation of Matter, the First Law of Thermodynamics, and the Second Law of Thermodynamics.

The first section is known as *stoichiometry*. It is concerned with the relations between the composition of the reactants, e.g. the fuel and air, and the composition of the products. It expresses some simple facts about the chemical constitution of matter, and involves accounting for each of the chemical elements as their atoms change partners in the reaction.

In Chapter 6 it was emphasized that the First Law is completely general. In later chapters however we imposed the restriction that the chemical state of the system should not change. Now this restriction must be removed, and we must study the interrelation between the heat, the work and the energy changes in processes undergone by systems which are not pure substances.

The Second Law gives two types of information about chemical reactions. First, by distinguishing between reversible and irreversible processes, it tells us whether a given reaction will go forwards or backwards; whether, for example, we can ever expect combustion products to decompose spontaneously into fuel and oxygen. Secondly, the Second Law makes it possible to determine how much work can be obtained by burning a fixed mass of fuel; this gives us a standard against which the performance of actual power plants can be measured. In the present book, this aspect of chemical thermodynamics is touched on only briefly.

Summary of the remainder of the chapter

The treatment will follow the sub-division of the subject which has just been indicated. In the first section, the composition of fuels will be discussed, together with methods of calculating the composition of the combustion products from the fuel-air ratio, and vice versa.

The second section introduces enthalpies of reaction and formation, and the calorific values of fuels. It is shown how the temperatures attained in burning mixtures can be calculated, and how much heat can be transferred from the products of combustion.

The final section deals with the entropy changes which occur during reaction, and explains their relation to the maximum possible work output. This makes it possible to define and discuss the efficiency of internal-combustion engines.

Symbols

c_p	Specific heat at constant pressure.	S'	Entropy of a chemical substance.
H'	Enthalpy of a chemical substance	T	Absolute temperature.
h_f'	Enthalpy of formation of a chemical species, per unit mass.	t	Temperature.
		U'	Internal energy of a chemical substance.
$\tilde{h}_f'$	Enthalpy of formation of a chemical species, per mole.	$[\Delta U']_{v,t}$	Increase in energy of a chemical substance in a constant-volume, isothermal reaction.
$[\Delta H']_{p,t}$	Increase in enthalpy of a chemical substance in a constant-pressure, isothermal reaction.	$\tilde{u}_{fg}$	Internal energy of evaporation of a pure substance, per mole.
$\tilde{h}_{fg}$	Enthalpy of evaporation of a pure substance, per mole.	u_{fg}	Internal energy of evaporation of a pure substance, per unit mass.
h_{fg}	Enthalpy of evaporation of a pure substance, per unit mass.	V'	Volume of a system comprising a chemical substance.
h	Specific enthalpy of a pure substance.	W	Net work done by a system.
		W_x	External work.
M	Relative molecular mass (molecular mass).	x	Air-fuel ratio by mass.
		x_{stoich}	Stoichiometric air-fuel ratio by mass.
m	Mass.	y	Air-fuel ratio by volume.
$\dot{m}$	Mass flow rate.	$\overline{CV}$	Calorific value of a fuel.
n	Number of moles.		
p	Pressure.	$\overline{LCV}$	Lower (or net) calorific value of a fuel.
Q	Heat transfer.		
$\mathscr{R}$	Molar (Universal) Gas Constant.	$\overline{HCV}$	Higher (or gross) calorific value of a fuel.

η_c	Combustion efficiency.	*Subscripts*	
η_h	Heating efficiency.	P	Products.
$\eta_{i.c.}$	Efficiency of an internal-	R	Reactants.
	combustion engine.	0, 1, 2 . . . States of a fluid.	

Stoichiometry

Chemical composition

We begin with a resumé of some facts about the constitution of matter which are normally treated in an elementary chemistry course.

The chemical elements. All matter is made up of a limited variety of elementary substances, the *chemical elements.* In combustion, the most common reacting elements are carbon, hydrogen, oxygen, sulphur and nitrogen; they are denoted by the capital letters, C, H, O, S and N, respectively.

Atoms. A quantity of a chemical element is not indefinitely divisible into smaller and smaller amounts, but must be regarded as consisting of a collection of tiny particles which cannot be split.* These particles are the *atoms.* They are so small that 1 kg of oxygen, for example, comprises 18.75×10^{24} of them. Atomic masses on the scale that makes the mass of the carbon-12 atom equal to 12, i.e. *relative atomic masses,* are given in Table 16.1 for elements of importance in combustion.

TABLE 16.1   *Relative Atomic Masses of Elements*

Element	H	C	N	O	S
Relative atomic masses	1·007 97	12·011 15	14·0067	15·9994	32·064

From now on we shall, for convenience, use the name atomic mass for relative atomic mass; it replaces the formerly-used name, "atomic weight".

Molecules. The individual atoms of a chemical element are rarely found in a "free" state. Even in a substance comprising only one chemical element, for example, hydrogen gas, the atoms are almost always found to be tied together in pairs; this fact is expressed by writing the formula for hydrogen gas as H_2. Oxygen and nitrogen gases also consist, as a rule, of multitudes of particles, each of which comprises two atoms; their formulae are accordingly O_2 and N_2 respectively.

Atoms need not "mate" with others of their own kind, however, but can form close bonds with atoms of different elements. Thus steam comprises particles each of which contains two hydrogen atoms and one oxygen atom:

---

* Modern knowledge of nuclear physics does not invalidate this statement. For when an atom is split it ceases to have the properties of the original chemical element. Such nuclear changes are not considered in this chapter.

its formula is accordingly H_2O. The groups of atoms are known as *molecules* and it is the constitution of these species that determines the chemical and other properties of the species in which they occur.

The inter-atomic forces which bind together the atoms in the molecule operate over only very small distances. It is therefore usual for each molecule to consist of only a small number of atoms; for larger groups of atoms would be but weakly bound and so would easily decompose. Carbon and hydrogen form the most notable exceptions to this general rule: their elements can jointly form molecules comprising some scores of atoms.

A consequence of the atomic and molecular construction of matter is that the ratio of the numbers of atoms of the elements in the molecules of a species must be expressible in terms of small whole numbers; in the case of hydrogen and oxygen in steam, for example, the ratio is 2 : 1. If the masses are expressed in molar units (e.g. kmol) the mass ratios are identical with the number-of-atom ratios; this is a major reason for using different mass units for the different species; the arithmetic is made simpler thereby.

Oxides. When oxygen gas takes part in a combustion reaction, the atoms of its molecules become parted; they then form new molecules jointly with atoms of carbon and hydrogen which were originally in the fuel molecules. The new molecules are known as *oxides*. Thus steam is an oxide of hydrogen.

Carbon forms two common oxides. The first, known as carbon monoxide, consists of molecules containing one atom of carbon and one of oxygen each; it has the chemical formula CO. The second, carbon dioxide, has two oxygen atoms and one carbon atom in its molecule; it has the formula CO_2. Both these substances are gaseous at normal temperatures and pressures; for most engineering purposes they may be regarded as Ideal Gases.

Fuels

It is convenient to classify fuels according to the phase in which they are normally handled. The gaseous fuels are chemically the simplest; liquid fuels contain more complex molecules; while solid fuels often have so complicated a molecular structure that full knowledge of their constitution is not yet available.

Gaseous fuels. Gaseous fuels are stored naturally beneath the earth in many parts of the world, often in the vicinity of oilfields. Where natural gas is not available, gaseous fuels may be manufactured, by the thermal treatment of coal or oil. Gaseous fuels may be conveniently transported through pipes over very large distances. Natural gas may also be carried in liquid form in ships and in land vehicles.

We will now list some of the more important chemical substances which occur in these fuels.

The simplest is *hydrogen*, H_2, which is a major constituent of manufactured gas derived from coal.

Carbon monoxide, CO, is a fuel, as well as being an oxide, because it can combine with more oxygen to form carbon dioxide, CO_2. Carbon monoxide is also found in gases manufactured from coal, both in that which is supplied in a town's gas system, and in so-called *producer gas*. The latter is manufactured for use in furnaces and engines by the simple process of burning coke with a limited supply of air.

The other gaseous fuels of chief importance are combinations of carbon and hydrogen, the so-called *hydrocarbons*. The simplest is *methane*, which has the chemical formula CH_4, signifying that its molecule contains one atom of carbon and four atoms of hydrogen. It is a major constituent of natural gas. A whole family of further fuels can be formed by adding to a methane molecule further carbon and hydrogen atoms in the ratio 1 : 2.

TABLE 16.2 *Volume Analyses of Some Fuel-Gas Mixtures* (%)

	CO	H_2	CH_4	C_2H_6	C_2H_6	C_4H_8	O_2	CO_2	N_2
Coal gas (Town gas)	9	53·6	25	—	—	3	0·4	3	6
Producer gas	29	12	2·6	0·4	—	—	—	4	52
Blast-furnace gas	27	2	—	—	—	—	—	11	60
Natural gas (U.K.)	1	—	93	—	3	—	—	—	3
Natural gas (U.S.A.)	—	—	80	—	18	—	—	—	2
Natural gas (U.S.S.R.)	—	1	93	—	3·5	—	—	2	0·5

Thus *ethane* has the formula C_2H_6, *propane* has the formula C_3H_8, *butane* is C_4H_{10}, and so on. This family is known as the *paraffins*. They, too, are found in natural gas.

It is not possible to add further hydrogen atoms to a paraffin molecule, and so form new chemical substances; paraffin hydrocarbons are therefore called *saturated*. However, hydrocarbon molecules can exist which have *fewer* hydrogen atoms combined with a given number of carbon atoms than has the corresponding paraffin; such hydrocarbons are called *unsaturated*. A simple example is the gas *acetylene* which has two carbon atoms and two hydrogen atoms in its molecule; its formula is accordingly C_2H_2. Others are *ethylene*, C_2H_4, *propylene*, C_3H_6, and so on. Many of them are found naturally, but some (acetylene is one) have to be manufactured.

Most gaseous fuels are mixtures of several gases, including some, such as nitrogen, which are not easily oxidized. Table 16.2 contains volume analyses of typical fuels used industrially. The components may be regarded as Ideal Gases under atmospheric conditions; the volume analysis is therefore the same as the molar analysis (see Chapter 15, p. 333).

Liquid fuels. There is no clear-cut boundary between gaseous and liquid fuels, because the one can be transformed into the other by changing the temperature and pressure. Most liquid fuels are hydrocarbons, either saturated or unsaturated. As the molecular mass increases, as a result of an increased number of carbon and hydrogen atoms in the molecule, the boiling point of the substance at atmospheric pressure rises. Roughly speaking, hydrocarbon molecules containing six or more carbon atoms in the molecule have atmospheric-pressure boiling points in excess of atmospheric temperatures; such hydrocarbons are therefore classified as liquid fuels.

Table 16.3 contains the names, chemical formulae, molecular masses and atmospheric-pressure boiling points of several hydrocarbon fuels. Both gaseous and liquid fuels are included so that the trend can be perceived.

TABLE 16.3 *Boiling Points of Hydrocarbon Fuels*

Name	Chemical formula	Molecular mass (rounded value)	Boiling point at $p = 1$ atm
Methane	CH_4	16	$-161 \cdot 4$ °C
Ethylene	C_2H_4	28	$-103 \cdot 9$
Ethane	C_2H_6	30	$-89 \cdot 0$
Propylene	C_3H_6	42	$-47 \cdot 6$
n-Butane	C_4H_{10}	58	$-0 \cdot 5$
n-Pentane	C_5H_{12}	72	$36 \cdot 0$
Benzene	C_6H_6	78	$80 \cdot 1$
Toluene	C_7H_8	92	$110 \cdot 7$
n-Octane	C_8H_{18}	114	$125 \cdot 6$

The fuels used in practice are invariably mixtures; often some hundreds of different varieties of molecule can be found in a single fuel sample. This does not, however, usually create difficulty for the combustion engineer, since all he needs to know about the fuel composition are the relative numbers or masses of carbon and hydrogen atoms in the sample. Table 16.4 gives some typical data. It will be seen that elements other than carbon and hydrogen are normally present, notably sulphur. There are also traces of incombustible impurities which are lumped together under the designation *ash*.

The *composition* of a fuel, expressed in terms of the percentage masses of the chemical elements, is known as the *mass analysis* of the fuel; sometimes it is called the *ultimate analysis*. It is most useful in the case of solid fuels, for which the molecular structure is too complex to be unravelled. For liquid hydrocarbon fuels, it is often sufficiently accurate to take the mass analysis as $85 \cdot 8\%$ C and $14 \cdot 2\%$ H; this corresponds to a $1 : 2$ ratio of number of carbon atoms to number of hydrogen atoms.

TABLE 16.4 *Liquid Fuels: Mass Analysis* (%)

Fuel	Carbon	Hydrogen	Sulphur	Ash etc.
Aviation gasoline (100 Octane)	85·1	14·9	0·01	—
Motor gasoline	85·5	14·4	0·1	—
Vaporizing oil	86·2	12·9	0·3	—
Motor benzole	91·7	8·0	0·3	—
Kerosine	86·3	13·6	0·1	—
Diesel oil, distilled (Gas oil)	86·3	12·8	0·9	—
Light fuel oil	86·2	12·4	1·4	—
Heavy fuel oil	86·1	11·8	2·1	—
Residual fuel oil (Bunker C)	88·3	9·5	1·2	1·0

Solid fuels. Pure carbon is one of the simplest solid fuels. Alone among the fuels mentioned so far, it burns in the solid phase; oxygen has to diffuse to the carbon surface and the gaseous oxides are formed there. The gaseous and liquid fuels, by contrast, burn in the gaseous phase, the liquids vaporizing

TABLE 16.5 *Some Properties of Solid Fuels*

Fuel	Moisture content of good commercial fuel: % by mass	Analysis of good commercial fuel: % by mass in dry fuel					Volatile matter: % by mass in dry fuel
		C	H	O	N + S	Ash	
Anthracite	1	90·27	3·00	2·32	1·44	2·97	4
Bituminous coal	2	81·93	4·87	5·98	2·32	4·90	25
Lignite	15	56·52	5·72	31·89	1·62	4·25	50
Peat	20	43·70	6·42	44·36	1·52	4·00	65
Wood	15	42·5	6·78	49·87	0·85	Trace	80

and mixing with the oxygen before burning. The boiling point of carbon at normal pressures is well above the temperatures attained in flames.

Most solid fuels are obtained by mining; they consist mainly of carbon, together with hydrogen, sulphur and some incombustible ash. According to their chemical and physical properties, these fuels are called anthracite, bituminous coal, brown coal, or peat. Some of their typical properties are indicated in Table 16.5. Anthracite is the most valuable, and peat the least. For comparison, wood is included also.

The chemical elements comprising industrial solid fuels are bound together, often fairly loosely, to form complex molecules of high molecular mass. The incombustible ash is interspersed in the fuel both in minute particles and also in larger lumps. In addition there is always a certain amount of water in the fuel, some of it chemically combined, some merely admixed; for a fuel from a given mine, the moisture content varies with the conditions of treatment and storage.

Engineers keep a check on the quality of solid fuel with which they are supplied by carrying out what is called a *proximate analysis*. By heating at atmospheric pressure to successively higher temperatures, with intermediate weighings, the percentages are determined of moisture, volatile matter, combustible solids and ash. The volatile matter mentioned is that driven off at temperatures above 100 °C, and so excludes the water; it consists chiefly of hydrocarbons of high molecular mass.

Standard testing methods have been laid down by the British Standards Institution and the American Society for Testing Materials (See Bibliography).

Exhaust gases. Fuel and air form the input streams to most combustion appliances. The output stream is invariably gaseous, apart from the ash and possibly some condensed water. The exhaust-gas stream therefore contains the products of combustion, consisting of: the oxides of hydrogen, carbon and, if present, sulphur; excess oxygen and unburned fuel; and nitrogen and other chemically inert gases flowing through the plant. Whether any of the steam condenses depends on the temperature of the exhaust gas and on the partial pressure of steam in it. The most usual oxide of sulphur is the dioxide, SO_2.

Chemical change

Chemical reaction is the process in which the interatomic bonds in the *reactant* molecules are broken, followed by the re-arrangement of the atoms thus set free, in new molecular combinations: the *product* molecules. Thus new chemical species appear while the old ones disappear, the actual atoms remaining the same. From this it follows that chemical reaction is not an interaction between systems, like heat and work; it is a change *within* a system. It is recognized by the change in the chemical properties of the system. Combustion is the particular class of chemical reaction in which the products are oxides.

Chemical equations. In order to keep track of chemical reactions, we make use of an accounting system which is expressed symbolically in the form of an equation. Thus the reaction in which methane burns with oxygen to carbon dioxide and steam is described by the equation:

$$CH_4 + 2O_2 = CO_2 + 2H_2O \qquad (16.1)$$

The convention adopted differs from that of ordinary algebra. The symbols for the reactant molecules appear on the left and those for the product molecules on the right. Each molecular symbol is multiplied by a numeral (which may be unity, and therefore omitted) signifying the relative number of molecules taking part. In the setting up of such an equation, usually the molecular symbols are written down first; the multiplying numerals are then prefixed in a way that conforms with the *Principle of Conservation of Matter*. This states that the number of atoms of any chemical element in the reactants is the same as the number of that element in the products.

Since, in eq. (16.1), the CH_4 symbol has the coefficient unity, so must the CO_2 symbol; otherwise the numbers of carbon atoms would not be equal on the two sides of the equation. Since hydrogen appears in the products only in the form H_2O, there must be two H_2O molecules to give the same number of H atoms (four) in both products and reactants. This fixes the number of oxygen atoms in the products as four (two in CO_2, two in $2H_2O$); thus two oxygen molecules must appear on the left-hand side.

By combining eq. (16.1) with the atomic-mass data* given in Table 16.1, we obtain the relative masses of the reactants and products taking part in the reaction; expressed in kg, they are:

$$16 \text{ kg } CH_4 + 64 \text{ kg } O_2 = 44 \text{ kg } CO_2 + 36 \text{ kg } H_2O$$

or on a molar basis:

$$1 \text{ mole } CH_4 + 2 \text{ moles } O_2 = 1 \text{ mole } CO_2 + 2 \text{ moles } H_2O.$$

Merely writing a chemical equation leaves unanswered important questions about the chemical reaction, for example whether it is even possible that it should proceed in the direction indicated, i.e. from left to right. Questions of possibility have to be answered by reference to the Second Law of Thermodynamics. For the present, however, we will take it that combustion reactions in which carbon- and hydrogen-bearing fuels are completely oxidized can always proceed spontaneously, *if sufficient oxygen is available* and if the appropriate combustion equipment is provided. CO_2 and H_2O are therefore regarded as the products of *complete combustion*. Only when the products are at extremely high temperatures is this assumption invalid.

Often, substances are present which take no part in the chemical reaction; for example, when most fuels burn with air initially at atmospheric temperature, the temperatures attained are too low for the nitrogen to be affected significantly. Such inert substances are usually omitted from the chemical equation; if it is found desirable to take note of them, the appropriate symbols are the same on both sides of the equation. Thus, with 100 moles of air taken to consist of a mixture of 21 moles of oxygen with 79 moles of nitrogen, the equation for the reaction of methane with sufficient air just to cause complete combustion could be written:

$$CH_4 + 2O_2 + 2 \times \tfrac{79}{21}N_2 = CO_2 + 2H_2O + 2 \times \tfrac{79}{21}N_2 \qquad (16.2$$

* It is usually adequate to use the rounded values of atomic masses.

The calculation of composition change

Identity of system analysis and control-volume analysis of composition change. In this section a number of commonly useful equations and methods will be set down relating to the composition changes occurring in combustion. The discussion will be in terms of a system, i.e. a fixed body of matter, contained, for example, in a reaction vessel. The procedure is identical however if steady flow has to be analysed by means of a control volume; then the masses within the system are merely replaced by the masses crossing the control-volume boundary.

Combustion with air: mass units. Consider a fuel of mass analysis: $100m_C$ % C, $100m_H$ % H, $100m_O$ % oxygen, $100m_A$ % ash. Let the system contain unit mass of fuel and mass x of air, the latter consisting of 0·232 kg O_2 per kg air and 0·768 kg N_2 per kg air. We will calculate the composition of the exhaust gases, assuming that *complete combustion* takes place.

The carbon burns according to the equation

$$C + O_2 = CO_2 \qquad (16.3)$$

The corresponding masses are

$$m_C \text{ C}, \qquad \tfrac{32}{12}m_C \text{ O}_2, \qquad \tfrac{44}{12}m_C \text{ CO}_2$$

since the molecular masses of C, O_2 and CO_2 are 12, 32 and 44 respectively.

For hydrogen (molecular mass = 2), the reaction and corresponding masses are

$$H_2 \qquad + \qquad \tfrac{1}{2}O_2 \qquad = \qquad H_2O \qquad (16.4)$$

$$m_H \qquad \qquad \tfrac{1}{2} \times \tfrac{32}{2}m_H \qquad \qquad \tfrac{18}{2}m_H$$

The species present at the end of reaction are CO_2, H_2O, O_2, N_2 and ash. We consider each in turn.

CO_2: the final mass is $\tfrac{44}{12}m_C$, as already stated.

H_2O: the final mass is $\tfrac{18}{2}m_H$, as already stated.

O_2: The initial mass is m_O supplied in the fuel, plus 0·232x supplied in the air. The quantity consumed in the reaction, as has been seen, is $\tfrac{32}{12}m_C + \tfrac{16}{2}m_H$.

The final mass is therefore:

$$m_O + 0.232x - \tfrac{32}{12}m_C - \tfrac{16}{2}m_H.$$

N_2: The mass supplied is 0·768x. Since nitrogen takes no part in the reaction, this is the final mass also.

Ash: The initial and final masses are both m_A.

The *exhaust gases*, if at sufficiently high temperature, consist of CO_2, H_2O (as steam), O_2 and N_2. If however the temperature is reduced sufficiently, the steam condenses. We will calculate the composition of the resulting *dry* gases; it will be remembered that it is the dry-gas analysis which is given by the Orsat apparatus (p. 347).

The total mass of dry exhaust gas (CO_2, O_2 and N_2) is

$$\underbrace{\tfrac{44}{12}m_C}_{CO_2} + \underbrace{m_O + 0{\cdot}232x - \tfrac{32}{12}m_C - \tfrac{16}{2}m_H}_{O_2} + \underbrace{0{\cdot}768x}_{N_2}$$

which may be written more simply as

$$m_C - \tfrac{16}{2}m_H + m_O + x.$$

This represents the original mass of $(1 + x)$, less the masses of the ash and condensed steam.

The *mass analysis* of the dry exhaust gas is therefore

$$(\tfrac{44}{12}m_C \times 100)/(m_C - \tfrac{16}{2}m_H + m_O + x) \qquad \% \; CO_2$$

$$\{(m_C + 0{\cdot}232x - \tfrac{32}{12}m_C - \tfrac{16}{2}m_H) \times 100\}/\{m_C - \tfrac{16}{2}m_H + m_O + x\} \; \% \; O_2$$

and $\qquad (0{\cdot}768x \times 100)/(m_C - \tfrac{16}{2}m_H + m_O + x) \qquad \% \; N_2.$

For comparison with measurements made with an Orsat apparatus, the *volume* (i.e. *molar*) *analysis* of the dry gases is required. The procedure is as follows:

The molecular masses of CO_2, O_2 and N_2 can be taken as 44, 32 and 28 respectively. Then the numbers of moles present are:

$$\tfrac{1}{44} \times \tfrac{44}{12}m_C, \qquad \text{i.e.} \quad \tfrac{1}{12}m_C \text{ mole } CO_2$$

$$\tfrac{1}{32}(m_O + 0{\cdot}232x - \tfrac{32}{12}m_C - \tfrac{16}{2}m_H) \text{ mole } O_2$$

and $\qquad \tfrac{1}{28} \times 0{\cdot}768x$ mole $N_2.$

The total number of moles is obtained by addition as:

$$\tfrac{1}{32}(m_O + 0{\cdot}232x - \tfrac{16}{2}m_H) + \tfrac{1}{28} \times 0{\cdot}768x.$$

The *molar analysis* is therefore:

$$\{\tfrac{1}{12}m_C \times 100\}/\{\tfrac{1}{32}(m_O + 0{\cdot}232x - \tfrac{16}{2}m_H) + (\tfrac{1}{28} \times 0{\cdot}768x)\} \qquad \% \; CO_2,$$

$$\{\tfrac{1}{32}(m_C + 0{\cdot}232x - \tfrac{32}{12}m_C - \tfrac{16}{2}m_H) \times 100\}/$$

$$\{\tfrac{1}{32}(m_O + 0{\cdot}232x - \tfrac{16}{2}m_H) + (\tfrac{1}{28} \times 0{\cdot}768x)\} \qquad \% \; O_2.$$

and $\quad (\tfrac{1}{28} \times 0{\cdot}768x \times 100)/$

$$\{\tfrac{1}{32}(m_O + 0{\cdot}232x - \tfrac{16}{2}m_H) + \tfrac{1}{28} \times 0{\cdot}768x\} \qquad \% \; N_2.$$

Deduction of air-fuel ratio from exhaust-gas analysis. The above results are often used in calculating the mass of air x supplied per unit mass of fuel from the exhaust-gas analysis, which is easily measured; for, if the percentage of CO_2 in the gas is known, together with the mass analysis of the fuel (i.e. m_C, m_H and m_O), x can be calculated.

The same can be done if the percentage of O_2 is measured. In general, the value of x calculated from the CO_2 percentage will differ slightly from that calculated from the O_2 percentage. This is a result of experimental inaccuracy. Normally the value based on CO_2 measurements is the more reliable.

Stoichiometric air-fuel ratio. It is of interest to calculate the minimum value of x which would supply sufficient oxygen for complete combustion. This is known as the *stoichiometric* or "theoretical" air-fuel ratio, x_{stoich}; it is calculated by equating the final mass of oxygen to zero. Thus

$$m_O + 0 \cdot 232 x_{stoich} - \tfrac{32}{12}m_C - \tfrac{16}{2}m_H = 0$$

that is,
$$x_{stoich} = \frac{\tfrac{32}{12}m_C + \tfrac{16}{2}m_H - m_O}{0 \cdot 232} \qquad (16.5)$$

For a hydrocarbon fuel, x_{stoich} is usually about 15; for a coal it is nearer 11.

Excess air. It is undesirable to supply more than the stoichiometric quantity of air when fuel is being burned for heating purposes, because the excess air merely lowers the gas temperature and so lowers the heat-transfer rate. However, since it is seldom possible to mix the fuel and air with complete uniformity, some *excess air* must be provided, in order to avoid the still less desirable phenomenon of incompletely-burned fuel passing out of the furnace in the exhaust gases. The quantity of excess air is usually expressed as a percentage, viz.

$$\text{per cent excess air} = \frac{x - x_{stoich}}{x_{stoich}} \times 100\%.$$

Often a value of 20 per cent excess air is regarded as acceptably low. Gas turbines however must operate with about 300 per cent excess air in order to avoid melting the turbine blades.

Mixture strength. For gasoline engines, a different term is adopted for comparing x and x_{stoich}. The reason for this is that, in gasoline engines, x can have values both above and below x_{stoich}. It is therefore usual to express the variation in air-fuel ratio in terms of the *mixture strength* defined by the relation:

$$\text{Mixture strength} = \frac{x_{stoich}}{x} \times 100 \text{ per cent} \qquad (16.6)$$

A typical range of mixture strength is from a *weak* mixture of 90 per cent to a *rich* mixture of 120 per cent. The terms "weak" and "rich" express respectively the deficiency and the excess of the fuel in a given quantity of air entering the engine, by comparison with the stoichiometric quantity of fuel.

The "theoretical", or stoichiometric, air-fuel ratio corresponds to a mixture strength of 100 per cent.

Consequences of deficient air supply. If the actual air-fuel ratio x is less than x_{stoich}, the assumption of complete combustion of fuel leads to the absurd conclusion that the concentration of oxygen in the exhaust gases is negative; the assumption is therefore wrong. What the exhaust gases actually truly consist of is hard to determine, and involves Second-Law considerations. An approximate rule however is to assume that, first, all the carbon is burned to the monoxide, CO, and that thereafter any remaining oxygen is equally shared between this gas, oxidizing to CO_2, and the hydrogen, oxidizing to H_2O. The exhaust gases then comprise CO_2, CO, H_2O, H_2 and N_2 if the fuel is a hydrocarbon.

Combustion with air: molar units. If the fuel has a known molecular constitution and mass, the calculation of the volume analysis of the exhaust-gas is more easily performed in molar units throughout. This is particularly the case when the fuel consists of a mixture of Ideal Gases. An example now follows:

EXAMPLE 16.1

Problem. The volume analysis of a producer gas is 26% CO, 12% H_2, 7% CO_2, and 55% N_2. The air-fuel ratio, by volume,* is y. What is the volume analysis of the dry exhaust gases?

Solution. We suppose 1 mole of gas and y mole of air to constitute the system, making use of the fact throughout that numbers of moles are proportional to standard volumes. The reactions are:

$$CO \quad + \quad \tfrac{1}{2}O_2 \quad = \quad CO_2 \qquad (16.7)$$
$$\text{1 mole} \quad \tfrac{1}{2} \text{ mole} \quad \text{1 mole}$$

$$H_2 \quad + \quad \tfrac{1}{2}O_2 \quad = \quad H_2O \qquad (16.8)$$
$$\text{1 mole} \quad \tfrac{1}{2} \text{ mole} \quad \text{1 mole}$$

Air consists, we assume, of 21% O_2 and 79% N_2 by volume. Treating each component of the exhaust gases in turn, we have:

CO_2: 0·07 moles are supplied. 0·26 moles result from the combustion of CO. Finally, therefore, we have 0·33 moles CO_2.

H_2O: None is supplied. 0·12 moles result from combustion of H_2.

O_2: 0·21 y moles are supplied. $\tfrac{1}{2} \times 0.26 + \tfrac{1}{2} \times 0.12$ moles are consumed in the oxidation of CO and H_2 respectively. Finally, we have $0.21y - 0.13 - 0.06$, i.e. $0.21y - 0.19$ moles O_2.

N_2: 0·55 moles are supplied in the fuel, and $0.79y$ moles in the air. Finally, therefore, we have $0.55 + 0.79y$ moles N_2.

The total number of moles of non-condensable gas (i.e. excluding H_2O) is therefore
$$(0.33) + (0.21y - 0.19) + (0.55 + 0.79y), \text{ i.e. } 0.69 + y \text{ moles.}$$

* The volumes are supposed to be measured at some standard condition, say: $p = $ 1 atm; $t = 15\ °C$.

The volume analysis of the dry exhaust gas is therefore

$$\frac{0 \cdot 33}{y + 0 \cdot 69} \times 100 \qquad \% \ CO_2,$$

$$\frac{0 \cdot 21y - 0 \cdot 19}{y + 0 \cdot 69} \times 100 \qquad \% \ O_2,$$

and

$$\frac{0 \cdot 79y + 0 \cdot 55}{y + 0 \cdot 69} \times 100 \qquad \% \ N_2. \qquad\qquad \textit{Answer}$$

Remarks. (i) The stoichiometric air-fuel ratio in terms of volumes is given by putting the oxygen concentration in the exhaust gases equal to zero; thus $y_{stoich} = 0 \cdot 19 \div 0 \cdot 21 = 0 \cdot 905$ m³ air/m³ producer gas.

(ii) With the stoichiometric air-fuel ratio, the CO_2 percentage has its highest possible value of $33 \div (0 \cdot 69 + 0 \cdot 905) = 20 \cdot 7 \%$. Measurement of the CO_2 percentage in the exhaust gases is often used as a control on the excess-air quantity; the operator of the furnace tries to keep it at as high a value as possible consistent with the non-appearance of CO and H_2 in the exhaust. Values of 18% are common for solid fuels, and for gases such as producer gas which are directly derived from them. Liquid fuels and gaseous hydrocarbons give CO_2 percentages of around 15% in stoichiometric combustion.

(iii) If the H_2O in the products condenses, the total number of gaseous moles changes from $1 + y$ in the reactants to $0 \cdot 69 + y$ in the products during the combustion process, a decrease of $0 \cdot 31$ moles. If the H_2O does not condense, the final number of gaseous moles is $0 \cdot 81 + y$, representing a decrease of $0 \cdot 19$ moles. At constant temperature and pressure therefore, the volume of gas suffers a decrease during this reaction.

Air-fuel ratio deduced from the carbon-nitrogen ratio. If the exhaust-gas analysis is being used solely for the calculation of the air-fuel ratio, the quickest procedure is usually to deduce the latter from the ratio of carbon to nitrogen in exhaust. This will be illustrated by an example.

EXAMPLE 16.2

Problem. A fuel contains 84% carbon by mass and no nitrogen. The composition of the exhaust gas produced by combustion with air is: $14 \% \ CO_2$, $0 \cdot 5 \% \ CO$, $5 \% \ O_2$ and $80 \cdot 5 \% \ N_2$ by volume. Determine the air-fuel ratio x.

Solution. 1 mole of exhaust gas contains $0 \cdot 14$ moles of (monatomic) carbon in the CO_2, $0 \cdot 005$ moles of (monatomic) carbon in the CO, and $0 \cdot 805$ moles of (diatomic) nitrogen. The mass ratio of carbon to nitrogen in the exhaust gas is therefore

$$\frac{12(0 \cdot 14 + 0 \cdot 005)}{28 \times 0 \cdot 805} = 0 \cdot 077 \ 2$$

Since 1 kg of air contains $0 \cdot 768$ kg of nitrogen, the carbon-nitrogen ratio in the reactants is

$$\frac{0 \cdot 84}{0 \cdot 768x}$$

Since the ratios before and after combustion must be equal, we have

$$x = \frac{0 \cdot 84}{0 \cdot 768 \times 0 \cdot 772} = 14 \cdot 2 \qquad\qquad \textit{Answer}$$

Condensation of moisture. We conclude this section by showing how to determine the phase of the H_2O present in the exhaust gases. In practice, it is undesirable for condensation to occur within the heating equipment; for oxides of sulphur, even if present in only small concentration in the gas, can dissolve in the droplets of water and seriously corrode metal surfaces.

EXAMPLE 16.3

Problem. A furnace is supplied with the producer gas of Example 16.1, together with 1·2 cubic metres of air per cubic metre of gas. What is the minimum temperature allowable for any surface in contact with the flue gas, if condensation is not to occur when the gas pressure is 98×10^3 N/m²?

Solution. We have to determine the dew-point of the gases. From the previous calculation (pp. 376, 377), 1 mole of fuel produces

$$0\cdot33 \text{ moles } CO_2$$

$$0\cdot12 \text{ moles } H_2O$$

$$0\cdot21 \times 1\cdot2 - 0\cdot19 = 0\cdot062 \text{ moles } O_2$$

and

$$0\cdot55 + 0\cdot79 \times 1\cdot2 = 1\cdot499 \text{ moles } N_2$$

$$\text{Total} \qquad 2\cdot011 \text{ moles}$$

If the steam behaves as an Ideal Gas, its partial pressure is therefore $98 \times 10^3 \times \dfrac{0\cdot12}{2\cdot011}$ $= 5\cdot85 \times 10^3$ N/m². From Steam Tables Appendix B, the temperature at which this partial pressure is also the saturation pressure is found to be 35·7 °C. Therefore no part of the metal surface should fall below 35·7 °C.   *Answer*

The First Law of Thermodynamics Applied to Combustion

Energy and enthalpy

All systems and processes obey the First Law of Thermodynamics; energy and enthalpy are concepts therefore of universal application. When dealing with a pure substance, we found it convenient to give special symbols to these quantities; we follow the same practice now that chemical reaction has to be allowed for. First a special sort of system will be defined; *the chemical substance.* The analogy with the pure substance should be clear.

Definition. A system which is homogeneous in composition will be called a chemical substance.

As with the pure substance, chemical substances will be considered in the absence of effects of gravity, motion, capillarity, electricity and magnetism. Energy terms expressing the first two of these influences can be added when required, as before.

Internal energy. The internal energy of a chemical substance, i.e. its energy in the absence of the above effects, will be denoted by the symbol U'. For a system of unit mass, the lower case symbol u' is used. The units are J for U', and J/kg for u', respectively.

Enthalpy. The enthalpy of a chemical substance is given the symbol H' and defined by

$$H' \equiv U' + pV' \qquad (16.9)$$

where p and V' are respectively the pressure and volume of the system. The lower case symbol h' is used for systems of unit mass. The units are as for U' and u'.

Relations between the zeros. Energy is defined by the First Law only as a difference (eq. 6.5); the state of zero internal energy can therefore be chosen arbitrarily. This choice fixes the zero of enthalpy also. Another relation between zeros, peculiar to chemical substances, will now be discussed.

Dependence of U' and H' on chemical aggregation. Whereas all the properties of a pure substance, including its internal energy and enthalpy, are fixed by the specification of two independent properties, e.g. pressure and specific volume, this is not true of a chemical substance; the state of chemical aggregation must be specified as well. Consider, for example, a system comprising hydrogen and oxygen. Let the two elements be present at first as hydrogen and oxygen gases (the reactant state), and finally in the form of steam molecules (the product state). If the pressures and specific volumes are the same in both states, it follows from the Ideal-Gas Rule that the final temperature must exceed the initial temperature; for the number of molecules in the system is reduced. U' and H' also alter.

Consequently we are not free, when chemical reactions are in question, to choose independently the zeros of the internal energies of chemical substances which may be transformed into each other by reaction. This will become clearer from the analysis which follows.

Energy and enthalpy changes in isothermal reactions

The changes of internal energy and enthalpy which result from chemical reaction depend on the initial and final states of the system. Attention will first be concentrated on changes in which the initial and final temperatures are the same. When such changes are considered, it is to be noted that we do not require that the temperature should remain constant during the whole course of the reaction. It seldom will. We may allow the temperature to change temporarily provided that it returns to the initial value. For brevity, however, we speak of constant-temperature or isothermal reactions.

Temperature alone does not determine the state, so in addition we specify that either the final volume or the final pressure must be equal to its initial value.

Constant-volume isothermal reaction: $[\Delta U']_{v,t}$. If a system comprising mass m of a chemical substance changes its composition from a reactant to a product state, by way of a chemical reaction at constant temperature and *volume*, the internal energy of the system in the final (product) state, U'_P, is related to the internal energy in the initial (reactant) state, U'_R, through the First Law.

From eq. (6.5), in the absence of shear work, we have

$$Q = \Delta E$$
$$= U'_\mathrm{P} - U'_\mathrm{R}$$
$$= [\Delta U']_{v,t} \qquad (16.10)$$

where $[\Delta U']_{v,t}$ signifies the increase in the internal energy of the chemical system, and the subscripts v and t denote that the reaction has taken place at constant volume and constant temperature. The magnitude and sign of the quantity, $[\Delta U']_{v,t}$ are obtained experimentally by measuring the heat transfer Q.

Constant-pressure isothermal reaction: $[\Delta H']_{p,t}$. For a chemical reaction carried out at constant temperature and *pressure*, we have, from the First Law, eq. (6.5), that

$$Q - W = U'_\mathrm{P} - U'_\mathrm{R}$$

Here the work W, in the absence of shear work, is given by

$$W = \int p \, dV$$

and so, for a constant-pressure reaction:

$$W = p \int dV$$
$$= p(V'_\mathrm{P} - V'_\mathrm{R})$$

where $V'_\mathrm{P} \equiv$ volume of the chemical system in the final (product) state, and $V'_\mathrm{R} \equiv$ volume of the chemical system in the initial (reactant) state. It follows that

$$Q - p(V'_\mathrm{P} - V'_\mathrm{R}) = U'_\mathrm{P} - U'_\mathrm{R}$$

or
$$Q = (U'_\mathrm{P} + pV'_\mathrm{P}) - (U'_\mathrm{R} + pV'_\mathrm{R})$$
$$= H'_\mathrm{P} - H'_\mathrm{R} \qquad \text{from eq. (16.9)}$$
$$= [\Delta H']_{p,t} \qquad (16.11)$$

where $[\Delta H']_{p,t} \equiv$ increase in enthalpy of the chemical system in a constant-pressure, constant-temperature reaction.

In this case it is $[\Delta H']_{p,t}$ which is obtained experimentally by measuring the heat transfer Q.

Relation between $[\Delta U']_{v,t}$ *and* $[\Delta H']_{p,t}$. From the derivation of eq. (16.11), it is evident that normally the internal-energy increase and the enthalpy increase have different values. They are identical in magnitude only when the volume change $\Delta V'$, i.e. $(V'_P - V'_R)$, is zero. In general, as has been seen above, a change in the number of moles, or a change of phase, will cause $\Delta V'$ to be finite.

Reactions in Ideal Gases. When the reactants and products are Ideal Gases, the lack of dependence of their internal energies and enthalpies on volume and pressure makes it permissible to drop one of the subscripts; we therefore write $[\Delta U']_t$ and $[\Delta H']_t$ simply*. However, the first quantity is still equal to the heat transfer to the surroundings only in a constant-volume reaction, while the second equals the heat transfer only in a constant-pressure reaction or in a steady-flow process with zero external work. The last result follows from the Steady-Flow Energy Equation.

These conclusions also hold fairly closely when solid or liquid phases are present in reactant or product; for U' and H' for solids and liquids depend only mildly on pressure. The Gibbs-Dalton Law implies that $[\Delta U']_t$ and $[\Delta H']_t$ are not influenced by the extent to which individual components of the system are mixed. The *phase* of the components is important however, as will be shown below.

For reactions between Ideal Gases, an algebraic relation between $[\Delta U']_t$ and $[\Delta H']_t$ may be developed as follows. From the definition of enthalpy, eq. (16.9), we have

$$[\Delta H']_t = [\Delta U']_t + \Delta[pV']_t.$$

Now from the Ideal-Gas Rule, on a molar basis, we may write

$$p_R V'_R = n_R \mathscr{R} T$$

and

$$p_P V'_P = n_P \mathscr{R} T$$

where n_R and n_P are the numbers of moles of gaseous reactants and gaseous products respectively, and T is the absolute temperature at which the isothermal reaction is carried out.

Hence

$$\Delta[pV']_t = [p_P V'_P - p_R V'_R]_t$$

$$= (n_P - n_R)\mathscr{R} T$$

It follows that

$$[\Delta H']_t = [\Delta U']_t + (n_P - n_R)\mathscr{R} T \qquad (16.12)$$

When the change in the number of moles is zero, $[\Delta H']_t$ and $[\Delta U']_t$ are equal.

* These quantities are often given the symbols ΔU°_T and ΔH°_T respectively.

382   ENGINEERING THERMODYNAMICS

EXAMPLE 16.4

Problem. A system comprising 1 kmol of CO and 0·5 kmol of O_2 reacts to form CO_2. The initial and final temperatures of the system are 25 °C. During the reaction process, the measured heat transfer from the system is 283·177 × 10^6 J. Assuming the reactants and products to be Ideal Gases, evaluate: (a) $[\Delta H']_{25\,°C}$ and (b) $[\Delta U']_{25\,°C}$ for the given system.

Solution (a). Since the gases are ideal, $[\Delta H']_t$ will have the same value for any (low) pressure; we may therefore use eq. (16.11) to obtain it. We note that Q is given as equal to −283·177 × 10^6 J and hence

$$[\Delta H']_{25\,°C} = -283\cdot177 \times 10^6 \text{ J/kmol} \qquad \text{Answer (a)}$$

Solution (b). To evaluate $[\Delta U']_{25\,°C}$ we make use of eq. (16.12). In this equation $[\Delta H']_{25\,°C}$, $\mathscr{R}$ and T are known quantities; $(n_P - n_R)$ has to be obtained from a knowledge of the reaction:

The reaction is: $$CO + \tfrac{1}{2}O_2 = CO_2$$

So
$$n_P - n_R = 1 - (1 + 0\cdot5)$$
$$= -0\cdot5 \text{ kmol}$$

Hence
$$(n_P - n_R)\mathscr{R}T = (-0\cdot5) \times 8\cdot314 \times 10^3 \times (273 + 25)$$
$$= -1\cdot239 \times 10^6 \text{ J}$$

Finally, from eq. (16.12) we obtain:

$$-283\cdot177 \times 10^6 = [\Delta U']_{25\,°C} - 1\cdot239 \times 10^6$$

i.e.
$$[\Delta U']_{25\,°C} = -281\cdot938 \times 10^6 \text{ J/kmol} \qquad \text{Answer (b)}$$

Remarks: 1. The increase in U', $[\Delta U']_t$, and the increase in H', $[\Delta H']_t$, are negative, i.e. U' and H' *decrease.* This is invariably the case in isothermal combustion reactions, which are thus given the name *exothermic*, because the direction of the heat transfer is *outwards* (see eq. (16.10)).

2. The results of this problem could have been expressed in terms of unit mass of either of the reactants; usually the "fuel" is chosen. Thus in the present case we can express $[\Delta U']_{25\,°C}$ as $(-281\cdot938 \times 10^6)/28$, i.e. $-10\cdot069 \times 10^6$ J/kg CO.

3. There is a reduction in the number of moles in this case; $(n_P - n_R)$ is negative. Note that n_P and n_R are the numbers of *gaseous* moles present. If any portions of the reactants or products are present as solids or liquids, they are ignored in computing $(n_P - n_R)$. For example, in the reaction

$$C + O_2 = CO_2$$

in which the carbon is present as a solid reactant, n_R equals 1, and n_P equals 1. It follows from eq. (16.12) that for this case $[\Delta H']_t$ equals $[\Delta U']_t$.

Final remark on the relation between the zeros. If the internal energy of a reactant system is arbitrarily defined to be zero, and then a constant-temperature, constant-volume reaction ensues, it is now clear that the internal energy of the resultant product system cannot be arbitrarily defined. The reason is that a definite heat transfer occurs in the process, so that the final internal energy is fixed; it is $[\Delta U']_{v,t}$, according to eq. (16.10). The enthalpies are determined, by reason of the definition, eq. (16.9).

Dependence of $[\Delta U']_{v,t}$ and $[\Delta H']_{p,t}$ on temperature and phase

The effect of temperature. The increases in internal energy and enthalpy associated with a given change of chemical state depend on the condition of the experiment, particularly the temperature. This may be illustrated by plotting the internal energy of a substance versus temperature for a fixed volume. (For Ideal Gases it does not matter what volume is chosen, of course.)

Fig. 16.1 shows such a plot for a system which consists initially of hydrogen and oxygen gases mixed in stoichiometric proportions. The

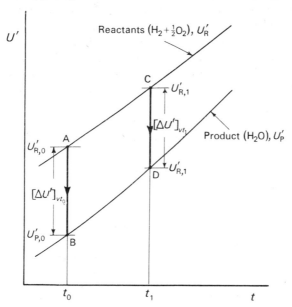

Fig. 16.1 The internal energy~temperature diagram
for the reaction $H_2 + \frac{1}{2}O_2 = H_2O$.

volume is supposed sufficiently large (and therefore the pressure sufficiently small) for all three substances, H_2, O_2 and H_2O, to act as Ideal Gases.

The upper curve gives the internal energy for the reactant gas (n moles of H_2 gas plus $n/2$ moles of O_2 gas); the lower gives that of the product (n moles of steam). It will be seen that the vertical spacing between them varies, being in fact less at the higher temperature than at the lower. Now this vertical distance measures the negative of $[\Delta U']_{v,t}$; for an isothermal reaction corresponds to a shift from the upper to the lower curve along a line of constant temperature. The distance changes with temperature because the c_v of the products differs from the c_v of reactants, c_v being the local slope of the internal-energy curve; there is, after all, no reason why the two c_v's should be the same. Only, if by chance, the c_v of the products were the same function of temperature as that of the reactants would the increase in internal energy be a constant at all temperatures.

Standard values of $[\Delta U']_{v,t}$ *and* $[\Delta H']_{p,t}$. It might be thought that the variation of $[\Delta U']_{v,t}$ with temperature would make it necessary for values for various temperatures to be tabulated in data books. Further thought shows however that, provided that data are available for the internal energies of the pure reactants and the pure products, a single value of $[\Delta U']_{v,t}$ will suffice.

Suppose that a measurement is made of the increase in the internal energy in a constant-volume reaction at a fixed temperature, t_0. For example, for the hydrogen-oxygen reaction discussed above, when 1 kmol of H_2 gas and 0·5 kmol of O_2 gas react at constant volume, at 25 °C, to form 1 kmol of steam, the value of $[\Delta U']_{v,t}$ may be measured and found to be equal to $-241·988 \times 10^6$ J. Then the two points A and B on Fig. 16.1 can be plotted, above same arbitrary base-line; the reactant state point A is above the corresponding product state point B because $[\Delta U']_{v,t}$ is a negative quantity. Internal-energy data for the pure gases H_2, O_2 and H_2O will then enable the two curves passing through A and B to be drawn. The increase in internal energy at any other temperature, say t_1, is then given as the intercept CD which these curves make on the vertical corresponding to t_1.

Normally it is not necessary to plot the curves. Instead an algebraic equivalent is used; thus, if number subscripts are used to denote temperatures, we may write:

$$(U'_{P,1} - U'_{R,1}) = (U'_{P,0} - U'_{R,0}) + (U'_{P,1} - U'_{P,0}) - (U'_{R,1} - U'_{R,0})$$

i.e. $$[\Delta U']_{v,t_1} = [\Delta U']_{v,t_0} + (U'_{P,1} - U'_{P,0}) - (U'_{R,1} - U'_{R,0})$$

$$(16.13)$$

In this equation, the terms in the last two pairs of brackets are respectively the internal-energy increases of the pure products and pure reactants for the temperature range $t_1 - t_0$; they are evaluated by reference to pure-substance property data. The truth of eq. (16.13) will be recognized by studying its geometrical equivalent, namely

$$\left.\frac{C}{D}\right| = \left.\frac{A}{B}\right| + \left.\frac{C}{A}\right| - \left.\frac{D}{B}\right|$$

where $\left.\dfrac{X}{Y}\right|$ is a symbol used to denote "vertical height of X above Y". A

sketch of Fig. 16.1 may be found to assist the correct ascription of positive and negative signs to the terms of eq. (16.13).

The enthalpy changes in constant-pressure isothermal reactions are similarly related as follows:

$$(H'_{P,1} - H'_{R,1}) = (H'_{P,0} - H'_{R,0}) + (H'_{P,1} - H'_{P,0}) - (H'_{R,1} - H'_{R,0})$$

i.e. $$[\Delta H']_{p,t_1} = [\Delta H']_{p,t_0} + (H'_{P,1} - H'_{P,0}) - (H'_{R,1} - H'_{R,0})$$

The value of $[\Delta U']_{v,t}$ at a standard temperature, t_0, often 25 °C (77 °F), is known as the *standard increase in internal energy at constant volume*. Correspondingly, $[\Delta H']_{v,t}$ at the standard temperature is known as the *standard increase in enthalpy at constant pressure*. In practice, it is customary to express these standard values in terms of a unit amount of the "fuel" component of the reactants. For example, in the hydrogen-oxygen reaction mentioned above, the H_2 gas in the reactants would be regarded as the fuel and $[\Delta U']_{25\,°C}$ would be given in combustion-data tables as either $-241\cdot988 \times 10^6$ J/kmol of H_2 gas or as $-120\cdot994 \times 10^6$ J/kg of H_2 gas; the latter figure of course equals the former divided by the molecular mass of hydrogen. It must be emphasized, however, that $[\Delta H']_{p,t}$ and $[\Delta U']_{v,t}$ relate to a *system* changing from a given reactant state to a known product state and not merely to a single substance. Values of $[\Delta H']_{p,t}$ and $[\Delta U']_{v,t}$ tabulated in data books are usually quoted for low pressure, i.e. large volume, under which conditions the substances may be taken as Ideal Gases. Correction for a change in pressure may be made, in the same way as for temperature, if $U'{\sim}p$ and $H'{\sim}p$ data are available for the substances in question; the correction is often negligible.

The First Law applied to non-isothermal reactions. Diagrams such as Fig. 16.1 are useful for illustrating the application of the First Law to non-isothermal processes also. We consider, as an example, the application of the Steady-Flow Energy Equation to a flow process in which a combustible

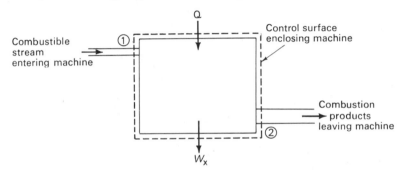

Fig. 16.2 Illustrating a steady-flow combustion process.

stream (reactants) of state 1 enters a machine while combustion products at state 2 flow out (Fig. 16.2). How are the heat transfer and shaft work related to the change of state?

Fig. 16.3 shows an $H'{-}t$ diagram for the reactant and product streams in question, which will be taken as mixtures of Ideal Gases so that no statement about pressure need be made. The S.F.E.E., eq. (8.43) gives

$$\frac{1}{m}\left[Q - W_x\right] - \Delta\left(\frac{V^2}{2g_c} + \frac{g}{g_c}z\right) = h'_{P,2} - h'_{R,1}$$

where
$$m = m_R$$
$$= m_P$$
$$h'_{P,2} = H'_{P,2}/m_P$$

and
$$h'_{R,1} = H'_{R,1}/m_R.$$

It follows that the above form of the S.F.E.E. may be written as:

$$Q - W_x - m\left[\Delta\left(\frac{V^2}{2g_c} + \frac{g}{g_c}z\right)\right] = H'_{P,2} - H'_{R,1} \qquad (16.14)$$

Let us suppose the quantities on the left-hand side of eq. (16.14) to be given and also the enthalpy of the reactant stream; the final state has to be evaluated. Knowledge of the initial state, of the enthalpy-temperature

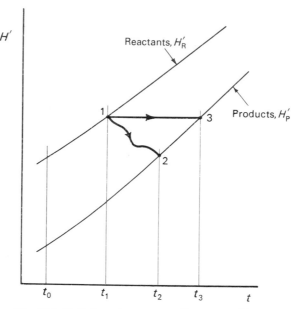

Fig. 16.3 Enthalpy~temperature diagram used in the discussion of steady-flow combustion processes.

relations of the substances and of the constant-pressure increase in enthalpy at temperature t_0, $[\Delta H']_{t_0}$, permits evaluation with the aid of an equation derived similarly to eq. (16.13); it is

$$H'_{P,2} - H'_{R,1} = (H'_{P,2} - H'_{P,0}) - (H'_{R,1} - H'_{R,0}) + (H'_{P,0} - H'_{R,0})$$

$$= (H'_{P,2} - H'_{P,0}) - (H'_{R,1} - H'_{R,0}) + [\Delta H']_{t_0}$$

$$(16.15)$$

The values of the first two brackets on the right-hand side are obtained directly from the enthalpy~temperature tables for the pure substances comprising the products and reactants, by way of the relations:

$$H'_{P,2} - H'_{P,0} = \sum_{products} m\,(h_2 - h_0)$$

and

$$H'_{R,1} - H'_{R,0} = \sum_{reactants} m\,(h_1 - h_0)$$

The term $[\Delta H']_{t_0}$ in eq. (16.15) is, of course, the standard increase in enthalpy for the reaction in question; it has to be obtained from tables.

Temperature rise in constant-enthalpy combustion. A case of particular interest is that in which the enthalpy change is zero. This arises in adiabatic steady-flow reaction in the absence of shaft work and of significant changes in kinetic and gravitational energy; a well-lagged gas-turbine combustion chamber is an example. Since H' does not change, the product state resulting from the initial state 1 must correspond to the point 3 in Fig. 16.3. The diagram therefore affords a simple means for the determination of the temperature rise.

In the absence of an accurately plotted $H' \sim t$ diagram, t_3 is calculated from eq. (16.15). With $H'_{P,2}$ replaced by $H'_{P,3}$ and with $H'_{P,3}$ put equal to $H'_{R,1}$, the useful form of this equation is

$$(H'_{P,3} - H'_{P,0}) = (H'_{R,1} - H'_{R,0}) - [\Delta H']_{t_0} \qquad (16.16)$$

The first term on the right-hand side is evaluated from enthalpy~temperature data for the pure reactants; the second term is evaluated from tables of combustion data giving standard increases in enthalpy; the bracket on the left-hand side is then used for the determination of t_3 from enthalpy~temperature data for the pure products.

Three alternative forms of eq. (16.16), in terms of specific heats, will now be given without comment. Their equivalence may be perceived by study of Fig. 16.3. They are:

$$\left(\sum_P m \bar{c}_{p_{03}}\right)(t_3 - t_0) = \left(\sum_R m \bar{c}_{p_{01}}\right)(t_1 - t_0) - [\Delta H']_{t_0} \qquad (16.17)$$

$$\left(\sum_P m \bar{c}_{p_{13}}\right)(t_3 - t_1) = -[\Delta H']_{t_1} \qquad (16.18)$$

$$0 = \left(\sum_R m \bar{c}_{p_{13}}\right)(t_1 - t_3) - [\Delta H']_{t_3} \qquad (16.19)$$

These equations are useful when mean-specific-heat tables are available instead of enthalpy tables. Since $\bar{c}_p$ depends on the temperature range (see p. 301), which is not known in each case before the calculation starts, a trial-and-error procedure has to be used.

It will be evident from comparison of eq. (16.18) and eq. (16.19) that, in the particular case in which the specific heats of products and reactants are equal, $[\Delta H']_t$ is independent of temperature.

The effects of the phase of the substances. So far we have assumed that the reactants and products are in the gaseous phase. However, both liquid and solid fuels have to be considered; and at least one product of common combustion reactions, H_2O, is easily condensable. The effects of the pressure of non-gaseous phases will be illustrated by two examples.

First, we consider once more the $U' \sim t$ diagram for the H_2-O_2-H_2O system; we shall assume that the reactants comprise n mole of H_2 gas plus

$n/2$ mole of O_2 gas, and that, correspondingly, the product will be n mole of H_2O. Fig. 16.4 shows the same curves as Fig. 16.1, but an additional curve has been added, namely that for the products in the liquid (water) phase. This curve lies below that for steam by a distance proportional to the quantity denoted earlier by u_{fg} (p. 185), multiplied by the molecular mass

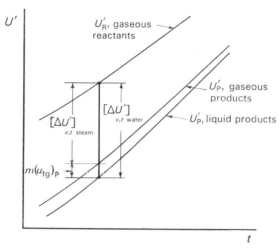

Fig. 16.4 Showing the effect of the phase of the products on $[\Delta U']_{v,t}$.

of steam. This is $\tilde{u}_{fg}$, the internal-energy increase in vaporization per mole of steam; it varies, of course, with temperature.*

Fig. 16.4 shows that the negative of $[\Delta U']_t$, i.e. the vertical distance between the reactant and product curves, is greater when the product is condensed than when it is not. The relations are:

$$- [\Delta U']_{t_{water}} = - [\Delta U']_{t_{steam}} + n\tilde{u}_{fg}$$

i.e.

$$[\Delta U']_{t_{water}} = [\Delta U']_{t_{steam}} - n\tilde{u}_{fg}$$

and

$$[\Delta U']_{t_{water}} = [\Delta U']_{t_{steam}} - mu_{fg}$$

where n is the number of moles of H_2O in the system and m is its mass. Calculations are carried out with the aid of Fig. 16.4, or of the corresponding tabulated data, in the same way as above.

Corresponding relations, in terms of enthalpy, may also be deduced. These are:

$$[\Delta H']_{t_{water}} = [\Delta H']_{t_{steam}} - n\tilde{h}_{fg}$$

$$[\Delta H']_{t_{water}} = [\Delta H']_{t_{steam}} - mh_{fg} \qquad (16.20)$$

where $\tilde{h}_{fg}$ is the enthalpy of vaporization of steam per mole.

* This variation implies that, when we are concerned with changes of phase in the products, the *pressure* of the latter has to enter our considerations; previously, when we were dealing with ideal-gas-product mixtures only, the product pressure could be ignored.

Secondly, we consider the $H'{\sim}t$ diagram of a system in which the reactants may be present in liquid form as well as the products. Fig. 16.5 might represent the enthalpy~temperature relations for a mixture of a condensable hydrocarbon with air. The upper reactant curve is valid if the hydrocarbon is in the vapour phase; the lower line holds if it is liquid. The two product curves correspond to the two possible phases of the H_2O produced. The vertical distance between the alternative curves is equal to

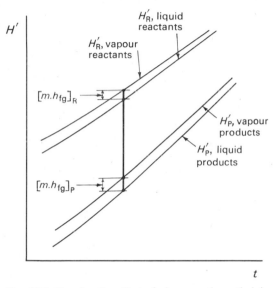

Fig. 16.5 Showing the effect of phase on the enthalpies of the reactants and products.

the enthalpy of evaporation of the condensable substance, multiplied by the proportion which that substance comprises of the mass of the system. It is evident that, in stating the values of a $[\Delta U']_t$ and $[\Delta H']_t$, the phase of both products and reactants must be completely specified.

Enthalpy of formation

In the last section it was shown how the necessity to tabulate the increases in internal energy and enthalpy of systems for various temperatures may be avoided, by reference to the values of these quantities at a standard temperature. We now discuss a similar device for avoiding tabulation of the standard increases in internal energy and enthalpy of all the multitudinous reactions which may actually occur. These matters involve the consideration of one reaction only for *each species*, together with the specification of arbitrary states of the elements to which are assigned zero values of enthalpy.

For each species, the reaction considered is one under specified standard conditions (usually 25 °C and 1 atm) in which the species is formed from its constituent chemical elements; it is a particular sort of standard reaction of the type mentioned earlier on p. 383.

Examples of datum states of the elements to which are assigned zero values of enthalpy are:

$$\text{Oxygen: diatomic* gas } (O_2) \text{ at } 25\,°C, 1 \text{ atm.}$$

$$\text{Hydrogen: diatomic* gas } (H_2) \text{ at } 25\,°C, 1 \text{ atm.}$$

$$\text{Carbon: graphite, solid } (C) \text{ at } 25\,°C, 1 \text{ atm.}$$

The data pertaining to such reactions are presented as *enthalpies of formation* which we define as follows:

The enthalpy of formation of a chemical species is the enthalpy of the product in a constant-pressure, isothermal reaction, in which the species is the only product and the reactants are the chemical elements in their datum states.

We use the symbol $\tilde{h}_f'$ for enthalpy of formation when this quantity is expressed on a molar basis and h_f' when it is given in terms of unit mass of material.† Some values of $\tilde{h}_f'$ are given in Table 16.6.

TABLE 16.6 *Enthalpy of Formation*

Species	Reaction	State	$\tilde{h}'_f$ J/kmol of species
O_2	—	gas at 25 °C, 1 atm	0, by definition
H_2	—	gas at 25 °C, 1 atm	0, by definition
C	—	solid at 25 °C, 1 atm	0, by definition
CO_2	$C + O_2 = CO_2$	gas at 25 °C, 1 atm	$-393 \cdot 776 \times 10^6$
CO	$C + \frac{1}{2}O_2 = CO$	gas at 25 °C, 1 atm	$-110 \cdot 599 \times 10^6$
H_2O	$H_2 + \frac{1}{2}O_2 = H_2O$	gas at 25 °C, 1 atm	$-241 \cdot 988 \times 10^6$
H_2O	$H_2 + \frac{1}{2}O_2 = H_2O$	liquid at 25 °C, 1 atm	$-286 \cdot 030 \times 10^6$
CH_4	$C + 2H_2 = CH_4$	gas at 25 °C, 1 atm	$-\ 74 \cdot 897 \times 10^6$
C_2H_6	$2C + 3H_2 = C_2H_6$	gas at 25 °C, 1 atm	$-\ 84 \cdot 725 \times 10^6$

The meaning of the definition and of the data given in Table 16.6 will be explained by reference to the reaction in which 1 mole of CO is produced from the reaction of carbon with oxygen:

$$C_{\text{solid}} + \tfrac{1}{2}O_{2\text{gas}} = CO_{\text{gas}}$$

Table 16.6 gives $\tilde{h}_f'$ for CO as $-110 \cdot 599 \times 10^6$ J/kmol CO.

For any reaction, we have the definition

$$[\Delta H']_{p,t} = H_P' - H_R'$$

* The monatomic gases at the same temperatures and pressures will have corresponding positive values of enthalpy of formation.

† Commonly the symbol $\Delta \tilde{h}_f'$ is used instead of $\tilde{h}_f'$

Now, since the reactants are chemical elements in their datum states, H'_R equals zero; on the other hand H'_P equals $\tilde{h}'_f$ for CO multiplied by 1 kmol.

Hence $\qquad [\Delta H']_{p,t} = -110 \cdot 599 \times 10^6 \text{ J}$

By writing the enthalpy of the product, H'_P, as $\tilde{h}'_f$, we emphasize the special nature and conditions of the reaction. If the zero values of enthalpy are not introduced, the data for the reaction would be presented as:

$$\text{Standard } [\Delta H']_{p,t} = -110 \cdot 599 \times 10^6 \text{ J}$$
$$= -110 \cdot 599 \times 10^6 \text{ J/kmol CO}$$
$$= -110 \cdot 599 \times 10^6 \text{ J/kmol C}$$

The enthalpy-of-formation concept, with its associated zero values of enthalpy, simplifies the calculations of $[\Delta H']_{p,t}$ for more complicated reactions. Example 16.5 below illustrates this.

Internal energy of formation. A procedure similar to that given above for enthalpy of formation could be adopted for internal energy of formation. However, since for Ideal Gases $[\Delta U']_t$ can be obtained from $[\Delta H']_t$ via eq. (16.12), it suffices to tabulate enthalpy of formation only.

Deduction of increase in enthalpy of reaction from enthalpy of formation.
The use of data for enthalpy of formation to obtain the standard increase in enthalpy in any reaction will now be illustrated by an example. The technique is to ascribe the appropriate values of enthalpy of formation to each of the reactants and products, to obtain H'_R and H'_P; the difference of these quantities then gives $[\Delta H']_{p,t}$.

EXAMPLE 16.5

Problem Steam reacts with solid carbon to form carbon monoxide and hydrogen (water-gas). Determine the increase in enthalpy at 25 °C and 1 atm.
Solution. The chemical equation is

$$C_{\text{solid}} + H_2O_{\text{gas}} = CO_{\text{gas}} + H_2{\text{ gas}}$$

The reactants (on the left-hand side of the equation) have enthalpy H'_R, given by

$$H'_R = n_C \cdot \tilde{h}'_{f,} + n_{\text{steam}} \cdot \tilde{h}'_{f \text{ steam}}$$

Inserting the values of $\tilde{h}'_f$ from Table 16.6, we obtain

$$H'_R = (1 \text{ kmol C}) \times (0) + (1 \text{ kmol H}_2O) \times (-241 \cdot 988 \times 10^6 \text{ J/kmol H}_2O)$$
$$= -241 \cdot 988 \times 10^6 \text{ J}.$$

Similarly, $H'_P = n_{CO} \cdot \tilde{h}'_{f,CO} + n_{H_2} \cdot \tilde{h}'_{f,H_2}$

$$= (1 \text{ kmol CO}) \times (-110 \cdot 599 \times 10^6 \text{ J/kmol CO}) + (1 \text{ kmol H}_2) \times (0)$$

$\therefore [\Delta H']_{p,t} = H'_P - H'_R$

$$= -110 \cdot 599 \times 10^6 - (-241 \cdot 988 \times 10^6)$$
$$= +131 \cdot 389 \times 10^6 \text{ J} \qquad\qquad\qquad \textit{Answer}$$

Comment. The reaction is endothermic; heat transfer to the system from the surroundings must occur. This may be seen from eq. (16.11), which states

$$Q = [\Delta H']_{p,t}$$

$$= +131 \cdot 389 \times 10^6 \text{ J in this case.}$$

Heat of reaction; heat of formation

So far we have considered the internal-energy and enthalpy changes which result from reaction, in a system comprising a chemical substance. The application of the First Law to such systems led to eq. (16.10) and eq. (16.11); these state how $[\Delta U']_{v,t}$ and $[\Delta H']_{p,t}$ are related to Q, the heat transfer *to the system*. We now consider the effects which the reaction in the system has *on the surroundings*.

Heat of reaction. In a constant-volume, isothermal reaction, the heat transfer *to the surroundings*, $-Q$, is, from eq. (16.10), equal to the *negative* of the increase in the internal energy of the system, $-[\Delta U']_{v,t}$; $-[\Delta U']_{v,t}$ is called the *constant-volume heat of reaction*. Similarly, for an isothermal reaction at constant pressure, the *negative* of the increase in enthalpy, $-[\Delta H']_{p,t}$, is called the *constant-pressure heat of reaction*. Clearly, since heats of reaction express the heat transfer to the surroundings of a chemically reacting system, they are of great interest in heating applications.

Standard heats of reaction. It will be recalled (p. 383) that in order to limit the amount of tabulated data on chemical reaction, standard values of $[\Delta U']_{v,t}$ and $[\Delta H']_{p,t}$ were used. Correspondingly, the negatives of these quantities become standard heats of reactions: the value of $-[\Delta U']_{v,t}$ at a standard temperature, often 25 °C (77 °F) is known as the *standard heat of reaction at constant volume*; $-[\Delta H']_{p,t}$ at the standard temperature is known as the *standard heat of reaction at constant pressure*. These standard heats of reaction may be regarded respectively as alternatives to the standard values of $[\Delta U']_{v,t}$ and $[\Delta H']_{p,t}$ for the presentation of tabulated data on chemical reaction; the corresponding quantities are equal in magnitude but of opposite sign.

Heat of formation. In parallel with the above pattern, the heat of formation of a substance is the negative of its enthalpy of formation, $-\tilde{h}'_f$. Again it is to be regarded as an alternative way of presenting data for reacting substances.

EXAMPLE 16.6

Problem. In the reaction $C + O_2 = CO_2$, the enthalpy of formation of CO_2 at 25 °C (77 °F) and 1 atm is $-393 \cdot 776 \times 10^6$ J/kmol (from Table 16.6).

Evaluate (*a*) the standard heat of reaction at constant pressure, i.e. at 25 °C and 1 atm;

(*b*) the heat of reaction when the reactants are at 50 °C and the products at 500 °C.

Take the following values of mean specific heat, $\bar{c}_p$: C, 833; O_2, 984; CO_2, 1025 J/kg K. The gaseous reactants and products may be assumed to be Ideal Gases.

Solution (*a*). We are given that

$$h_f' = -393.776 \times 10^6 \text{ J/kmol CO}_2$$

Therefore, for a chemical system comprising 1 kmol of C plus 1 kmol of O_2, the increase in enthalpy of reaction at 1 atm, 25 °C is given by

$$[\Delta H']_{25 \text{ °C}} = -393.776 \times 10^6 \text{ J}$$

since h_f' assumes zero values for the enthalpy of C and O_2 at 25 °C.

Hence the standard heat of reaction at constant pressure is given by:

$$-[\Delta H']_{25 \text{ °C}} = +393.776 \times 10^6 \text{ J}$$

i.e.
$$= 393.776 \times 10^6 \text{ J/kmol CO}_2$$

or
$$= 393.776 \times 10^6 \text{ J/kmol C}$$

or
$$= \frac{393.776 \times 10^6}{12} = 32.815 \times 10^6 \text{ J/kg C} \quad \textit{Answer (a)}$$

Solution (*b*)

Using subscripts 1 and 2 to denote the temperatures 50 °C and 500 °C respectively, and subscript 0 to denote the standard temperature, we have, from eq. (16.15), for the $C-O_2-CO_2$ system:

$$H_{P,2}' - H_{R,1}' = (H_{P,0}' - H_{R,0}') + (H_{P,2}' - H_{P,0}') - (H_{R,1}' - H_{R,0}')$$

Now

$$(H_{P,0}' - H_{R,0}') = [\Delta H']_t = -393.776 \times 10^6 \text{ J}$$

$$(H_{P,2}' - H_{P,0}') = m\bar{c}_p(t_2 - t_0) \quad \text{for the only product, namely, CO}_2.$$

$$= 44 \times 1025 \times (500 - 25) = 21.423 \times 10^6 \text{ J}$$

$$(H_{R,1}' - H_{R,0}') = [m\bar{c}_p(t_1 - t_0)]_C + [m\bar{c}_p(t_1 - t_0)]_{O_2}$$

$$= 12 \times 833 \times (50 - 25) + 32 \times 984 \times (50 - 25)$$

$$= 1.037 \times 10^6 \text{ J}.$$

Hence

$$H_{P,2}' - H_{R,1}' = (-393.776 + 21.423 - 1.037) \times 10^6$$

$$= -373.390 \times 10^6 \text{ J}$$

Finally, since the heat of reaction is equal to the heat transfer *to* the surroundings, and is therefore equal to the *negative* of the enthalpy increase of the *system*, we have

$$\text{Heat of reaction} = +373.390 \times 10^6 \text{ J}. \quad \textit{Answer (b)}$$

Calorific value

"Heat of reaction" and "heat of formation" are terms originating in chemical thermodynamics. The mechanical engineer who is more exclusively concerned with combustion reactions uses a different terminology. We begin this section with a definition:

The calorific value of a fuel is the standard heat of reaction at constant pressure of the reaction in which the fuel burns completely with oxygen.

By "completely" is meant that all the hydrogen in the products is contained in H_2O molecules and all the carbon in CO_2 molecules; if sulphur is present in the fuel, the appropriate oxide is the dioxide, SO_2. Other oxidizable elements, such as vanadium, are normally present in such small quantities as to make it immaterial what oxides are formed.

The calorific value, also called *heating value*, is expressed as J/kg of fuel, J/kmol of fuel, or J/standard m^3 of fuel according to convenience. Reference to a unit of fuel makes it unnecessary to know how much air happens to be mixed with the fuel, though of course this must exceed the stoichiometric quantity if combustion is to be complete. It is likewise immaterial whether air or pure oxygen is the oxidant.

In accordance with the foregoing analysis, the calorific value of a fuel, $\overline{CV}$, is related to the increase in enthalpy in the reaction, $[\Delta H']_{p,t}$ by the definition:

$$\overline{CV} \equiv -[\Delta H']_{p,t}$$

Higher and lower calorific values. It has been seen above that the phases of the reactants and products influence the magnitudes of the internal-energy and enthalpy increases. It follows, therefore, that the calorific value of a fuel will be also so influenced. The terms *higher calorific value*, $\overline{HCV}$, and *lower calorific value*, $\overline{LCV}$, are used, respectively, to distinguish the cases in which any H_2O formed is in the liquid or the gaseous phase. The same distinction is sometimes rendered by the terms "gross" and "net". The two calorific values are related, through eq. (16.20), as follows:

$$\overline{HCV} = \overline{LCV} + mh_{fg}$$

where m is the mass of H_2O produced per unit mass of fuel and h_{fg} is the enthalpy of evaporation of water at the standard temperature. In practice h_{fg} is taken as 2442×10^3 J/kg although it actually depends somewhat on the pressure and the temperature.

The phase of the fuel, if in question, must be stated separately.

Some typical calorific values of fuels are given in Table 16.7.

Mixtures of fuels. It is a consequence of the Gibbs-Dalton Law that the calorific value of a mixed fuel can be determined by simple addition. We give as an example the calculation of the calorific value of the producer gas of example 16.1 on p. 376.

EXAMPLE 16.7

Problem. Determine the higher calorific value of a producer gas with volume analysis: 26% CO, 12% H_2, 7% CO_2 and 55% N_2.

Solution. Using the data in Table 16.7, we draw up the following table:

Gas	$\overline{HCV}$ of pure gas MJ/m³ at 15 °C, 1 atm	Volume fraction	Contribution to $\overline{HCV}$ of mixture. MJ/m³ at 15 °C, 1 atm.
CO	11·84	0·26	0·26 × 11·84 = 3·08
H_2	11·92	0·12	0·12 × 11·92 = 1·43
		Total	= 4·51

The answer is therefore: 4·51 MJ/m³ at 15 °C, 1 atm. The CO_2 and N_2 of course make no contribution to the calorific value of the mixture.

TABLE 16.7 *Typical Calorific Values of Fuels*

Solid fuels

Fuel	Calorific value, MJ/kg (Bomb calorimeter, 15 °C)	
	Higher	Lower
Anthracite	34·583	33·913
Bituminous coal	33·494	32·406
Lignite	21·646	20·390
Peat	15·910	14·486
Wood	15·826	14·319
Coke	30·731	30·480

Liquid fuels

Fuel	Calorific value, MJ/kg (Bomb calorimeter, 15 °C)	
	Higher	Lower
Aviation gasoline (100 octane)	47·311	44·003
Motor gasoline	46·892	43·710
Vaporizing oil	46·055	43·210
Motor benzole	41·973	40·193
Kerosine	46·180	43·166
Diesel oil, distillate (Gas oil)	45·971	43·166
Light fuel oil	44·799	42·077
Heavy fuel oil	43·961	41·366
Residual fuel oil (Bunker C)	42·054	39·961

Gaseous fuels

Fuel	Calorific value MJ/m³ at 15 °C, 1 atm	
	Higher	Lower
Coal gas (Town gas)	20·11	17·95
Producer gas	6·06	6·02
Blast-furnace gas	3·44	3·40
Natural gas (U.K.)	36·38	32·75
Carbon monoxide gas, CO	11·84	11·84
Hydrogen gas, H_2	11·92	10·05

Measurement of calorific values. The engineer needs to have simple means of measuring the calorific values of the fuels with which he is supplied. *Calorimeters* are used for this purpose. They are of two types: *steady-flow* calorimeters, and *bomb* (or constant-volume) calorimeters. The first type is used for measurements on gaseous fuels, the second for liquid and solid fuels.

Fig. 16.6 shows a gas calorimeter. The entering fuel is saturated with water vapour; it is mixed with sufficient air to ensure complete combustion and the condensation of most of the steam in the exhaust gases. Since the process is one of steady flow at low velocity, in the absence of external work, the enthalpy decrease of the gas stream is equal to the heat transferred to the surroundings; the latter is measured by the temperature rise of the cooling water. The gas calorimeter therefore measures the higher calorific value of the fuel gas; the lower calorific value can then be calculated, from the measured quantity of steam condensed in the calorimeter and the enthalpy of evaporation of steam.

Steady-flow techniques are more difficult to devise for liquid and solid fuels; it is easier to carry out experiments at *constant volume*. Fig. 16.7 shows a bomb calorimeter used for solid and liquid fuels. A sample of fuel is placed in the bomb, into which pure oxygen is then forced under pressure. Ignition is effected by an electric current passed through a wire in contact with the fuel. When combustion is complete, the temperature rise of the bomb and its surrounding water bath is measured; since this temperature rise amounts only to a few degrees, the reaction can be regarded as isothermal, although a correction can be applied if great accuracy is required.

Two other corrections are needed if the bomb calorimeter is to yield the calorific value of the fuel. First we note that, since the experiment is at constant volume rather than constant pressure, it measures $-[\Delta U']_{v,t}$ rather than $-[\Delta H']_{p,t}$; these quantities differ because of the change in the number of gaseous moles during the reaction; eq. (16.12) enables $-[\Delta H']_{p,t}$ to be deduced from the measured $-[\Delta U']_{v,t}$. Secondly, the conditions of the experiment are usually such that most of the steam condenses; the measurement therefore gives more nearly the higher calorific value than the lower. However a small but calculable proportion of the steam remains in the vapour phase, so a small correction has to be applied. Often the experimental accuracy, or the use to which the data are to be put, renders it unnecessary to apply these corrections; then the heat transfer from the fuel and oxygen to their surroundings is simply taken as the higher calorific value of the fuel.

Standard methods of measuring calorific values have been laid down by the British Standards Institution and the American Society for Testing Materials (see bibliography).

The efficiency of combustion appliances

The discussion of the application of the First Law of Thermodynamics to combustion processes will be concluded by mention of two ways in which the performance of combustion equipment may be compared with

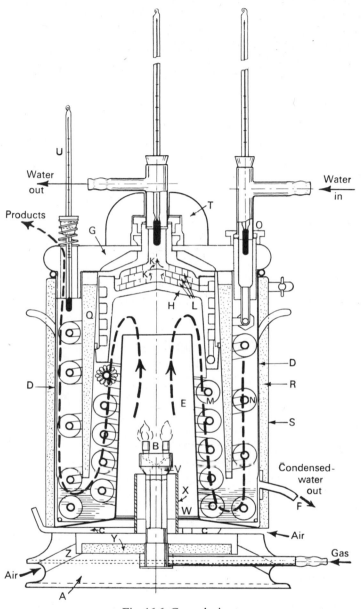

Fig. 16.6 Gas calorimeter.

A. Wooden base
B. Gas burners
C. Metal plate
D. Metal vessel
E. Copper chimney
F. Condensed-water outlet

G. Wooden lid
H. Mixing chamber
K. Dished plates
M. Inner coil
N. Outer coil
O. Inlet-water box
Q. Baffle

R. Felt jacket
S. Polished sheet metal
T. Wooden shield
W. Ferrule
X. Fibre tube
Y. Felt disc
Z. Air holes in base

an absolute standard. These involve the concepts of combustion efficiency, η_c, and heating efficiency, η_h.

Combustion efficiency. In the foregoing sections we have considered reactions which go to completion, i.e. in which all the carbon and hydrogen

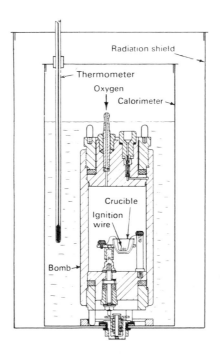

Fig. 16.7 Bomb calorimeter (Griffin-Sutton) (stirrer omitted).

in the fuel leave the equipment in the form of CO_2 and H_2O. This desirable result is sometimes difficult to achieve, the main causes of failure being incomplete mixing of fuel and air in the combustion chamber and inadequate time of residence in the chamber for the reaction to be completed. The extent to which combustion is completed is measured by the *combustion efficiency*, η_c, defined in such a way that η_c is 100 per cent if combustion is complete and 0 per cent if no combustion has taken place.

Various definitions of η_c are in use; each has its own advantages and fields of convenient application. One which is susceptible of easy measurement is the so-called "CO_2 efficiency", defined by:

$$\eta_c \equiv \frac{\text{actual volume fraction of } CO_2 \text{ in the exhaust gases}}{\text{volume fraction of } CO_2 \text{ in the exhaust gases from the same}} \qquad (16.21)$$
$$\text{fuel and air streams if combustion is complete.}$$

It has the disadvantage that, if all the carbon were completely burned while the hydrogen escaped combustion, η_c would appear as 100 per cent.* This situation rarely arises in practice.

Another definition of η_c may be used if the combustion chamber is adiabatic. It is the "temperature-rise efficiency", defined by:

$$\eta_c \equiv \frac{\text{actual temperature rise of gases in the chamber}}{\text{temperature rise which would occur}} \qquad (16.22)$$
$$\text{if combustion were complete.}$$

This avoids the disadvantage of the former definition, but is more difficult to measure and is restricted to adiabatic combustion chambers.

Heating efficiency. The most important measure of performance from the point of view of the designer of heating equipment is the "heating efficiency", η_h. This is defined as

$$\eta_h \equiv \frac{\text{heat transferred from gases to material to be heated}}{\text{maximum possible heat transfer}} \qquad (16.23)$$
$$\text{in isothermal reaction}$$

The denominator is the calorific value of the fuel. The numerator is less than this in practice because (*a*) the exhaust gases rarely leave the equipment at as low a temperature as the entering fuel and air streams; (*b*) wasteful heat transfers occur to the surroundings of the equipment because of inadequate insulation; (*c*) the combustion efficiency may not be 100 per cent. The heating efficiency of a well-designed steam boiler is of the order of 85 per cent; the heating efficiency of an open coal fire on the other hand, if the room is regarded as the system to be heated, is nearer 15 per cent.

Opinion varies as to whether the higher or the lower calorific value should be used as the denominator of η_h. On the one hand it may be argued that the steam in the exhaust gases could be condensed; then the higher calorific value is appropriate. On the other hand, such condensation has to be avoided as a rule, lest oxides of sulphur should dissolve in the water to form acid and cause corrosion; in this case it may be thought more "fair" to the heat-transfer equipment to use the lower calorific value. However the important thing is to make known on what basis any quoted heating efficiency has been calculated.

* Or very nearly. The number of moles in the exhaust gas would differ somewhat from that for the products of complete combustion; so CO_2 efficiency would not be quite 100%.

EXAMPLE 16.8

Problem. A boiler generates 11 kg of steam per kg of fuel burned, when fired with fuel oil having a higher calorific value of $42 \cdot 0 \times 10^6$ J/kg. The feed water is supplied at a temperature of 60 °C and the steam is produced at a pressure of 4×10^6 N/m² and a temperature of 500 °C. Calculate the heating efficiency of the boiler.

Solution. We apply the S.F.E.E. to the H_2O flowing steadily through the boiler, with neglect of the kinetic and potential-energy terms, to obtain the heat transfer to H_2O in the boiler.

Thus
$$\frac{Q}{m} = h_{steam} - h_{feed}.$$

Now, from the Steam Tables, Appendix B,
$$h_{steam} = 3445 \cdot 0 \times 10^3 \text{ J/kg}$$
$$h_{feed} = 251 \cdot 1 \times 10^3 \text{ J/kg}$$

Hence
$$\frac{Q}{m} = 3193 \cdot 9 \times 10^3 \text{ J/kg}$$

The heating efficiency, often called the *boiler efficiency*, is then obtained from eq. (16.23) as
$$\eta_h = \frac{11 \times 3193 \cdot 9 \times 10^3}{42 \cdot 0 \times 10^6} \times 100\%$$
$$= 83 \cdot 6\% \qquad \text{based on the } \overline{HCV}. \qquad \textit{Answer}$$

The Second Law of Thermodynamics Applied to Combustion

The entropy of a chemical substance

Just as values can be ascribed to the internal energy and enthalpy of a chemical substance, each substance in a definite state has a definite value of entropy. The symbol S' will be used for this quantity. The remarks made earlier about the restrictions on the choice of arbitrary zeros of U' and H' apply to S' also; if the entropy of a given chemical substance in a reactant state is arbitrarily put equal to zero, the entropy of the system in any product state is fixed.

Entropy change in adiabatic reaction. It is a consequence of the Second Law that the entropy of an adiabatic system can only increase (eq. (13.33), p. 263). Consequently, any chemical substance which can react spontaneously under adiabatic conditions is found to have a higher entropy after reaction than before it.

The *possibility* of any reaction can therefore be determined by comparing the entropy of the reactants with the entropy of the (supposed) products, while bearing in mind that the temperatures of the products and reactants will in general differ because the reaction is adiabatic. To do this it is necessary to have data available for the entropies of the materials in question. We shall not discuss here how these are determined.

Entropy change in isothermal reaction. The changes of U' and H' in chemical reactions have been discussed above largely for isothermal reactions; accordingly we now re-formulate the above statement about the possibility of reaction in the corresponding terms.

In Chapter 13 we derived* the inequality:

$$dQ \leqslant T \, dS' \tag{16.24}$$

where dQ is an infinitesimal heat transfer, T is the absolute temperature of the system and dS' is the change in entropy of the (chemical) system.

Now, from the application of the First Law to processes in which shear work, electrical work, gravity, motion, capillarity, etc. are absent, we can write:

$$dQ = dU' \text{ if the process is at constant volume} \tag{16.25}$$

and $\quad dQ = dH'$ if the process is at constant pressure $\qquad$ (16.26)

Combination of eq. (16.25) and eq. (16.26), in turn, with the Second-Law statement (16.24), yields:

$$dU' - T \, dS' \leqslant 0 \text{ for a constant-volume process} \tag{16.27}$$

and $\quad dH' - T \, dS' \leqslant 0$ for a constant-pressure process $\qquad$ (16.28)

Since we are restricting attention to changes in which T is constant,† there is no need to consider only infinitesimal changes. Eq. (16.27) and eq. (16.28) can therefore be integrated, giving

$$\Delta U' - T(\Delta S') \leqslant 0\ddagger$$

or $(U' - TS')_\mathrm{P} - (U' - TS')_\mathrm{R} \leqslant 0$ for a constant-volume process,

$$\tag{16.29}$$

and $\qquad \Delta H' - T(\Delta S') \leqslant 0\ddagger$

or $(II' - TS')_\mathrm{P} - (H' - TS')_\mathrm{R} \leqslant 0$ for a constant-pressure process.

$$\tag{16.30}$$

Our criterion for the possibility of isothermal chemical reaction can now be formulated. It is:

An isothermal chemical reaction is possible at constant volume only if $U' - TS'$ for the products is less than $U' - TS'$ for the reactants. Isothermal reaction is possible at constant pressure, only if $H' - TS'$ for the products is less than $H' - TS'$ for the reactants.

The possibility of combustion reactions. Table 16.8 contains data for the increases in $H' - TS'$ for reactions important in combustion, at 25 °C and 1 atm.

---

* In Chapter 13, S had no prime; but the change of symbol in no way affects the validity of the inequality.

† i.e. to isothermal reactions. See p. 379.

‡ To conform with the nomenclature used earlier in this chapter we should write $[\Delta U']_{v,t}$ and $[\Delta H']_{p,t}$. We omit the brackets here for clarity.

Examination of Table 16.8 reveals two features. The first is that $\Delta(H' - TS')$ is indeed negative; the reactions are therefore possible, as

TABLE 16.8

Reaction	$\Delta H'$ J/kmol	$\Delta(H' - TS')$ J/kmol
H_2 gas $+ \frac{1}{2}O_2$ gas $= H_2O$ gas	$-241 \cdot 988 \times 10^6$	$-227 \cdot 748 \times 10^6$
C solid $+ O_2$ gas $= CO_2$ gas	$-393 \cdot 776 \times 10^6$	$-393 \cdot 427 \times 10^6$
C solid $+ \frac{1}{2}O_2$ gas $= CO$ gas	$-110 \cdot 599 \times 10^6$	$-113 \cdot 884 \times 10^6$

common experience asserts.* The second is that $\Delta(H' - TS')$ has very nearly the same value as $\Delta H'$; this implies that the contribution of the entropy change at constant temperature is small. Closer study shows that S' increases in the course of reactions leading to H_2O and CO_2, but decreases in that leading to CO.

Work done by a chemically-reacting system

In the introduction to this chapter, it was stated that the Second Law furnishes (a) a criterion for the possibility of chemical reaction, and (b) a means of determining the maximum attainable work. Having discussed the first of these, we turn now to the second. As a preliminary, we first consider how it is that any of the reactions of Table 16.8 can ever be reversed; we know that H_2O can be split into H_2 and O_2, yet $\Delta(H' - TS')$ for this reaction is *positive* (i.e. >0). How can this be?

Reflexion about how H_2O is decomposed in practice, namely by means of an electrolytic cell, reveals the answer to the query; for a flow of electricity counts as external work, which was specifically excluded in deriving eq. (16.30). External work, W_x, here means work other than pdV work at the (constant-pressure) system boundary. (cf. p. 140 for the corresponding definition for a control volume). We now remedy this restriction, considering only constant-pressure isothermal processes.

The First Law yields $\mathrm{d}Q - \mathrm{d}W_x = \mathrm{d}H'$ (16.31)

---

* This conclusion requires qualification. What if there is an intermediate state, occurring before complete reaction, in which $H' - TS'$ is even lower than in the completely reacted state? Then the reaction will proceed to this intermediate state and no farther. This is not merely an academic point; there are indeed always such intermediate states, although at low temperatures they are so close to the state of complete reaction as to be indistinguishable for practical purposes. At the temperatures occurring in gasoline engines and rocket motors on the other hand, the intermediate equilibrium states are far indeed from those of complete combustion. The theory of this phenomenon is found in more advanced texts than the present, usually under the heading of "dissociation". This word implies that reaction goes to completion and then partly retraces its steps; i.e. the products "dissociate". This is not the way things really happen, but it is quite a useful way of imagining them.

Combination of this equation with eq. (16.24), followed by integration, yields

$$W_x \leqslant -(\Delta H' - T\Delta S') \tag{16.32}$$

In electrolysis, W_x is negative. This enables the condition eq. (16.32) to be satisfied even though $(\Delta H' - T\Delta S')$ is positive.

The maximum work in chemical reaction. Eq. (16.32) enables an important practical question to be answered: given a fuel-air mixture, how much work can be obtained from isothermal constant-pressure reaction? The answer is obtained by replacing the inequality by an equality, namely:

$$W_{x,\text{max}} = -(\Delta H' - T\Delta S') \tag{16.33}$$

Examination of Table 16.8 shows that, since the $T\Delta S'$ terms are small, the maximum work quantity is very nearly equal to $-\Delta H'$ and therefore to the calorific value, $\overline{CV}$, for combustion of both hydrogen and carbon. This result holds for all the fuels used in practice. Symbolically we write

$$W_{x,\text{max}} \approx \overline{CV} \tag{16.34}$$

Efficiency of internal-combustion engines. Although much of the world's power is produced by gasoline and diesel engines, and by "open-cycle" gas turbines, these were specifically excluded from the consideration of efficiency advanced in Chapter 10. Eq. (16.33) makes it possible to remedy this omission by defining an efficiency $\eta_{\text{i.c.}}$ for internal-combustion engines, as

$$\eta_{\text{i.c.}} \equiv \frac{W_x}{W_{x,\text{max}}} \tag{16.35}$$

wherein W_x is the shaft work produced by the engine and $W_{x,\text{max}}$ is equal to $-(\Delta H' - T\Delta S')$ for the fuel used, evaluated at the constant pressure and temperature of the surrounding atmosphere.

However, since the term $T\Delta S'$ is so small, and since actual engines anyway deliver much less than the maximum possible work, it is sufficient in practice to use an approximate expression, obtained by combining eq. (16.34) and eq. (16.35), namely

$$\eta_{\text{i.c.}} \approx \frac{W_x}{\overline{CV}} \tag{16.36}$$

Sometimes the higher calorific value is inserted in this relation, and sometimes the lower. There is no decisive reason for using one rather than the other; but which is used should always be stated explicitly.

Values of $\eta_{\text{i.c.}}$ which are achieved in practice vary from about 20 per cent for the simplest gas-turbine engines, to about 40 per cent for Diesel engines. Gasoline engines have efficiencies between 30 per cent and 35 per cent at full load.

EXAMPLE 16.9

Problem. An industrial gas-turbine engine develops 4700 kW and has a specific fuel consumption of 0·46 kg of fuel/kWh. The higher calorific value of the fuel used is $42·5 \times 10^6$ J/kg of fuel. Calculate the efficiency, $\eta_{1.c.}$ of the engine.

Solution. The shaft power, $\dot{W}_x$, is given by

$$\dot{W}_x = 4700 \text{ kJ/s}$$

From equation (16·34) we have

$$\dot{W}_{x,\text{max}} = \overline{CV}$$

Therefore the maximum power obtainable from the fuel is given by:

$$\dot{W}_{x,\text{max}} = \overline{CV} \times \dot{m}_{\text{fuel}}$$

where $\dot{m}_{\text{fuel}} = $ mass flow rate of fuel.

In this case, $$\dot{m}_{\text{fuel}} = \frac{0·46 \times 4700 \times 10^3}{3600 \times 10^3} = 0·60 \text{ kg fuel/s}$$

Hence $$\dot{W}_{x,\text{max}} \approx 0·60 \times 42·5 \times 10^6 = 25\,500 \text{ kJ/s}$$

Finally, from equation (16.35) we have

$$\eta_{1.c.} = \frac{4700 \times 10^3}{25·5 \times 10^6} \times 100\% = 18·4\% \qquad \textit{Answer}$$

EXAMPLE 16.10

Problem. An automobile gasoline engine develops a brake power of 50 kW. The mixture strength is 110% and the engine air consumption is 3·12 kg/min. The stoichiometric air-fuel ratio of the fuel used is 14·5 and the higher calorific value of the fuel is $44·0 \times 10^6$ J/kg. Determine (*a*) the brake specific fuel consumption in kg/kWh, and (*b*) the efficiency, $\eta_{1.c.}$.

Solution (*a*). First we determine the air-fuel ratio, x, of the mixture entering the engine by the use of eq. (16.6), as follows:

$$x = 14·5 \times \frac{100}{110} = 13·18$$

Next, from the air consumption and x, we deduce the fuel mass flow rate:

$$\dot{m}_{\text{fuel}} = \frac{3·12}{60} \times \frac{1}{13·18}$$

$$= 3·95 \times 10^{-3} \text{ kg fuel/s}$$

Finally, we obtain the specific fuel consumption

$$\frac{\dot{m}}{\dot{W}_x} = \frac{3·95 \times 10^{-3} \times 3600 \times 10^3}{50 \times 10^3}$$

$$= 0·284 \text{ kg fuel/kWh} \qquad \textit{Answer (a)}$$

Solution (*b*). The shaft power, $\dot{W}_x$, $= 50 \times 10^3$ J/s

Further, from eq. (16.34), the maximum power obtainable from the fuel is given by:

$$\dot{W}_{x,\text{max}} = \overline{CV} \times \dot{m}_{\text{fuel}}$$

$$= 44·0 \times 10^6 \times 3·95 \times 10^{-3}$$

$$= 173·8 \times 10^3 \text{ J/s}$$

The value of $\eta_{i.c.}$ then follows from eq. (16.35):

$$\eta_{i.c.} = \frac{50 \times 10^3}{173 \cdot 8 \times 10^3}$$

$$= 0 \cdot 288 \qquad\qquad \textit{Answer (b)}$$

The upper limit of efficiency of internal-combustion engines. A major difference must be emphasized between $\eta_{i.c.}$ and the heat-engine efficiency of Chapter 10. Whereas the latter is limited by the temperature of the available "reservoirs", no such limitation is imposed on the efficiency of an internal-combustion engine. According to the Second Law, $\eta_{i.c.}$, can reach 100 per cent.

Internal-combustion engines should therefore be judged more strictly than, for example, steam power plants. A Diesel engine giving 40 per cent efficiency is wasting 60 per cent of the available power. Engineers cannot be satisfied until a much closer approach to the permissible limit has been achieved.

One way of doing this has been mentioned already in Chapter 11 (pp. 222, 230). This is the fuel cell. It is currently the most hopeful development in the effort materially to increase the efficiency of man's usage of fuel.

BIBLIOGRAPHY

A.S.T.M. STANDARDS, Parts 17 and 18. *Petroleum Products*; Part 19, *Gaseous Fuels; Coal and Coke.* 1966.

BS 1016: 1957–1971. *Methods for the Analysis and Testing of Coal and Coke.* British Standards Institution.

BS 1017: 1960. *The Sampling of Coal and Coke.* British Standards Institution.

BS 2869: 1970. *Petroleum Fuels for Oil Engines and Burners.* British Standards Institution.

BS 526: 1961. *Definitions of the Calorific Values of Fuels.* British Standards Institution.

BS 1756: 1971. *Methods for the Sampling and Analysis of Flue Gases.* British Standards Institution.

MAXWELL, J. B. *Data Book on Hydrocarbons.* Van Nostrand, New York, 1950.

PERRY, J. H., CHILTON, C. H., KIRKPATRICK, S. D. *Chemical Engineers Handbook.* 4th Ed. McGraw-Hill, New York, 1963.

Selected Values of Properties of Hydrocarbons. Circular C 461, National Bureau of Standards, 1947.

SPIERS, H. B. *Technical Data on Fuel.* 6th Ed. The British National Committee, World Power Conference, 1962.

I. P. Standards for Petroleum and its Products. Institute of Petroleum, 1965.

CHAPTER 16—PROBLEMS

Notes.

1. Assume air contains $23 \cdot 2\%$ of oxygen and $76 \cdot 8\%$ of nitrogen, by mass; $21 \cdot 0\%$ of oxygen and 79% of nitrogen, by volume.

2. Assume all gaseous reactants and non-condensable products to be Ideal Gases.

16.1 (*a*) A solid fuel, having a composition by mass of 84% carbon, 14% hydrogen and 2% sulphur, is burned completely with oxygen. Determine the minimum quantity of oxygen required, and the mass composition and the molar composition of the products of combustion.

(b) The fuel of (a) is burned completely with air. Determine the minimum quantity of air required and the mass composition and molar composition of the products of combustion.

16.2 (a) A stoichiometric mixture of n-octane (C_8H_{18}) and air burns completely. Determine the stoichiometric air-fuel ratio by mass, and the molar composition of the products of combustion.

(b) A six-cylinder, four-stroke reciprocating engine uses n-octane as fuel. The engine develops 120 kW when running at a speed of 3600 rev/min at a mixture strength of 100 per cent. The specific fuel consumption is 0·29 kg/kWh. Evaluate the volumes (in cubic centimetres) of liquid octane and of atmospheric air, consumed per working cycle per cylinder, assuming combustion to be complete. The density of liquid octane is 709 kg/m³ and the atmospheric pressure and temperature are 100×10^3 N/m² and 20 °C respectively; R for air is 287 J/kg K.

16.3 A liquid fuel has a composition, by mass of 84% carbon and 16% hydrogen. The fuel used in an engine at a mixture strength of 115 per cent. On the assumption that all the hydrogen is burned completely and that there is no "free" oxygen in the exhaust gas estimate:

(i) the molar composition of the exhaust gas;
(ii) the "Orsat analysis" of the exhaust gas.

16.4 A boiler burns fuel having a composition by mass of 82% carbon, 5% hydrogen, 6% oxygen and 7% ash. The composition by volume, of the dry flue gas is 13·2% carbon dioxide, 1·5% carbon monoxide, 6·8% oxygen, 78·5% nitrogen.

Evaluate (i) the air-fuel ratio;
(ii) the percentage excess air supplied;
(iii) the volume of air handled by the boiler fans in m³/s, given that the hourly fuel consumption of the boiler is 1000 kg. Assume that the air enters the fans at a pressure of 100×10^3 N/m² and a temperature of 40 °C; R for air is 287 J/kg K.

16.5 The fuel of problem **16.1** is burned in a boiler with 40 per cent excess air. The fuel, as fired, contains 3 per cent by mass of moisture, and the air supplied has a relative humidity of 50 per cent. The temperature and pressure of the air supplied are 30 °C and 1 atm; the pressure of the flue gas is 98×10^3 N/m². Assuming the fuel to be burned completely, evaluate:

(i) the molar composition of the flue gas;
(ii) the mass of H_2O in the flue gas per kilogram of *dry* fuel burned;
(iii) the dew point of the flue gas.

16.6 A fuel gas has a composition, by volume, of 30% carbon monoxide, 10% hydrogen, 3% methane, 5% carbon dioxide and 52% nitrogen. The gas is burned in a furnace, the air-fuel ratio by volume being 2. Assuming combustion to be complete, determine the volume analysis of the dry exhaust gas.

16.7 A stoichiometric mixture of carbon monoxide and oxygen is contained in a rigid vessel of volume 1·6 dm³ at a pressure of 1 atm and a temperature of 25 °C.

When the mixture is ignited and burns completely, it is found that a heat transfer of 12·3 kJ is required from the contents of the vessel to restore their temperature to 25 °C. Assuming carbon monoxide and oxygen to be Ideal Gases evaluate:

(i) the number of kmol of carbon monoxide and oxygen in the vessel.
(ii) the internal-energy increase of reaction at 25 °C in J/kmol of CO;
(iii) the final pressure of the contents of the vessel;
(iv) the enthalpy increase of reaction at 1 atm, 25 °C.

16.8 For the combustion of 1 kmol of carbon to carbon monoxide at 25 °C, 1 atm, increase in enthalpy of reaction is $-110 \cdot 599 \times 10^6$ J.

(a) For the reaction $2C + O_2 = 2CO$, plot to scale, the $H' \sim t$ diagram, for the combustion of 1 kmol of carbon, using the following data for mean specific heat at constant pressure.

$t\,°C \rightarrow$	100	200	300	400	500	600
Substance $\downarrow$	$\bar{c}_p$ in J/kg K for the range of temperature from 0 °C to t °C					
C	816	925	1034	1138	1230	1300
O_2	924	936	950	964	978	993
CO	1034	1042	1052	1062	1073	1085

(b) Evaluate the heat of reaction at 600 °C.

(c) Determine the product temperature in a constant-pressure reaction in which the heat transfer to the surroundings is 7937 kJ/kg of carbon. The temperature of the reactants is 0 °C.

16.9 The enthalpy increase of reaction of liquid octane (C_8H_{18}) with oxygen at a pressure of 1 atm and a temperature of 25 °C is $-44\cdot426 \times 10^6$ J/kg of octane when the H_2O product is in the vapour state.

(i) Sketch the $H'{\sim}t$ diagram, showing the reactant curves corresponding to the liquid and gaseous states of octane and the product curves corresponding to the liquid and vapour states of H_2O.

(ii) Evaluate the internal-energy increase of reaction for these conditions.

(iii) Calculate the enthalpy increase of reaction and the internal-energy increase of reaction of liquid octane-oxygen gas system at a temperature of 130 °C when the H_2O product is in the vapour state. Use the following values of mean specific heat at constant pressure: liquid octane, 1760 J/kg K; CO_2, 900 J/kg K; steam, 2010 J/kg K; O_2, 920 J/kg K.

(iv) The enthalpy of evaporation of octane at a temperature of 130 °C is 298×10^3 J/kg; the enthalpy of evaporation of H_2O at this temperature may be taken as 2174 kJ/kg. Calculate the enthalpy increase of reaction and the internal-energy increase of reaction of gaseous octane-oxygen system at a temperature of 130 °C when the H_2O product is in the liquid state.

16.10 For the reaction of the carbon-oxygen system to carbon dioxide only, at a temperature of 25 °C and a pressure of 1 atm, the enthalpy increase is $-32\cdot813 \times 10^6$ J/kg of carbon; at a temperature of 800 °C and a pressure of 1 atm, the enthalpy increase is $-33\cdot287 \times 10^6$ J/kg of C. The mean specific heat of carbon between 25 °C and 800 °C is 1385 J/kg K, and of oxygen is 1022 J/kg. Evaluate the mean specific heat of CO_2 for this temperature range.

16.11 Carbon monoxide is burned adiabatically in steady flow at atmospheric pressure with 100 per cent excess air. The CO is supplied to the gas burner at a temperature of 150 °C and the air at a temperature of 40 °C. The standard enthalpy increase of reaction at 1 atm, 25 °C is $-283\cdot177 \times 10^6$ J/kmol of carbon monoxide. Sketch the $H'{\sim}t$ diagram and hence calculate the temperature of the combustion products assuming combustion to be complete. Use the following values of mean specific heat at constant pressure: CO, 1038 J/kg K; O_2, 1120 J/kg K; N_2, 1171 J/kg K; CO_2, 1185 J/kg K; air, 1000 J/kg K.

16.12 Calculate the calorific value of carbon monoxide, in J/kg using the enthalpy-of-formation data given in Table 16.6

16.13 (a) Ethane gas, C_2H_6, is burned completely with oxygen at a constant pressure of 1 atm. The reactants and products are both at a temperature of 25 °C and the H_2O product is in the liquid state. Using the data given in Table 16.6, evaluate:

(i) the enthalpy increase of reaction;

(ii) the heat of reaction;

(iii) the higher and lower calorific values.

(b) Ethane gas is burned completely in adiabatic steady flow with 50 per cent excess of pure oxygen. The reactants are at a temperature of 25 °C. Determine the temperature of the combustion products using the following values of mean specific heat at constant pressure: CO_2, 1294 J/kg K; O_2, 1134 J/kg K; steam, 2595 J/kg K.

16.14 (a) Recalculate problem **16.13** (a), assuming the reaction to take place in air.

(b) Recalculate problem **16.13** (b), assuming 50% excess air to be supplied. The specific heat of nitrogen at constant pressure may be taken as 1177 J/kg K.

16.15 Determine the higher and lower calorific values of the fuel gas of problem **16.6** in J/m^3 at 15 °C when the gas is (a) dry, (b) saturated with water vapour. Use the data given in Table 16.7. Assume that the calorific values of methane are the same as those of natural gas.

16.16 Kerosine having a composition by mass of 86·5% carbon, 13·5% hydrogen, is burned adiabatically in the combustion chamber of an open-cycle gas-turbine engine; the air-fuel ratio is 63 : 1 by mass. The fuel is supplied to the chamber at a temperature of 15 °C and the air at a temperature of 240 °C. The measured CO_2 content of the dry products of combustion is 3·18 per cent by volume and the measured products temperature is 830 °C. The higher calorific value of kerosine is $46·180 \times 10^6$ J/kg.

Evaluate (i) the "CO_2 efficiency";

(ii) the "temperature-rise efficiency";

Use the following values of mean specific heat at constant pressure (J/kg K): O_2, 1022; N_2, 1103; CO_2, 1097: steam, 2148.

16.17 (a) By applying the Steady-Flow Energy Equation to the fuel-air stream as it flows through a steam boiler, show that the so-called "energy account" may be written

$$\overline{LCV} = q_1 + q_u + m\bar{c}_p (t - t_0)$$

where q_1 is the heat transfer to the H_2O in the boiler per unit mass of fuel burned,

q_u is the unmeasured ("unaccounted-for") heat transfer (to the surroundings) per unit mass of fuel burned,

m is the mass of flue gas per unit mass of fuel burned,

$\bar{c}_p$ is the mean specific heat at constant pressure of the flue gas,

t is the flue-gas temperature,

t_0 is the $\overline{LCV}$ datum temperature.

Assume that the fuel and air are supplied to the boiler at t_0.

(b) A boiler generates steam at a pressure of 2×10^6 N/m^2 and a temperature of 300 °C, from feed water at a temperature of 65 °C; the quantity of steam produced is 11·3 kg per kg of oil fuel burned. The oil fuel has a composition, by mass, of 88% carbon, and 12% hydrogen, and a $\overline{LCV}$ of 40×10^6 J/kg (datum 15 °C). The Orsat analysis of the flue gas is 10·6% CO_2, 7·4% O_2, 82% N_2; the CO content is negligible. The flue gas temperature is 342 °C.

(i) Draw up the "energy account" given that $\bar{c}_p = 1025$ J/kg K.

(ii) Evaluate the heating efficiency of the boiler based on the $\overline{LCV}$.

16.18 (a) The $\overline{LCV}$ of light fuel oil is $44·80 \times 10^6$ J/kg. What is the maximum power which could be produced by the reaction of this fuel with air at constant pressure at the rate of 0·01 kg/s. (Neglect the entropy change occurring during the reaction.)

(b) An open-cycle gas-turbine engine burns the fuel oil of (a). The specific fuel consumption of the engine is 0·425 kg fuel/kWh. Evaluate $\eta_{l.c.}$, the efficiency of the engine.

16.19 In the formation of steam by the reaction of hydrogen with oxygen at a temperature of 25 °C and a pressure of 1 atm, the enthalpy increase of reaction is $-241·988 \times 10^6$ J/kmol of steam. Correspondingly the entropy of the product steam is $22·2 \times 10^3$ J/K less than the entropy of the reactants, per kg of hydrogen present.

Evaluate the maximum work which could be produced by the isothermal reaction of hydrogen to steam at 25 °C, 1 atm.

Appendix A Unit Abbreviations and Tables of Conversion Factors

These tables are extracts from BS 1991: 1961 *Recommendations for Letter Symbols, Signs and Abbreviations*, and from BS 350: Part 1: 1959 *Conversion Factors and Tables*. They are reproduced by permission of the British Standards Institution, 2 Park Street, London W1A 2BS, from whom copies of the complete standards may be obtained.

TABLE A.1 *Recommended Unit Abbreviations*

	SI	*Imperial System*
Acceleration	m/s²	ft/s²
Area	m²	ft²
Constant in Newton's Second Law	kg m/N s²*	lb ft/lbf s²*
Current	A	A
Density	kg/m³	lb/ft³
Displacement	m	ft
Electric charge	C	C
Energy	Nm; J	ft lbf; Btu
Enthalpy	J	Btu
Entropy	J/K	Btu/°R
Force	N	lbf
Gas constant	J/kg K	ft lbf/lb °R
Heat transfer	J	Btu
Heat transfer rate	J/s; W	Btu/h
Internal energy	J	ft lbf; Btu
Length	m	ft
Mass	kg	lb
Mechanical equivalent of heat	Nm/J*	ft lbf/Btu
Molar (Universal) Gas Constant	J/kmol K	Btu/lb-mole °R
Molecular mass	kg/kmol	lb/lb-mole*
Potential difference	V	V
Power	W	Btu/s; hp
Pressure, stress	N/m²; bar	lbf/ft²
Rotational speed	rev/s	rev/min
Shear stress	N/m²	lbf/ft²
Specific energy	J/kg	ft lbf/lb; Btu/lb
Specific enthalpy	J/kg	Btu/lb
Specific entropy	J/kg K	Btu/lb °R
Specific heat	J/kg K	Btu/lb degF
Specific internal energy	J/kg	Btu/lb
Specific volume	m³/kg	ft³/lb
Temperature	°C; K	°R; °F
Temperature difference	K	deg R
Time	s	s; min; h
Torque	Nm	lbf ft
Velocity, speed	m/s	ft/s
Volume	m³	ft³
Weight-force	N	lbf
Work	Nm; kWh	ft lbf; hp h

* Not specified in the Standard.

TABLE A.2 *Prefixes for SI Units*

Prefix	Multiplying factor	Symbol
atto	10^{-18}	a
femto	10^{-15}	f
pico	10^{-12}	p
nano	10^{-9}	n
micro	10^{-6}	μ
milli	10^{-3}	m
centi	10^{-2}	c
deci	10^{-1}	d
deka	10	da
hecto	10^{2}	h
kilo	10^{3}	k
mega	10^{6}	M
giga	10^{9}	G
tera	10^{12}	T

TABLE A.3 *Units of Length**

	metre (m)	inch (in)	foot (ft)	kilometre (km)	mile
1 metre (m) =	**1**	39·370 1	3·280 84	**0·001**	6·213 71 $\times 10^{-4}$
1 inch (in) =	**0·025 4**	1	0·083 333 3	**25·4** $\times 10^{-6}$	1·578 28 $\times 10^{-5}$
1 foot (ft) =	**0·304 8**	12	1	**304·8** $\times 10^{-3}$	1·893 94 $\times 10^{-4}$
1 kilometre (km) =	**1000**	3·937 01 $\times 10^{4}$	3·280 84 $\times 10^{3}$	**1**	0·621 371
1 mile =	**1 609·344**	63 360	5280	**1·609 344**	1

* Figures given in bold type are *exact*.

TABLE A.4 *Units of Area**

	square metre (m²)	square inch (in²)	square foot (ft²)	square yard (yd²)
1 square metre = (m)	1	1 550·00	10·7639	1·195 99
1 square inch = (in²)	6·4516 × 10⁻⁴	1	6·944 44 × 10⁻³	7·716 05 × 10⁻⁴
1 square foot = (ft²)	0·092 903 0	144	1	0·111 111
1 square yard = (yd²)	0·836 127	1296	9	1

TABLE A.5 *Units of Volume**

	cubic metre (m³)	litre (l)	cubic inch (in³)	cubic foot (ft³)
1 cubic metre = (m³)	1	1000	61 023·8	35·3147
1 litre = (l)	1 × 10⁻³	1	61·023 8	0·035 314 7
1 cubic inch = (in³)	1·638 71 × 10⁻⁵	0·016 387 0	1	5·787 04 × 10⁻⁴
1 cubic foot = (ft³)	0·028 316 8	28·3168	1728	1

TABLE A.6 *Units of Velocity**

	metre per second (m/s)	foot per second (ft/s)	mile per hour (mile/h)	international knot (kn)
1 metre per second = (m/s)	1	3·280 84	2·236 94	1·943 84
1 foot per second = (ft/s)	0·3048	1	0·681 818	0·592 484
1 mile per hour = (mile/h)	0·447 04	1·466 67	1	0·868 976
1 international knot = (kn)	0·514 444	1·687 81	1·150 78	1

* Figures given in bold type are *exact*.

TABLE A.7 *Units of Mass**

	kilogram (kg)	pound (lb$_m$)†	slug
1 kilogram = (kg)	**1**	2·204 62	0·068 521 8
1 pound = (lb$_m$)	**0·453 592 37**	**1**	0·031 081 0
1 slug =	14·593 9	32·174 0	**1**

Note also that the tonne is equivalent to 1000 kg; the UK ton ≡ **2240** lb$_m$ ≡ 984·207 kg; the US ton ≡ **2000** lb$_m$ ≡ **907·184 74** kg.

TABLE A.8 *Units of Specific Volume**

	m^3/kg	ft^3/lb$_m$†	in^3/lb$_m$
1 m^3/kg =	**1**	16·0185	27 679·9
1 ft^3/lb$_m$ =	0·062 428 0	**1**	**1728**
1 m^3/lb$_m$ =	3·612 73 × 10^{-5}	5·787 04 × 10^{-4}	**1**

TABLE A.9 *Units of Force**

	N	lbf	pdl	kgf
1 newton = (N)	**1**	0·224 809	7·233 01	0·101 972
1 pound-force = (lbf)	4·448 22	**1**	32·1740	0·453 592
1 poundal = (pdl)	0·138 255	0·031 081 0	**1**	0·014 098 1
1 kilogram-force = (kgf)	**9·806 65**	2·204 62	70·9316	**1**

* Figures given in bold type are *exact*.
† The BS unit symbol for pound is lb.

TABLE A.10 *Units of Pressure**

	N/m²	lbf/in²	lbf/ft²	atm	kgf/cm²
1 N/m² =	**1**	$1{\cdot}450\ 38 \times 10^{-4}$	$2{\cdot}088\ 54 \times 10^{-2}$	$9{\cdot}869\ 23 \times 10^{-6}$	$1{\cdot}019\ 72 \times 10^{-5}$
1 lbf/in² =	$6{\cdot}894\ 76 \times 10^{3}$	**1**	**144**	$6{\cdot}804\ 60 \times 10^{-2}$	$7{\cdot}030\ 70 \times 10^{-2}$
1 lbf/ft² =	$47{\cdot}8807$	$9{\cdot}944 \times 10^{-3}$	**1**	$4{\cdot}725 \times 10^{-4}$	$4{\cdot}882 \times 10^{-4}$
1 atm =	$1{\cdot}013\ 250 \times 10^{5}$	$14{\cdot}6959$	$2{\cdot}116\ 215 \times 10^{3}$	**1**	$1{\cdot}033\ 23$
1 kgf/cm² =	$9{\cdot}806\ 65 \times 10^{4}$	$14{\cdot}2233$	$2{\cdot}048\ 160 \times 10^{3}$	$0{\cdot}967\ 841$	**1**

TABLE A.11 *Additional Conversion Factors for Units of Pressure**

$$1 \text{ bar (bar)} = \mathbf{10^5} \ N/m^2$$
$$1 \text{ pascal (Pa)} = \mathbf{1} \ N/m^2$$
$$1 \text{ standard atmosphere (atm)} = \mathbf{760} \text{ barometric millimetre of mercury, mmHg}$$
$$= 29{\cdot}9213 \text{ barometric inch of mercury, inHg}$$

* Figures given in bold type are *exact*.

TABLE A.12 *Units of Work**

	N m	ft lbf	kWh	hp h	m kgf
1 N m (≡ 1 joule (J)) =	**1**	0·737 562	$0.277\ 778 \times 10^{-6}$	$0.372\ 506 \times 10^{-6}$	0·101 972
1 ft lbf =	1·355 82	**1**	$0.376\ 616 \times 10^{-6}$	$0.505\ 051 \times 10^{-6}$	0·138 255
1 kWh =	$\mathbf{3.6 \times 10^{6}}$	$2.655\ 22 \times 10^{6}$	**1**	1·341 02	$0.367\ 098 \times 10^{6}$
1 hp h =	$2.684\ 52 \times 10^{6}$	$\mathbf{1.98 \times 10^{6}}$	0·745 700	**1**	$0.273\ 745 \times 10^{6}$
1 m kgf =	**9·806 65**	7·233 01	$2.724\ 07 \times 10^{-6}$	$3.653\ 04 \times 10^{-6}$	**1**

TABLE A.13 *Units of Power**

	N m/s	ft lbf/s	metric hp	hp	kgf m/s
1 N m/s (≡ 1 watt (W)) =	**1**	0·737 562	$1.359\ 62 \times 10^{-3}$	$1.341\ 02 \times 10^{-3}$	0·101 972
1 ft lbf/s =	1·355 82	**1**	1.843×10^{-3}	$1.818\ 18 \times 10^{-3}$	0·138 255
1 metric hp =	735·499	542·476	**1**	0·986 320	**75**
1 hp =	745·7	**550**	1·013 87	**1**	76·040 2
1 kgf m/s =	**9·806 65**	7·233 01	0·013 333 3	0·013 150 9	**1**

* Figures given in bold type are *exact.*

TABLE A.14 *Units of Heat, Work and Energy**

	newton metre (N m)	foot pound-force (ft lbf)	kilocalorie (kcal)	British thermal unit (Btu)	kilowatt hour (kWh)	horsepower hour (hp h)
1 newton metre (N m) (≡ 1 Joule (J))	1	0·737 562	$0·238\ 846 \times 10^{-3}$	$0·947\ 817 \times 10^{-3}$	$0·277\ 778 \times 10^{-6}$	$0·372\ 506 \times 10^{-6}$
1 foot pound-force (ft lbf)	1·355 82	1	$0·323\ 832 \times 10^{-3}$	$1·285\ 07 \times 10^{-3}$	$0·376\ 616 \times 10^{-6}$	$0·505\ 051 \times 10^{-6}$
1 kilocalorie (kcal)	**4186·8**	3088·03	1	3·968 32	0·001 163	$1·559\ 61 \times 10^{-3}$
1 British thermal unit (Btu)	1055·06	778·170	0·251 996	1	$0·293\ 071 \times 10^{-3}$	$0·393\ 015 \times 10^{-3}$
1 kilowatt hour (kWh)	$\mathbf{3·6 \times 10^{6}}$	$2·655\ 22 \times 10^{6}$	859·845	3412·14	1	1·341 02
1 horsepower hour (hp h)	$2·684\ 52 \times 10^{6}$	$\mathbf{1·98 \times 10^{6}}$	641·186	2544·43	0·745 700	1

TABLE A.15 *Additional Conversion Factors for Units of Heat, Work and Energy*

1 joule = 0·101 972 metre kilogram-force, m kgf
       = $9·868\ 96 \times 10^{-3}$ litre atmosphere
       = 23·7304 foot poundal, ft pdl
       = $2·389\ 20 \times 10^{-7}$ thermie, th

* Figures given in bold type are *exact*.

Appendix B
Steam Tables

The tables given in this Appendix are extracts, reproduced by permission, from *"U.K. Steam Tables in SI Units"*, Edward Arnold (Publishers) Ltd, London, 1970. Further data on the properties of steam may be obtained from the references given at the end of Chapter 9 (page 196).

The datum state used in the following tables is that fixed by the International Steam-Table Conference of 1956, namely, zero values for the internal energy and entropy of liquid water at the triple point. The data tabulated for 0 °C correspond to a fictitious state point and have been obtained by extrapolating the triple-point data to 0 °C.

TABLE I

TRIPLE-POINT DATA

TEMPERATURE: 0·01 °C; PRESSURE: 0·006 112 bar

Phase	Specific Volume m³/kg	Specific internal energy kJ/kg	Specific enthalpy kJ/kg	Specific entropy kJ/kg K
Ice	0·001 090 5	−333·5	−333·5	−1·2209
Water	0 001 000 2	0	0·000 611 3	0
Steam	206·1629	2375·6	2501·6	9·1575

TABLE II *Properties of Saturated Water and Saturated Steam. Temperature Table*

Temp. °C t	Abs. Press. bar p	Specific Volume m³/kg			Enthalpy kJ/kg			Entropy kJ/kg K			Temp. °C t
		Sat. Liq. v_f	Evap. v_{fg}	Sat. Vap. v_g	Sat. Liq. h_f	Evap. h_{fg}	Sat. Vap. h_g	Sat. Liq. s_f	Evap. s_{fg}	Sat. Vap. s_g	
0	0.006 017	0.001 000 2	206.298	206.299	−0.0	2501.6	2501.6	−0.0	9.1578	9.1578	0
0.01	0.006 112	0.001 000 1	206.162	206.163	+0.0	2501.6	2501.6	0	9.1575	9.1575	0.01
2	0.007 055	0.001 000 0	179.922	179.923	8.4	2496.8	2505.2	0.0306	9.0741	9.1047	2
4	0.008 129	0.001 000 0	157.271	157.272	16.8	2492.1	2508.9	0.0611	8.9915	9.0526	4
6	0.009 345	0.001 000 0	137.779	137.780	25.2	2487.4	2512.6	0.0913	8.9102	9.0015	6
8	0.010 720	0.001 000 1	120.965	120.966	33.6	2482.6	2516.2	0.1213	8.8300	8.9513	8
10	0.012 270	0.001 000 3	106.429	106.430	42.0	2477.9	2519.9	0.1510	8.7510	8.9020	10
12	0.014 014	0.001 000 4	93.834	93.835	50.4	2473.2	2523.6	0.1805	8.6731	8.8536	12
14	0.015 973	0.001 000 7	82.899	82.900	58.8	2468.5	2527.2	0.2098	8.5963	8.8060	14
15	0.017 039	0.001 000 8	77.977	77.978	62.9	2466.1	2529.1	0.2243	8.5582	8.7826	15
16	0.018 168	0.001 001 0	73.383	73.384	67.1	2463.8	2530.9	0.2388	8.5205	8.7593	16
18	0.020 624	0.001 001 3	65.086	65.087	75.5	2459.0	2534.5	0.2677	8.4458	8.7135	18
20	0.023 366	0.001 001 7	57.837	57.838	83.9	2454.3	2538.2	0.2963	8.3721	8.6684	20
25	0.031 660	0.001 002 9	43.401	43.402	104.8	2442.5	2547.3	0.3670	8.1922	8.5592	25
30	0.042 415	0.001 004 3	32.928	32.929	125.7	2430.7	2556.4	0.4365	8.0181	8.4546	30
35	0.056 216	0.001 006 0	25.244	25.245	146.6	2418.8	2565.4	0.5049	7.8495	8.3543	35
40	0.073 750	0.001 007 8	19.545	19.546	167.5	2406.9	2574.4	0.5721	7.6861	8.2583	40
45	0.095 820	0.001 009 9	15.275	15.276	188.4	2394.9	2583.3	0.6383	7.5277	8.1661	45
50	0.123 35	0.001 012 1	12.045	12.046	209.3	2382.9	2592.2	0.7035	7.3741	8.0776	50
55	0.157 41	0.001 014 5	9.577 9	9.578 9	230.2	2370.8	2601.0	0.7677	7.2248	7.9925	55
60	0.199 20	0.001 017 1	7.677 5	7.678 5	251.1	2358.6	2609.7	0.8310	7.0798	7.9108	60
65	0.250 09	0.001 019 9	6.201 3	6.202 3	272.0	2346.3	2618.4	0.8933	6.9388	7.8321	65
70	0.311 62	0.001 022 8	5.045 3	5.046 3	293.0	2334.0	2626.9	0.9548	6.8017	7.7565	70
75	0.385 49	0.001 025 9	4.133 1	4.134 1	313.9	2321.5	2635.4	1.0154	6.6681	7.6835	75
80	0.473 60	0.001 029 2	3.408 1	3.409 1	334.9	2308.8	2643.8	1.0753	6.5380	7.6133	80
85	0.578 03	0.001 032 6	2.827 8	2.828 8	355.9	2296.1	2652.0	1.1343	6.4111	7.5454	85
90	0.701 09	0.001 036 1	2.360 3	2.361 3	376.9	2283.2	2660.1	1.1925	6.2873	7.4798	90
95	0.845 26	0.001 039 9	1.981 2	1.982 2	398.0	2270.2	2668.1	1.2501	6.1665	7.4166	95
100	1.013 25	0.001 043 7	1.672 0	1.673 0	419.1	2256.0	2676.0	1.3069	6.0485	7.3554	100
105	1.208 0	0.001 047 7	1.418 3	1.419 3	440.2	2243.6	2683.7	1.3630	5.9331	7.2962	105
110	1.432 7	0.001 051 9	1.208 9	1.209 9	461.3	2230.0	2691.3	1.4185	5.8203	7.2388	110
115	1.690 6	0.001 056 2	1.035 3	1.036 3	482.5	2216.2	2698.7	1.4733	5.7099	7.1832	115
120	1.985 4	0.001 060 6	0.890 46	0.891 52	503.7	2202.2	2706.0	1.5276	5.6017	7.1293	120
125	2.321 0	0.001 065 2	0.769 17	0.770 23	525.0	2188.0	2713.0	1.5813	5.4957	7.0769	125
130	2.701 3	0.001 070 0	0.667 07	0.668 14	546.3	2173.6	2719.9	1.6344	5.3917	7.0261	130
135	3.130 8	0.001 075 0	0.580 74	0.581 81	567.7	2158.9	2726.6	1.6869	5.2897	6.9766	135
140	3.613 8	0.001 080 1	0.507 41	0.508 49	589.1	2144.0	2733.1	1.7390	5.1894	6.9284	140
145	4.155 2	0.001 085 3	0.444 89	0.445 97	610.6	2128.7	2739.3	1.7906	5.0910	6.8815	145

t	p	v_f	v_{fg}	v_g	h_f	h_{fg}	h_g	s_f	s_{fg}	s_g	t
150	4·760 0	0·001 090 8	0·391 36	0·392 45	632·1	2113·2	2745·4	1·8416	4·9941	6·8358	150
155	5·433 3	0·001 094 6	0·345 55	0·346 44	653·8	2097·4	2751·2	1·8923	4·8989	6·7911	155
160	6·180 6	0·001 102 2	0·305 66	0·306 76	675·5	2081·3	2756·7	1·9425	4·8050	6·7475	160
165	7·007 7	0·001 108 2	0·271 29	0·272 40	697·3	2064·8	2762·0	1·9923	4·7126	6·7048	165
170	7·920 2	0·001 114 5	0·241 44	0·242 55	719·1	2047·9	2767·1	2·0416	4·6214	6·6630	170
175	8·924 4	0·001 120 9	0·215 54	0·216 54	741·1	2030·7	2771·8	2·0906	4·5314	6·6221	175
180	10·027	0·001 127 5	0·192 67	0·193 80	763·1	2013·2	2776·3	2·1393	4·4426	6·5819	180
185	11·233	0·001 134 4	0·172 73	0·173 86	785·3	1995·2	2780·4	2·1876	4·3548	6·5424	185
190	12·551	0·001 141 5	0·155 18	0·156 32	807·5	1976·7	2784·3	2·2356	4·2680	6·5036	190
195	13·987	0·001 148 9	0·139 69	0·140 84	829·9	1957·9	2787·8	2·2833	4·1821	6·4654	195
200	15·549	0·001 156 5	0·126 01	0·127 16	852·4	1938·6	2790·9	2·3307	4·0971	6·4278	200
205	17·243	0·001 164 4	0·113 87	0·115 03	875·0	1918·8	2793·8	2·3778	4·0128	6·3906	205
210	19·077	0·001 172 6	0·103 07	0·104 24	897·7	1898·5	2796·2	2·4247	3·9293	6·3539	210
215	21·060	0·001 181 1	0·093 45	0·094 625	920·6	1877·6	2798·3	2·4713	3·8463	6·3176	215
220	23·198	0·001 190 0	0·084 85	0·086 038	943·7	1856·2	2799·9	2·5178	3·7639	6·2817	220
225	25·501	0·001 199 2	0·077 15	0·078 349	966·9	1834·3	2801·2	2·5641	3·6820	6·2461	225
230	27·976	0·001 208 7	0·070 24	0·071 450	990·3	1811·7	2802·0	2·6102	3·6006	6·2107	230
235	30·632	0·001 218 7	0·064 03	0·065 245	1013·8	1788·5	2802·3	2·6561	3·5194	6·1756	235
240	33·478	0·001 229 1	0·058 42	0·059 645	1037·6	1764·6	2802·2	2·7020	3·4386	6·1406	240
245	36·523	0·001 239 9	0·053 37	0·054 606	1061·6	1740·0	2801·6	2·7478	3·3579	6·1057	245
250	39·776	0·001 251 3	0·048 79	0·050 037	1085·8	1714·7	2800·4	2·7935	3·2773	6·0708	250
255	43·246	0·001 263 2	0·044 64	0·045 896	1110·2	1688·5	2798·7	2·8392	3·1968	6·0359	255
260	46·943	0·001 275 6	0·040 86	0·042 130	1134·9	1661·5	2796·4	2·8848	3·1161	6·0010	260
265	50·877	0·001 288 7	0·037 43	0·038 710	1159·9	1633·5	2793·5	2·9306	3·0353	5·9658	265
270	55·058	0·001 302 5	0·034 29	0·035 588	1185·2	1604·6	2789·9	2·9763	2·9541	5·9304	270
275	59·496	0·001 317 0	0·031 42	0·032 736	1210·9	1574·7	2785·5	3·0222	2·8725	5·8947	275
280	64·202	0·001 332 4	0·028 80	0·030 126	1236·8	1543·6	2780·4	3·0683	2·7903	5·8586	280
285	69·186	0·001 348 7	0·026 38	0·027 733	1263·2	1511·3	2774·5	3·1146	2·7074	5·8220	285
290	74·461	0·001 365 9	0·024 17	0·025 535	1290·0	1477·6	2767·6	3·1611	2·6237	5·7848	290
295	80·037	0·001 384 4	0·022 13	0·023 513	1317·3	1442·6	2759·8	3·2079	2·5389	5·7469	295
300	85·927	0·001 404 1	0·020 25	0·021 649	1345·1	1406·0	2751·0	3·2552	2·4529	5·7081	300
305	92·144	0·001 425 2	0·018 50	0·019 927	1373·4	1367·7	2741·1	3·3029	2·3656	5·6685	305
310	98·700	0·001 448 0	0·016 89	0·018 334	1402·4	1327·6	2730·0	3·3512	2·2766	5·6278	310
315	105·61	0·001 472 6	0·015 38	0·016 856	1432·1	1285·5	2717·6	3·4002	2·1856	5·5858	315
320	112·89	0·001 499 5	0·013 98	0·015 480	1462·6	1241·1	2703·7	3·4500	2·0923	5·5423	320
325	120·56	0·001 528 9	0·012 67	0·014 195	1494·0	1194·0	2688·0	3·5008	1·9961	5·4969	325
330	128·63	0·001 561 5	0·011 43	0·012 989	1526·5	1143·6	2670·2	3·5528	1·8962	5·4490	330
335	137·12	0·001 597 8	0·010 26	0·011 854	1560·3	1089·5	2649·7	3·6063	1·7916	5·3977	335
340	146·05	0·001 638 7	0·009 14	0·010 780	1595·5	1030·7	2626·2	3·6616	1·6811	5·3427	340
345	155·45	0·001 685 8	0·008 08	0·009 763 1	1632·5	966·4	2598·9	3·7193	1·5636	5·2828	345
350	165·35	0·001 741 1	0·007 06	0·008 799 1	1671·9	895·7	2567·7	3·7800	1·4376	5·2177	350
355	175·77	0·001 808 9	0·006 04	0·007 859 2	1716·6	813·8	2530·4	3·8489	1·2953	5·1442	355
360	186·75	0·001 895 9	0·005 04	0·006 939 8	1764·2	721·3	2485·4	3·9210	1·1390	5·0600	360
365	198·33	0·002 016 0	0·003 99	0·006 011 6	1818·0	610·0	2428·0	4·0021	0·9558	4·9579	365
370	210·54	0·002 213 6	0·002 76	0·004 972 8	1890·2	452·6	2342·8	4·1108	0·7036	4·8144	370
374	220·81	0·002 842 7	0·000 63	0·003 465 9	2046·7	109·5	2156·2	4·3493	0·1692	4·5185	374
374·15	221·20	0·003 17	0	0·003 17	2107·4	0	2107·4	4·4429	0	4·4429	374·15

TABLE III Properties of Saturated Water and Saturated Steam. Pressure Table

Abs. Press. bar p	Temp. °C t	Specific volume m³/kg Sat. Liq. v_f	Specific volume Sat. Vap. v_g	Enthalpy kJ/kg Sat. Liq. h_f	Enthalpy Evap. h_{fg}	Enthalpy Sat. Vap. h_g	Internal Energy kJ/kg Sat. Liq. u_f	Internal Energy Sat. Vap. u_g	Entropy kJ/kg K Sat. Liq. s_f	Entropy Evap. s_{fg}	Entropy Sat. Vap. s_g	Abs. Press. bar p
0·006 02	0	0·001 000 2	206·298 7	−0·0	2501·6	2501·6	−0·0	2375·6	−0·0	9·1578	9·1578	0·006 02
0·006 11	0·01	0·001 000 2	206·162 9	+0·0	2501·6	2501·6	0	2375·6	0	9·1575	9·1575	0·006 11
0·010	6·98	0·001 001	129·210 7	29·3	2485·0	2514·4	29·3	2385·2	0·1060	8·8706	8·9767	0·010
0·020	17·51	0·001 001 2	67·011 6	73·5	2460·2	2533·6	73·5	2399·6	0·2606	8·4640	8·7246	0·020
0·030	24·10	0·001 002 7	45·670 0	101·0	2444·6	2545·6	101·0	2408·6	0·3543	8·2242	8·5785	0·030
0·040	28·98	0·001 004 0	34·803 3	121·4	2433·1	2554·5	121·4	2415·3	0·4225	8·0530	8·4755	0·040
0·050	32·90	0·001 005 2	28·194 5	137·8	2423·8	2561·6	137·8	2420·6	0·4763	7·9197	8·3960	0·050
0·060	36·18	0·001 006 4	23·740 6	151·5	2416·0	2567·5	151·5	2425·1	0·5209	7·8103	8·3312	0·060
0·070	39·03	0·001 007 4	20·530 4	163·4	2409·2	2572·6	163·4	2428·9	0·5591	7·7176	8·2767	0·070
0·080	41·54	0·001 008 4	18·103 8	173·9	2403·2	2577·1	173·9	2432·3	0·5926	7·6370	8·2295	0·080
0·090	43·79	0·001 009 4	16·203 4	183·3	2397·9	2581·1	183·3	2435·3	0·6224	7·5657	8·1881	0·090
0·10	45·83	0·001 010 2	14·673 7	191·8	2392·9	2584·8	191·8	2438·1	0·6493	7·5018	8·1511	0·10
0·15	54·00	0·001 014	10·022 1	226·0	2373·2	2599·2	226·0	2448·9	0·7549	7·2544	8·0093	0·15
0·20	60·09	0·001 017 2	7·649 2	251·5	2358·4	2609·9	251·5	2456·9	0·8321	7·0773	7·9094	0·20
0·25	64·99	0·001 019 9	6·204 0	272·0	2346·4	2618·3	272·0	2463·2	0·8933	6·9390	7·8323	0·25
0·30	69·13	0·001 022 3	5·229 0	289·3	2336·1	2625·4	289·6	2468·2	0·9441	6·8254	7·7695	0·30
0·35	72·71	0·001 024 5	4·525 5	304·3	2327·2	2631·5	304·3	2473·1	0·9878	6·7288	7·7166	0·35
0·40	75·89	0·001 026 5	3·993 2	317·7	2319·2	2636·9	317·7	2477·2	1·0261	6·6448	7·6709	0·40
0·45	78·74	0·001 028 4	3·576 1	329·6	2312·0	2641·7	329·6	2480·8	1·0603	6·5703	7·6306	0·45
0·50	81·35	0·001 030 1	3·240 1	340·6	2305·4	2646·0	340·5	2484·0	1·0912	6·5035	7·5947	0·50
0·60	85·95	0·001 033 3	2·731 7	359·9	2293·6	2653·6	359·8	2489·7	1·1455	6·3872	7·5327	0·60
0·70	89·96	0·001 036 1	2·364 7	376·8	2283·3	2660·1	376·3	2494·6	1·1921	6·2883	7·4804	0·70
0·80	93·51	0·001 038 7	2·086 9	391·7	2274·0	2665·8	391·6	2498·8	1·2330	6·2022	7·4352	0·80
0·90	96·71	0·001 041 2	1·869 1	405·2	2265·6	2670·9	405·1	2502·7	1·2696	6·1258	7·3954	0·90
1·00	99·63	0·001 043 4	1·693 7	417·5	2257·9	2675·4	417·4	2505·0	1·3027	6·0571	7·3598	1·00
1·013 25	100·00	0·001 043 7	1·673 0	419·1	2256·9	2676·0	419·0	2505·5	1·3069	6·0485	7·3554	1·013 25
1·20	104·81	0·001 047 6	1·428 1	439·4	2244·1	2683·4	439·3	2512·0	1·3609	5·9375	7·2984	1·20
1·40	109·32	0·001 051 3	1·236 3	458·4	2231·9	2690·3	458·3	2517·2	1·4109	5·8356	7·2465	1·40
1·60	113·32	0·001 054 7	1·091 1	475·4	2220·9	2696·2	475·2	2521·6	1·4550	5·7467	7·2017	1·60
1·80	116·93	0·001 057 9	0·977 18	490·7	2210·8	2701·5	490·5	2525·6	1·4944	5·6677	7·1622	1·80
2·00	120·23	0·001 060 8	0·885 40	504·7	2201·6	2706·3	504·5	2529·2	1·5301	5·5967	7·1268	2·00
2·50	127·43	0·001 067 6	0·718 40	535·4	2181·0	2716·4	535·1	2536·8	1·6072	5·4448	7·0520	2·50
3·00	133·54	0·001 073 5	0·605 53	561·4	2163·2	2724·7	561·1	2543·0	1·6717	5·3192	6·9909	3·00
3·50	138·88	0·001 078 9	0·523 97	584·3	2147·3	2731·6	583·9	2548·2	1·7273	5·2118	6·9392	3·50
4·00	143·63	0·001 083 9	0·462 20	604·7	2132·9	2737·6	604·3	2552·7	1·7764	5·1179	6·8943	4·00
4·50	147·92	0·001 088 5	0·413 73	623·2	2119·7	2742·9	622·7	2555·7	1·8204	5·0342	6·8547	4·50
5·00	151·85	0·001 092 8	0·374 66	640·1	2107·4	2747·5	639·6	2563·2	1·8604	4·9588	6·8192	5·00
6·00	158·84	0·001 100 9	0·315 46	670·4	2085·0	2755·5	669·7	2565·2	1·9308	4·8267	6·7575	6·00
7·00	164·96	0·001 108 2	0·272 68	697·1	2064·9	2762·0	696·3	2571·1	1·9918	4·7134	6·7052	7·00
8·00	170·41	0·001 115 0	0·240 26	720·9	2046·5	2767·5	720·0	2573·3	2·0457	4·6139	6·6596	8·00
9·00		0·001 …	0·211 …		2029 …	2772·1	741·6	2578·8	2·0941	4·5251	6·6192	9·00

p	t	v_f	v_g	h_f	h_fg	h_g	u_f	u_g	s_f	s_fg	s_g	p
10.00	179.88	0.001 127 4	0.194 30	762.6	2013.6	2776.2	761.5	2581.9	2.1382	4.4447	6.5828	10.00
11.00	184.06	0.001 133 1	0.177 39	781.1	1998.6	2779.7	779.9	2584.6	2.1786	4.3712	6.5498	11.00
12.00	187.96	0.001 138 6	0.163 21	798.4	1984.3	2782.7	797.0	2586.8	2.2160	4.3034	6.5194	12.00
13.00	191.60	0.001 143 8	0.151 14	814.7	1970.7	2785.4	813.2	2588.9	2.2509	4.2404	6.4913	13.00
14.00	195.04	0.001 148 9	0.140 73	830.1	1957.7	2787.8	828.5	2590.8	2.2836	4.1815	6.4651	14.00
15.00	198.28	0.001 153 8	0.131 67	844.6	1945.3	2789.9	842.7	2592.4	2.3144	4.1262	6.4406	15.00
16.00	201.37	0.001 158 6	0.123 70	858.5	1933.2	2791.7	856.6	2593.8	2.3436	4.0740	6.4176	16.00
17.00	204.30	0.001 163 3	0.116 64	871.8	1921.6	2793.4	869.8	2595.1	2.3712	4.0246	6.3958	17.00
18.00	207.11	0.001 167 8	0.110 33	884.5	1910.3	2794.8	882.4	2596.2	2.3976	3.9776	6.3751	18.00
19.00	209.79	0.001 172 3	0.104 67	896.8	1899.3	2796.1	894.6	2597.2	2.4227	3.9327	6.3555	19.00
20.00	212.37	0.001 176 6	0.099 549	908.6	1888.7	2797.2	906.2	2598.1	2.4468	3.8899	6.3367	20.00
25.00	223.94	0.001 197 2	0.079 915	961.9	1839.0	2800.9	958.9	2601.1	2.5542	3.6994	6.2537	25.00
30.00	233.84	0.001 216 3	0.065 632	1008.3	1794.0	2802.3	1004.7	2602.4	2.6455	3.5383	6.1838	30.00
35.00	242.54	0.001 234 5	0.057 028	1049.7	1752.2	2802.0	1045.4	2602.4	2.7252	3.3976	6.1229	35.00
40.00	250.33	0.001 252 1	0.049 749	1087.4	1712.9	2800.3	1082.4	2601.3	2.7965	3.2720	6.0685	40.00
45.00	257.41	0.001 269 1	0.044 035	1122.1	1675.6	2797.7	1116.4	2599.5	2.8612	3.1579	6.0191	45.00
50.00	263.92	0.001 285 8	0.039 425	1154.5	1639.7	2794.2	1148.1	2597.1	2.9207	3.0528	5.9735	50.00
55.00	269.94	0.001 302 3	0.035 624	1184.9	1605.0	2789.9	1177.7	2594.0	2.9758	2.9551	5.9309	55.00
60.00	275.56	0.001 318 7	0.032 433	1213.7	1571.3	2785.0	1205.8	2590.4	3.0274	2.8633	5.8907	60.00
65.00	280.83	0.001 335 0	0.029 714	1241.2	1538.3	2779.5	1232.5	2586.4	3.0760	2.7766	5.8526	65.00
70.00	285.80	0.001 351 4	0.027 368	1267.5	1506.0	2773.4	1258.0	2581.8	3.1220	2.6941	5.8161	70.00
75.00	290.51	0.001 367 8	0.025 323	1292.7	1474.1	2766.9	1282.4	2577.0	3.1658	2.6152	5.7810	75.00
80.00	294.98	0.001 384 3	0.023 521	1317.2	1442.7	2759.9	1306.1	2571.7	3.2077	2.5393	5.7470	80.00
85.00	299.24	0.001 401 0	0.021 923	1340.8	1411.6	2752.4	1328.9	2566.1	3.2480	2.4661	5.7141	85.00
90.00	303.31	0.001 417 9	0.020 493	1363.8	1380.8	2744.6	1351.0	2560.2	3.2867	2.3952	5.6820	90.00
95.00	307.22	0.001 435 1	0.019 206	1386.2	1350.2	2736.3	1372.6	2553.8	3.3242	2.3264	5.6506	95.00
100.00	310.96	0.001 452 6	0.018 041	1408.1	1319.7	2727.7	1393.6	2547.3	3.3606	2.2592	5.6198	100.00
110.00	318.04	0.001 488 7	0.016 007	1450.6	1258.8	2709.3	1434.2	2533.2	3.4304	2.1292	5.5596	110.00
120.00	324.64	0.001 526 7	0.014 285	1491.7	1197.5	2689.2	1473.4	2517.8	3.4971	2.0032	5.5003	120.00
130.00	330.81	0.001 567 1	0.012 800	1531.9	1135.1	2667.0	1511.5	2500.6	3.5614	1.8795	5.4409	130.00
140.00	336.63	0.001 610 5	0.011 498	1571.5	1070.9	2642.4	1549.0	2481.4	3.6241	1.7564	5.3804	140.00
150.00	342.12	0.001 657 8	0.010 343	1610.9	1004.2	2615.1	1586.0	2460.0	3.6857	1.6323	5.3180	150.00
160.00	347.32	0.001 710 2	0.009 309 9	1650.4	934.5	2584.9	1623.0	2435.9	3.7470	1.5063	5.2533	160.00
170.00	352.26	0.001 769 5	0.008 372 1	1691.6	860.0	2551.6	1661.5	2409.3	3.8106	1.3749	5.1856	170.00
180.00	356.96	0.001 839 6	0.007 497 3	1734.8	779.0	2513.9	1701.7	2378.9	3.8766	1.2362	5.1127	180.00
190.00	361.44	0.001 926 2	0.006 675 9	1778.7	691.8	2470.5	1742.1	2343.7	3.9430	1.0900	5.0330	190.00
200.00	365.71	0.002 037 4	0.005 874 5	1826.6	591.6	2418.2	1785.9	2300.7	4.0151	0.9259	4.9410	200.00
210.00	369.79	0.002 201 8	0.005 022 5	1886.3	461.2	2347.5	1840.1	2242.0	4.1049	0.7172	4.8222	210.00
220.00	373.68	0.002 667 5	0.003 693 7	2010.3	186.3	2196.6	1951.6	2144.4	4.2934	0.2881	4.5814	220.00
221.20	374.15	0.003 170 0	0.003 170 0	2107.4	0	2107.4	2037.3	2037.3	4.4429	0	4.4429	221.20

TABLE IV Properties of Superheated Steam
[Specific Volume, v, m^3/kg; Enthalpy, h, kJ/kg; Entropy, s, kJ/kg K]

Temperature—degrees Celsius

Abs. Press. bar (Sat. Temp. °C)		50	100	150	200	250	300	350	400	500	600	700	800
0·02 (17·5)	v	74·524	86·080	97·628	109·171	120·711	132·251	143·790	155·329	178·405	201·482	224·558	247·634
	h	2594·4	2688·5	2783·7	2880·0	2977·7	3076·8	3177·7	3279·7	3489·2	3705·6	3928·8	4158·7
	s	8·9226	9·1934	9·4327	9·6479	9·8441	10·0251	10·1934	10·3512	10·6413	10·9044	11·1464	11·3712
0·04 (29·0)	v	37·240	43·027	48·806	54·580	60·351	66·122	71·892	77·662	89·201	100·740	112·278	123·816
	h	2593·9	2688·3	2783·5	2879·9	2977·6	3076·8	3177·4	3279·7	3489·2	3705·6	3928·8	4158·7
	s	8·6016	8·8730	9·1125	9·3279	9·5241	9·7051	9·8735	10·0313	10·3214	10·5845	10·8265	11·0513
0·06 (36·2)	v	24·812	28·676	32·532	37·383	40·232	44·079	47·927	51·773	59·467	67·159	74·852	82·544
	h	2593·5	2688·0	2783·4	2879·8	2977·6	3076·7	3177·4	3279·6	3489·2	3705·6	3928·8	4158·7
	s	8·4135	8·6854	8·9251	9·1406	9·3369	9·5179	9·6863	9·8441	10·1342	10·3973	10·6394	10·8642
0·08 (41·5)	v	18·598	21·501	24·395	27·284	30·172	33·058	35·944	38·829	44·599	50·369	56·138	61·908
	h	2593·1	2687·8	2783·2	2879·7	2977·5	3076·7	3177·3	3279·6	3489·1	3705·5	3928·8	4158·7
	s	8·2797	8·5521	8·7921	9·0077	9·2041	9·3851	9·5535	9·7113	10·0014	10·2646	10·5066	10·7314
0·10 (45·8)	v	14·869	17·195	19·512	21·825	24·136	26·445	28·754	31·062	35·679	40·295	44·910	49·526
	h	2592·7	2687·5	2783·1	2879·6	2977·4	3076·6	3177·3	3279·6	3489·1	3705·5	3928·8	4158·7
	s	8·1757	8·4486	8·6888	8·9045	9·1010	9·2820	9·4504	9·6083	9·8984	10·1616	10·4036	10·6284
0·50 (81·3)	v	—	3·4181	3·8893	4·3560	4·8205	5·2839	5·7467	6·2091	7·1335	8·0574	8·9810	9·9044
	h	—	2682·6	2780·1	2877·7	2976·1	3075·7	3176·6	3279·0	3488·7	3705·2	3928·6	4158·5
	s	—	7·6953	7·9406	8·1587	8·3564	8·5380	8·7068	8·8649	9·1552	9·4185	9·6606	9·8855
1·00 (99·6)	v	—	1·6955	1·9363	2·1723	2·4061	2·6387	2·8708	3·1025	3·5653	4·0277	4·4898	4·9517
	h	—	2676·2	2776·3	2875·4	2974·5	3074·5	3175·6	3278·2	3488·1	3704·8	3928·2	4158·3
	s	—	7·3618	7·6137	7·8349	8·0342	8·2166	8·3858	8·5442	8·8348	9·0982	9·3405	9·5654
2·00 (120·2)	v			0·95954	1·0804	1·1989	1·3162	1·4328	1·5492	1·7812	2·0129	2·2442	2·4754
	h			2768·5	2870·5	2971·2	3072·1	3173·8	3276·7	3487·0	3704·0	3927·6	4157·8
	s			7·2794	7·5072	7·7096	7·8937	8·0638	8·2226	8·5139	8·7776	9·0201	9·2452
3·00 (133·5)	v			0·63374	0·71635	0·79644	0·87529	0·95352	1·0314	1·1865	1·3412	1·4957	1·6499
	h			2760·4	2865·5	2967·9	3069·7	3171·9	3275·2	3486·0	3703·2	3927·0	4157·3
	s			7·0771	7·3119	7·5176	7·7034	7·8744	8·0338	8·3257	8·5898	8·8325	9·0577
4·00 (143·6)	v			0·47066	0·53426	0·59519	0·65485	0·71385	0·77250	0·88919	1·0054	1·1214	1·2372
	h			2752·0	2860·4	2964·5	3067·2	3170·0	3273·6	3484·9	3702·3	3926·4	4156·9
	s			6·9285	7·1708	7·3800	7·5675	7·7395	7·8994	8·1919	8·4563	8·6992	8·9246

P (sat. T)		C1	C2	C3	C4	C5	C6	C7	C8	C9			
5·00 (151·8)	v	0·989 56	0·896 85	0·803 95	0·710 78	0·617 16	0·570 05	0·522 58	0·474 43	0·424 96	—	—	—
	h	4156·4	3925·8	3701·5	3483·8	3272·1	3168·1	3064·8	2961·1	2855·1	—	—	—
	s	8·8213	8·5957	8·3626	8·0879	7·7948	7·6343	7·4614	7·2721	7·0592	—	—	—
6·00 (158·8)	v	0·824 47	0·747 14	0·669 63	0·591 84	0·513 61	0·474 19	0·434 39	0·393 91	0·352 04	—	—	—
	h	4155·9	3925·1	3700·7	3482·7	3270·6	3166·2	3062·3	2951·6	2849·7	—	—	—
	s	8·7368	8·5111	8·2678	8·0027	7·7090	7·5479	7·3740	7·1829	6·9662	—	—	—
7·00 (165·0)	v	0·706 55	0·640 21	0·573 68	0·506 89	0·439 64	0·405 71	0·371 39	0·336 37	0·299 92	—	—	—
	h	4155·5	3924·5	3699·9	3481·6	3269·0	3164·3	3059·8	2954·0	2844·2	—	—	—
	s	8·6653	8·4395	8·1959	7·9305	7·6362	7·4745	7·2997	7·1066	6·8859	—	—	—
8·00 (170·4)	v	0·618 11	0·560 01	0·501 72	0·443 17	0·384 16	0·354 34	0·324 14	0·293 21	0·260 79	—	—	—
	h	4155·0	3923·9	3699·1	3480·5	3267·5	3162·4	3057·3	2950·4	2838·6	—	—	—
	s	8·6033	8·3773	8·1336	7·8678	7·5729	7·4107	7·2348	7·0397	6·8148	—	—	—
9·00 (175·4)	v	0·549 33	0·497 63	0·445 76	0·393 61	0·341 01	0·314 40	0·287 39	0·259 63	0·230 32	—	—	—
	h	4154·5	3923·3	3698·2	3479·4	3266·0	3160·5	3054·7	2946·8	2832·7	—	—	—
	s	8·5486	8·3225	8·0785	7·8124	7·5169	7·3540	7·1771	6·9800	6·7508	—	—	—
10·00 (179·9)	v	0·494 30	0·447 73	0·400 98	0·353 96	0·306 49	0·282 43	0·257 98	0·232 75	0·205 92	—	—	—
	h	4154·1	3922·7	3697·4	3478·3	3264·4	3158·5	3052·1	2943·0	2826·8	—	—	—
	s	8·4997	8·2734	8·0292	7·7627	7·4665	7·3031	7·1251	6·9259	6·6922	—	—	—
15·00 (198·3)	v	0·329 21	0·298 03	0·266 66	0·235 03	0·202 92	0·186 53	0·169 70	0·151 99	0·132 38	—	—	—
	h	4151·7	3919·6	3693·3	3472·8	3256·6	3148·7	3038·9	2923·5	2794·7	—	—	—
	s	8·3108	8·0838	7·8385	7·5703	7·2709	7·1044	6·9207	6·7099	6·4508	—	—	—
20·00 (212·4)	v	0·246 66	0·223 17	0·199 50	0·175 55	0·151 13	0·138 66	0·125 50	0·111 45	—	—	—	—
	h	4149·4	3916·5	3689·2	3467·3	3248·7	3138·6	3025·0	2902·4	—	—	—	—
	s	8·1763	7·9485	7·7022	7·4323	7·1296	6·9596	6·7696	6·5454	—	—	—	—
25·00 (223·9)	v	0·197 14	0·178 26	0·159 21	0·139 87	0·120 04	0·109 75	0·098 925	0·086 985	—	—	—	—
	h	4147·0	3913·4	3685·1	3461·7	3240·7	3128·2	3010·4	2879·5	—	—	—	—
	s	8·0716	7·8431	7·5956	7·3240	7·0178	6·8442	6·6470	6·4077	—	—	—	—
30·00 (233·8)	v	0·164 12	0·148 32	0·132 34	0·116 08	0·099 310	0·090 526	0·081 159	0·070 551	—	—	—	—
	h	4144·7	3910·3	3681·0	3456·5	3232·5	3117·5	2995·1	2854·8	—	—	—	—
	s	7·9857	7·7564	7·5079	7·2345	6·9246	6·7471	6·5422	6·2857	—	—	—	—
35·00 (242·5)	v	0·140 54	0·126 94	0·113 15	0·099 088	0·084 494	0·076 776	0·068 424	0·058 693	—	—	—	—
	h	4142·4	3907·2	3676·9	3450·6	3224·2	3106·5	2979·0	2828·1	—	—	—	—
	s	7·9128	7·6828	7·4332	7·1580	6·8443	6·6626	6·4491	5·1732	—	—	—	—
40·00 (250·3)	v	0·122 85	0·110 90	0·098 763	0·086 341	0·073 376	0·066 446	0·058 833	—	—	—	—	—
	h	4140·0	3904·1	3672·8	3445·0	3215·7	3095·1	2962·0	—	—	—	—	—
	s	7·8495	7·6187	7·3680	7·0909	6·7733	6·5870	6·3642	—	—	—	—	—
45·00 (257·4)	v	0·109 10	0·098 425	0·087 570	0·076 427	0·064 721	0·058 696	0·051 336	—	—	—	—	—
	h	4137·7	3901·0	3668·6	3439·3	3207·1	3083·3	2944·2	—	—	—	—	—
	s	7·7934	7·5619	7·3100	7·0311	6·7093	6·5182	6·2852	—	—	—	—	—

TABLE VI (Cont.) Properties of Superheated Steam

[Specific Volume, v, m³/kg; Enthalpy, h, kJ/kg; Entropy, s, kJ/kg K]

Abs. Press. bar (Sat. Temp. °C)		50	100	150	200	250	300	350	400	500	600	700	800
50·00 (263·9)	v	—	—	—	—	—	0·045 301	0·051 941	0·057 791	0·068 494	0·078 616	0·088 446	0·098 093
	h						2925·5	3071·2	3198·3	3433·7	3664·5	3897·9	4135·3
	s						6·2105	6·4545	6·6508	6·9770	7·2578	7·5108	7·7431
60·00 (275·6)	v	—	—	—	—	—	0·036 145	0·042 222	0·047 379	0·056 592	0·065 184	0·073 478	0·081 587
	h						2885·0	3045·8	3180·1	3422·2	3656·2	3891·7	4130·7
	s						6·0692	6·3386	6·5462	6·8818	7·1664	7·4217	7·6554
70·00 (285·8)	v	—	—	—	—	—	0·029 457	0·035 233	0·039 922	0·048 086	0·055 590	0·062 787	0·069 798
	h						2839·4	3018·7	3161·2	3410·6	3647·9	3885·4	4126·0
	s						5·9327	6·2333	6·4536	6·7993	7·0880	7·3456	7·5808
80·00 (295·0)	v	—	—	—	—	—	0·024 264	0·029 948	0·034 310	0·041 704	0·048 394	0·054 770	0·060 956
	h						2786·8	2989·9	3141·6	3398·8	3639·5	3879·2	4121·5
	s						5·7942	6·1349	6·3694	6·7262	7·0191	7·2790	7·5158
90·00 (303·3)	v	—	—	—	—	—	—	0·025 792	0·029 29	0·036 737	0·042 798	0·048 534	0·054 080
	h							2959·0	3121·2	3386·8	3631·1	3873·0	4116·7
	s							6·0408	6·2915	6·6600	6·9574	7·2196	7·4579
100·00 (311·0)	v	—	—	—	—	—	—	0·022 421	0·026 408	0·032 760	0·038 320	0·043 546	0·048 580
	h							2925·8	3099·9	3374·6	3622·7	3866·8	4112·0
	s							5·9489	6·2182	6·5994	6·9013	7·1660	7·4058
125·00 (327·8)	v	—	—	—	—	—	—	0·016 122	0·020 010	0·025 590	0·030 259	0·034 510	0·038 682
	h							2828·0	3042·9	3343·3	3601·4	3851·1	4100·3
	s							5·7155	6·0481	6·4654	6·7796	7·0504	7·2942
150·00 (342·1)	v	—	—	—	—	—	—	0·011 462	0·015 661	0·020 795	0·024 884	0·028 587	0·032 086
	h							2694·8	2979·1	3310·6	3579·8	3835·4	4088·6
	s							5·4467	5·8876	6·3487	6·6764	6·9536	7·2013
175·00 (354·6)	v	—	—	—	—	—	—	—	0·012 460	0·017 359	0·021 043	0·024 314	0·027 376
	h								2906·3	3276·5	3557·8	3819·7	4077·0
	s								5·7274	6·2432	6·5858	6·8698	7·1215
200·00 (365·7)	v	—	—	—	—	—	—	—	0·009 947 0	0·014 771	0·018 161	0·021 111	0·023 845
	h								28·205	3241·1	3535·5	3803·8	4065·3
	s								5·5585	6·1456	6·5043	6·7953	7·0511

300·00	v	0·002 830 6	0·008 680 8	0·011 436	0·013 647	0·015 619
	h	2161·8	3085·0	3443·0	3739·7	4018·5
	s	4·4896	5·7972	6·2340	6·5560	6·8288
400·00	v	0·001 909 1	0·005 615 6	0·008 088 4	0·009 930 2	0·011 521
	h	1934·1	2906·8	3346·4	3674·8	3971·7
	s	4·1190	5·4762	6·0135	6·3701	6·6606
500·00	v	0·001 729 1	0·003 882 2	0·006 111 3	0·007 719 7	0·009 075 9
	h	1877·7	2723·0	3248·3	3610·2	3925·3
	s	4·0083	5·1782	5·8207	6·2138	6·5222
600·00	v	0·001 632 4	0·002 951 5	0·004 835 0	0·006 269 0	0·007 460 3
	h	1847·3	2570·6	3151·6	3547·0	3879·6
	s	3·9383	4·9374	5·6477	6·0775	6·4031
700·00	v	0·001 567 1	0·002 466 8	0·003 971 9	0·005 256 6	0·006 320 8
	h	1827·8	2467·1	3060·4	3486·3	3835·3
	s	3·8855	4·7688	5·4931	5·9562	6·2979
800·00	v	0·001 518 0	0·002 188 1	0·003 379 2	0·004 519 3	0·005 480 5
	h	1814·2	2397·4	2980·3	3428·7	3792·8
	s	3·8425	4·6488	5·3595	5·8470	6·2034
900·00	v	0·001 478 8	0·002 012 9	0·002 966 8	0·003 964 2	0·004 840 7
	h	1804·6	2349·9	2913·5	3374·6	3752·4
	s	3·8059	4·5602	5·2468	5·7479	6·1179
1000·00	v	0·001 446 4	0·001 893 4	0·002 668 1	0·003 535 6	0·004 341 1
	h	1797·6	2316·1	2857·5	3324·4	3714·3
	s	3·7738	4·4913	5·1505	5·6579	6·0397

Appendix C
Property Data Relating to Fuels and Combustion

TABLE C.1 *Relative Atomic Masses of Elements*

Element	H	C	N	O	S
Relative atomic masses	1·007 97	12·011 15	14·0067	15·9994	32·064

TABLE C.2 *Volume Analyses of Some Fuel-Gas Mixtures* (%)

	CO	H_2	CH_4	C_2H_4	C_2H_6	C_4H_8	O_2	CO_2	N_2
Coal gas (Town gas)	9	53·6	25	—	—	3	0·4	3	6
Producer gas	29	12	2·6	0·4	—	—	—	4	52
Blast-furnace gas	27	2	—	—	—	—	—	11	60
Natural gas (U.K.)	1	—	93	—	3	—	—	—	3
Natural gas (U.S.A.)	—	—	80	—	18	—	—	—	2
Natural gas (U.S.S.R.)	—	1	93	—	3·5	—	—	2	0·5

TABLE C.3 *Boiling Points of Hydrocarbon Fuels*

Name	Chemical formula	Molecular mass (rounded value)	Boiling point at $p = 1$ atm
Methane	CH_4	16	−161·4 °C
Ethylene	C_2H_4	28	−103·9
Ethane	C_2H_6	30	−89·0
Propylene	C_3H_6	42	−47·6
n-Butane	C_4H_{10}	58	−0·5
n-Pentane	C_5H_{12}	72	36·0
Benzene	C_6H_6	78	80·1
Toluene	C_7H_8	92	110·7
n-Octane	C_8H_{18}	114	125·6

TABLE C.4 *Liquid Fuels: Mass Analysis* (%)

Fuel	Carbon	Hydrogen	Sulphur	Ash etc.
Aviation gasoline (100 Octane)	85·1	14·9	0·01	—
Motor gasoline	85·5	14·4	0·1	—
Vaporizing oil	86·2	12·9	0·3	—
Motor benzole	91·7	8·0	0·3	—
Kerosine	86·3	13·6	0·1	—
Diesel oil distillate (Gas oil)	86·3	12·8	0·9	—
Light fuel oil	86·2	12·4	1·4	—
Heavy fuel oil	86·1	11·8	2·1	—
Residual fuel oil (Bunker C)	88·3	9·5	1·2	1·0

TABLE C.5 *Some Properties of Solid Fuels*

Fuel	Moisture content of good commercial fuel: % by mass	Analysis of good commercial fuel: % by mass in dry fuel					Volatile matter: % by mass in dry fuel
		C	H	O	N + S	Ash	
Anthracite	1	90·27	3·00	2·32	1·44	2·97	4
Bituminous coal	2	81·93	4·87	5·98	2·32	4·90	25
Lignite	15	56·52	5·72	31·89	1·62	4·25	50
Peat	20	43·70	6·42	44·36	1·52	4·00	65
Wood	15	42·5	6·78	49·87	0·85	Trace	80

TABLE C.6 *Enthalpy of Formation*

Species	Reaction	State	$\tilde{h}_f$ J/kmol of species
O_2	—	gas at 25 °C, 1 atm	0, by definition
H_2	—	gas at 25 °C, 1 atm	0, by definition
C	—	solid at 25 °C, 1 atm	0, by definition
CO_2	$C + O_2 = CO_2$	gas at 25 °C, 1 atm	$-393\cdot776 \times 10^6$
CO	$C + \frac{1}{2}O_2 = CO$	gas at 25 °C, 1 atm	$-110\cdot599 \times 10^6$
H_2O	$H_2 + \frac{1}{2}O_2 = H_2O$	gas at 25 °C, 1 atm	$-241\cdot988 \times 10^6$
H_2O	$H_2 + \frac{1}{2}O_2 = H_2O$	liquid at 25 °C, 1 atm	$-286\cdot030 \times 10^6$
CH_4	$C + 2H_2 = CH_4$	gas at 25 °C, 1 atm	$-74\cdot897 \times 10^6$
C_2H_6	$2C + 3H_2 = C_2H_6$	gas at 25 °C, 1 atm	$-84\cdot725 \times 10^6$

TABLE C.7 *Typical Calorific Values of Fuels*

Solid fuels

Fuel	Calorific value, MJ/kg (Bomb calorimeter, 15 °C)	
	Higher	Lower
Anthracite	34·583	33·913
Bituminous coal	33·494	32·406
Lignite	21·646	20·390
Peat	15·910	14·486
Wood	15·826	14·319
Coke	30·731	30·480

Liquid fuels

Fuel	Calorific value, MJ/kg (Bomb calorimeter, 15 °C)	
	Higher	Lower
Aviation gasoline (100 octane)	47·311	44·003
Motor gasoline	46·892	43·710
Vaporizing oil	46·055	43·210
Motor benzole	41·973	40·193
Kerosine	46·180	43·166
Diesel oil, distillate (Gas oil)	45·971	43·166
Light fuel oil	44·799	42·077
Heavy fuel oil	43·961	41·366
Residual fuel oil (Bunker C)	42·054	39·961

Gaseous fuels

Fuel	Calorific value MJ/m³ at 15 °C, 1 atm	
	Higher	Lower
Coal gas (Town gas)	20·11	17·95
Producer gas	6·06	6·02
Blast-furnace gas	3·44	3·40
Natural gas (U.K.)	36·38	32·75
Carbon monoxide gas, CO	11·84	11·84
Hydrogen gas, H_2	11·92	10·05

TABLE C.8

Reaction	$\Delta H'$ J/kmol	$\Delta(H' - TS')$ J/kmol
$H_{2\ gas} + \frac{1}{2}O_{2\ gas} = H_2O_{\ gas}$	$-241{\cdot}988 \times 10^6$	$-227{\cdot}748 \times 10^6$
$C_{\ solid} + O_{2\ gas} = CO_{2\ gas}$	$-393{\cdot}776 \times 10^6$	$-393{\cdot}427 \times 10^6$
$C_{\ solid} + \frac{1}{2}O_{2\ gas} = CO_{gas}$	$-110{\cdot}599 \times 10^6$	$-113{\cdot}884 \times 10^6$

Answers to Problems

Chapter 1

1.1   (a) 3.6×10^6 N m; 2.68×10^6 N m.
1.2   (a) 2.43 kg/kWh; 3.99 lb_m/hp h; (b) 18.1 kg/kWh; 29.8 lb_m/hp h;
     (c) 2.89 kg/kWh; 4.75 lb_m/hp h; (d) 1.27 kg/kWh; 2.08 lb_m/hp h.
1.3   (a) 0.250 kg/kWh; (b) 1.42 kg/kWh; (c) 0.498 kg/kWh; (d) 0.357 kg/kWh;
     (e) 0.429 kg/kWh; (f) 13.9×10^{-6} kg/kWh.
1.4   (a) 0.960 kg/h N; (b) 0.0955 kg/h N; (c) 0.0552 kg/h N.
1.5   (a) 4 kW; (b) 3.795 kW; (c) 0.435 kg/kWh.

Chapter 2

2.1   2 400 N.
2.2   (a) 8 000 cm/s²; (b) 17.6×10^3 cm/s².
2.3   (a) 100 kg m/c s²; (b) 1c = 100 N; (c) numerically the mass is 10.92 times the
     weightforce.
2.4   (a) 8.91 m/s²; (b) 766 N.
2.5   14.8 lbf/in².
2.6   (a) 1.99 kN/m²; (b) 169 m.
2.7   1 503 kN/m²; 8.27 kN/m².
2.8   12.22 in; 76.8%.

Chapter 3

3.1   45 N m.   **3.2**   (a) 3.5 N m; (b) 1800 N m; −1800 N m;
3.3   10 N m.
3.4   (a) 490.5 N m; −490.5 N m; (b) 1 800 N m; −1 800 N m;
     (c) 120 N m; −120 N m; (d) 0; (e) 0.
3.5   −2760 N m; −39 240 N m; 0; 0.
3.6   (a) −6.0 N m; (b) 20 N m; (c) −14 N m.
3.7   (a) 990 N m; (b) 990 N m.
3.8   7069 N m; 70.7 kW.
3.9   (a) 25×10^6 N/m³; (b) 566 mmm⁰.
3.10 367 kW.
3.11 (i) Sketch;

$$(ii) \ W = \frac{n}{n-1} p_1 V_{sw} \left[1 - \frac{c}{100} \left\{ \left(\frac{p_2}{p_1}\right)^{\frac{1}{n}} - 1 \right\} \right] \left[1 - \left(\frac{p_2}{p_1}\right)^{\frac{n-1}{n}} \right];$$

     (iii) −415 N m/cycle; (iv) −2.70 kW.
3.12 1443 N m.

$$\text{(i) Sketch; (ii) } p_m = \frac{p_1}{100} \left[a + (a+c) \ln \frac{100+c}{a+c} \right] - p_e$$

3.13
     (iii) 276×10^3 N/m²; 3 220 N m/cycle; (iv) 14.15 kW.
3.14 1225 J.
3.15 (a) Sketch; (b) (i) −300 N m; (ii) 0; (iii) −82 N m; (iv) 0; (v) 382 N m.
3.16 $−22.8 \times 10^3$ N m.
3.17 58.2×10^3 N m.
3.18 81.7 kW.
3.19 (a) 99.5×10^3 N m; (b) 14 520 kW; (c) $−480 \times 10^3$ N m/s.
3.20 0.079 N m; 6.96 W.
3.21 (a) 5040 N m; (b) −870 N m.

Chapter 4
4.1   (*a*) 636·8 °F;   (*b*) −15 °C;   (*c*) 30·2 °F;   (*d*) 565·6 °C;   (*e*) 270 K.
4.2   22·8 °C.
4.3   (*a*) 24·27 °C;   (*b*) See page 81.
4.4   (*a*) $t_c = 0·273\,p − 273$;   (*b*) 20·5 °C.

Chapter 5
5.1   (i) $0·239 × 10^{-3}$ kcal;   (ii) $1·055 × 10^{-3}$ Btu.
5.2   (i) 5·02 kJ;   (ii) −5·02 kJ;   (iii) 0.
5.3   (i) 100·5 kJ;   (ii) 5·02 kJ;   (iii) −105·52 kJ;   (iv) 0.
5.4   (i) 5·02 kJ;   (ii) −105·52 kJ;   (iii) 100·5 kJ;   (iv) 0.
5.5   0.
5.6

	(*a*)	(*b*)	(*c*)	(*d*)	(*e*)	(*f*)	(*g*)	(*h*)
Q	−	0	+	+	0	0	0	0
W	0	+	0	−	−	−	0	−

5.7

	(*a*)	(*b*)
Q	0	−0·251 kJ
W	0	0

5.8

	(*a*)	(*b*)	(*c*)
Q	0	−	−
W	0	0	0

Chapter 6
6.1   (*a*) −23 350 ft lbf;   (*b*) 778 ft lbf/Btu;   (*c*) +30 Btu in each vessel.
6.2   (*a*) +0·560 kN m;   −0·640 kN m;   (*b*) −0·560 kN m;   (*c*) (i) +0·640 kJ;   (ii) −0·640 kJ;   (*d*) 0.
6.3   1·0.
6.4   −6 kJ.
6.5   (*a*) −;   (*b*) −;   (*c*) +;   (*d*) +;   (*e*) +;   (*f*) +;(*g*) 0;   (*h*) +.
6.6   (*a*) 0;   (*b*) −.
6.7   (*a*) 0;   (*b*) −;   (*c*) −.
6.8   (i) −4000 N m;   (ii) +4000 N m;   (iii) 0.
6.9   0.
6.10   0.
6.11   (*a*) −5 kJ;   (*b*) +30 kJ.
6.12   (*a*) +54 kJ;   (*b*) −4 kJ; (*c*) 0.
6.13

	Q	W	ΔE
(i)	0	−	+
(ii)	0	+	1

6.14

	Q	W	ΔE
(i)	0	−	+
(ii)	0	0	0

6.15

	Q	W	ΔE
(*a*)	0	+	−
(*b*)	0	+	−
(*c*)	−	0	−

6.16   (*a*) $−392 × 10^3$ N m;   0; $861 × 10^3$ N m;   0;   (*b*) 371 kJ;   (*c*) +392 kJ;   840 kJ;   −861 kJ;   −371 kJ.
6.17   (*a*) 0;   (*b*) −45 kJ;   (*c*) 30 kJ;   −15 kJ.
6.18   −4·16 kJ.
6.19   (*a*) 13 kJ;   13 kJ;   −19 kJ;   (*b*) −3 kJ;   −1 kJ. The process is not cyclic; the system is not restored to its initial state because chemical reaction has occurred.

Chapter 7

7.1 (*a*) 200 °C; 0; (*b*) 12 kJ.
7.2 (*a*) 16 kJ; (*b*) −10 kJ.
7.3 (*a*) 16·0 kJ/kg; (*b*) −584·1 kJ; (*c*) 600·6 kJ.
7.4 137·3 kJ/kg.
7.5 Proof.
7.6 Proof.
7.7 Proof.
7.8 (*a*) 418·6 kJ; (*b*) 83·7 kJ; (*c*) (i) −118·5 kJ; (ii) −165·8 kJ. No. Only ($Q - W$) may be evaluated.
7.9 (*a*) 0·658 kJ/kg K; 0·918 kJ/kg K; (*b*) Proof;
(*c*) 310·5 °C; 20·1 kJ; 633 × 10³ N/m².
7.10 2·03 kJ/kg K; 2·20 kJ/kg K.
7.11 (*a*) −9·3 °C; (*b*) 4·8 kN m.
7.12 (*a*) 36·02 kJ; (*b*) +2·30 kJ.

Chapter 8

8.1 0·144 kg/s.
8.2 −49·8 kJ/kg.
8.3 518 m/s.
8.4 (*a*) 143·3 kJ/kg; (*b*) 152·9 kJ/kg.
8.5 (*a*) 111 m/s; (*b*) 2057·8 kJ/kg.
8.6 (*a*) 2770 kJ/kg; (*b*) 1·04 m/s; 10·19 m/s.
8.7 +7·96 kJ/kg.
8.8 14·97 m/s; 155·6 m/s; 1·39 kg/s.
8.9 (*a*) 490 m/s, (*b*) 169 m/s, (*c*) 0; (*d*) 105·8 kJ/kg; (*e*) 4·23 kW;
(*f*) 2600 kJ/kg; 2600 kJ/kg; 2494·2 kJ/kg.
8.10 0·436 kg/s.
8.11 2462 kJ/kg.
8.12 43500 kJ/kg.
8.13 −1877 kJ/kg mixture.

Chapter 9

9.1 (*a*) $v_f = 0\cdot001\ 000\ 4$ m³/kg; (*b*) $v_g = 0\cdot127\ 16$ m³/kg; (*c*) $t_f = 100$ °C;
(*d*) $t_g = 100$ °C; (*e*) $t = 100$ °C; (*f*) $h_{fg} = 2079\cdot0$ kJ/kg; (*g*) $h_g = 2676\cdot0$ kJ/kg;
(*h*) $u_f = 1063\cdot9$ kJ/kg;
(*i*) $v = 0\cdot772\ 50$ m³/kg; $h = 3273\cdot6$ kJ/kg; $u = 2964\cdot6$ kJ/kg;
(*j*) $t = 317\cdot7$ °C; $v = 0\cdot031\ 506$ m³/kg; (*k*) $p = 945$ kN/m²; $u = 2876\cdot1$ kJ/kg.
9.2 Sketch.
9.3 (*a*) 2592·0 kJ/kg; 2414·6 kJ/kg; 0·126 77 m³/kg;
(*b*) 2628·9 kJ/kg; 2449·3 kJ/kg; 0·045 158 m³/kg.
9.4 3·347 8 MN/m²; 8·506 kg; 0·004 70 m³/kg; 9317·9 kJ; 9583·0 kJ.
9.5 (*a*) 0·347 kg; 912 kJ; 851 kJ;
(*b*) 112·7 kJ; 92·3 kJ; 20·4 kJ; 112·7 kJ.
9.6 (*a*) 0·58 m³/kg; 133·54 °C; 0·958; 2459·3 kJ/kg; 2633·3 kJ/kg;
(*b*) 342·1 kN/m²; 144·8 kJ/kg; 120·4 kJ/kg; 120·4 kJ/kg; 315·75 kN/m².
9.7 (*a*) 0·003 17 m³; (*b*) $m_f = 0\cdot998\ 28$ kg; $m_g = 0\cdot001\ 72$ kg; (*c*) 405·8 °C;
(*d*) 803 kJ.
9.8 8·62 kg.
9.9 0·971.
9.10 0·916; no ($h_1 < h_g$ at 100 kN/m²).
9.11 (*a*) 0·961; no ($h_1 < h_g$ at 1 atm).

9.12   (*a*) 1·24;   (*b*) 4·69 kJ;   (*c*) −1·83 kJ.
9.13   0·901.
9.14   (*a*) −62·85 kJ/kg;   (*b*) −62·9 kJ/kg.
9.15   (*a*) 830 m/s;   0·951 kg/s;   (*b*) 38·7 kg/s.
9.16   0·871;   0·266 m².
9.17   (*a*) 400 kN/m²;   (*b*) 4·61 kg.
9.18   482 kg.
9.19   +6·47 kJ.

Chapter 10

10.1   23·9%;   68·5 kJ.
10.2   500 kW;   400 kW.
10.3   (*a*) 54·1%;   (*b*) 6968 kW;   3768 kW.
10.4   (*a*) 2614·0 kJ/kg;   (*b*) 248·1 kJ/kg;   (*c*) 2212·8 J/kg;
(*d*) 153·1 kJ/kg;   (*e*) 9·49%.
10.5   (*a*) 824 °C;   (*b*) 493 kJ/kg;   (*c*) 149 kJ/kg;   (*d*) 23·2%.
10.6   295 kJ;   2·89;   3·89.
10.7   (*a*) 300 kW;   250 kW;   (*b*) 3·59 kg/s.
10.8   4·26;   222·4 Btu/min.
10.9   (*a*) 11·1 × 10⁻³ kg/s;   (*b*) 1·19 × 10³ J/s;   (*c*) 64·6 kJ/kg;
(*d*) 171·7 kJ/kg;   0·941;   (*e*) 3·84.
10.10 1·81.
10.11 Proof.

Chapter 11

11.1   (*a*) 45%;   (*b*) (i) 2·22;   (ii) 3·375 kW;   (*c*) 1·22.
11.2   195·9, 1938·6;   707·8, 0;   −81·0, −1282·9;   −167·2, 0 kJ/kg;   ΣW = 655·5
kJ/kg;   33·8%.
11.3   0, 1938·6;   789·7, 0; 0, −1282·9;   −134·2, 0 kJ/kg;   ΣW = 655·5 kJ/kg;   33·8%.
11.4   Possible according to the First Law (ΣQ = ΣW). Impossible according to the
Second Law (PMM2).
11.5   Proof.

Chapter 12

12.1   (*a*) 33·1%;   (*b*) 6·04 kJ, 4·04 kJ;   (*c*) 3·02, 19·8 kW.
12.2   16·67%;   87 °C;   27 °C.
12.3   14·23 kW.
12.4   63·6%.
12.5   6·11 kWh.
12.6   (*a*) 40 kW;   4·0;   (*b*) 12·2;   3·28 kW.
12.7   166·7 kW.
12.8   Claim not valid unless a cold system at a very low temperature is available. Even
with no thermal irreversibility and an internally reversible engine a cold system
at −11 °C is required.
12.9   (*a*) 3934 kJ;   5584 kJ;   (*b*) 275 kJ;   1925 kJ.

Chapter 13

13.1   (*a*) 0·360 kJ/K;   (*b*) 0·156 kJ/kg K;   (*c*) 0·156 kJ/kg K.
13.2 s_g = 6·8358 kJ/kg K;   (*b*) s_f = 0·5591 kJ/kg K;
(*c*) *s* = 6·2339 kJ/kg K;   (*d*) *s* = 8·4563 kJ/kg K.

13.3   (a) 0·875; 133·59 °C; 6·3260 kJ/kg K;
(b) 918·5 kN/m²; 0·254 66 m³/kg; 2713·6 kJ/kg;
(c) 3·9776 MN/m²; 0·043 939 m³/kg; 2411·3 kJ/kg.

13.4   Sketch.

13.5   24·717 kJ/K.

13.6   4·6138 kJ/kg K; 1855·2 kJ/kg.

13.7   (a) 1·4391 kJ/kg K; 704·3 kJ/kg; (b) 235·5 kJ/kg; 273·5 kJ/kg;
(c) 468·8 kJ/kg; 468·8 kJ/kg; (d) 430·8 kJ/kg; 430·8 kJ/kg.

13.8   (a) 179·88 °C; 135·41 °C; (b) 16·09 kJ; (c) 0·0457 kJ/K; (d) 19·48 kJ.

13.9   (a) 254·8 °C; (b) 362·0 kJ/kg; (c) 428·2 kJ/kg; (d) 362·0 kJ/kg; (e) 428·2 kJ/kg.

13.10   (a) −41·2 kJ, (b) +0·1193 kJ/kg K.

13.11   (a) 341·5 kJ/kg; (b) 462·9 kJ/kg; (c) 73·8%.

13.12   Check.

13.13   (a) +1957·7 kJ/kg, 0; 0, +445·9 kJ/kg; −1558·6 kJ/kg, 0; 0, −46·8 kJ/kg.
(b) Σw_x = 399·1 kJ/kg = Σq; 20·4%, 20·4%.

13.14   (a) +2370·3 kJ/kg, 0; 0, +445·9 kJ/kg; −1924·4 kJ/kg, 0; ≈0.
(b) Σw_x = 445·9 kJ/kg = Σq; (c) $\eta_{Rankine}$ = 18·8%; η_R = 20·4%.

13.15   (a) Proof; (b) −3·02 kJ/kg.

13.16   (a) 0·929; 715·7 kJ/kg; (b) −6 kJ/kg; (c) 24·1%; 69·7%; 24·2%; 70·0%.

13.17   (a) 0·895; 68·1 kJ/kg; (b) 0·185; (c) 302·0 kJ/kg; (d) 4·43, 5·06.

13.18   (a) 0·978; 751 m/s; (b) 713 m/s; 0·989; +0·0689 kJ/kg K;
(c) A_a/A_b = 0·937.

13.19   (a) 2·3307 kJ/kg K; 6·4278 kJ/kg K; 6·7541 kJ/kg K; 2·2112 kJ/kg K.

(b) Irreversible; Δs exceeds $\int \dfrac{dQ}{T}$ in each case.

(c) +1938·6 kJ/kg; 0; −1693·4 kJ/kg; 0; 12·65%.
(d) −0·446 kJ/kg K; (e) 0; 21·25%.

13.20   (a) Proof; (b) +0·0579 kg J/K; (c) +0·0700 kJ/K; (d) +0·0874 kJ/K.

13.21   (a) +0·2686 kJ/kg K; 0; (b) +0·0689 kJ/kg K; 0; (c) +0·0579 kJ/kg K; 0;
(d) +0·0700 kJ/K; 0; (e) +0·0874 kJ/K; 0.

Chapter 14

14.1   Units: p, N/m²; v, m³/kg; T, K.

(i)
pv	195·3 × 10³	193·6 × 10³	188·3 × 10³
$\dfrac{pv}{T}$	462	458	445

(ii)
pv	310·7 × 10³	310·3 × 10³	309·0 × 10³	289·0 × 10³	198·9 × 10³
$\dfrac{pv}{T}$	462	461	459	429	296

(iii)
pv	495·3 × 10³	495·2 × 10³	494·9 × 10³	490·5 × 10³	476·9 × 10³
$\dfrac{pv}{T}$	462	461	461	457	444

14.2   (i) <0·2 kJ/kg; (ii) 79·0 kJ/kg.

14.3   (i) −0·09 K; (ii) −39·5 K.

14.4   Check.

14.5   0·0778 kg; 0·002 68 kmol.

14.6   (i) 1·647 kg; (ii) 1·665 kg. Answer (ii); steam is not an Ideal Gas.

14.7   (a) 37·5 kg; 1·163 MN/m²; 0·1067 m³/kg; 0·3326 kJ/kg K;
(b) 1·208 kJ/kg K; 0·875 kJ/kg K; (c) 898 kN/m²;
(d) −2789 kJ; −3847 kJ; −2789 kJ.

14.8   Sketch.

14.9   628 kN/m²; 489 °C; +100·5 kJ; +351·6 kJ.

14.10   (i) 0·841 m³/kg;   10 °C; 0;   (ii) 0; 2390 kW.
14.11   1·123 kg;   2·40 MN/m²; 0;   −258 kJ;   −258 kJ;   −0·523 kJ/K.
14.12   (a) (i) 0·1071 m³/kg;   49 °C +0·316 kJ/kg K.
             (ii) +146·4 kJ/kg;   +109·8 kJ/kg.
             (b) (i) 0·1071 m³/kg;   49 °C;   +0·316 kJ/kg K.
             (ii) +36·6 kJ/kg;   0.
14.13   (i) 35 °C;   82 °C;   (ii) 1·179;   (iii) −0·167 kJ;   −0·093 kJ.
14.14   (a) −230 kJ/kg;   (b) −193 kJ/kg;   (c) 83·9%.
14.15   (a) 245 °C;   96·0 kW.
14.16   (a) (i) 76·8 °C; 0·1245 m²;   (b) −13 440 kW.
14.17   (a) 9·7 °C;   +0·006 29 kJ/K;   (b) −6 °C;   +10·4 kJ;   (c) −6 °C;   +8·9 kJ.
14.18   (i) 0·100 kg;   (ii) −2·57 kJ;   (iii) 62·4 kN/m².
14.19   (i) 127·2 kN/m²;   0·014 64 kg.
14.20   (i) 7·0;   (ii) −0·0993 kJ;   −0·0053 kJ.
14.21   (i) 0·489 kJ/kg K;   1·712 kJ/kg K;   (ii) −2 °C;   418 m/s.
             (iii) 18·52 cm²;   7·90 cm².
14.22   (a)

$$V = \sqrt{2g_c R \frac{\gamma}{\gamma - 1} \cdot T_1 \left[1 - \left(\frac{p}{p_1} \right)^{\frac{\gamma - 1}{\gamma}} \right]}$$

$$A = \dot{m} \sqrt{\frac{\gamma - 1}{\gamma} \cdot \frac{RT_1}{2g_c}} \cdot \frac{(r)^{\frac{\gamma - 1}{\gamma}}}{r p_1 \sqrt{1 - (r)^{\frac{\gamma - 1}{\gamma}}}} \quad \text{where } r = \frac{p}{p_1}$$

(b) Plot.

$$(c) \frac{p}{p_1} = \left(\frac{2}{\gamma + 1} \right)^{\frac{\gamma}{\gamma - 1}}; \quad V = \sqrt{2g_c \frac{\gamma}{\gamma + 1} RT_1}.$$

(d) Proof.

14.23   (i)

	1	2	3	4
p, (kN/m²)	600	300	55·5	111
T, (K)	523	523	323	323
V, (m³)	0·1	0·2	0·667	0·334
S, (kJ/K)	S_1	$S_1 + 0.0795$	$S_1 + 0.0795$	S_1

(ii) 38·24%;   15·9 kJ;   (iii) 25·4 kW.

14.24   (i) Sketch;   (ii) $\eta = 1 - \left(\frac{1}{r} \right)^{\gamma - 1} \cdot \frac{\lambda^\gamma - 1}{\gamma(\lambda - 1)}$;

             (iii) 59·5%;   (iv) $T_2 = 805$ K; $T_3 = 1208$ K; $T_4 = 526$ K;
             (v) 241 kJ/kg.

14.25   (i) Sketch;   (ii) $\eta = 1 - \left(\frac{1}{r} \right)^{\frac{\gamma - 1}{\gamma}}$;   (iii) 36·9%;

             (iv) $W_x = R \cdot \frac{\gamma}{\gamma - 1} [(r)^{\frac{\gamma - 1}{\gamma}} - 1][T_3(r)^{\frac{1 - \gamma}{\gamma}} - T_1]$;

             (v) $W_x = 199$ kJ/kg;   (vi) Answers would be the same.
14.26   (i) Sketch;   (ii) $t_1 = 40$ °C; $t_2 = 260$ °C; $t_3 = 760$ °C; $t_4 = 429$ °C;
             (iii) 22·4%;   (iv) 112·6 kJ/kg.
14.27   (a) (i) 0·590 kJ/kg K;   (ii) 1·443;   (iii) 188 kJ/kg;   −32 kJ/kg;
             (b) (i) 1·252;   (ii) 0·860 m³/kg;   725 kJ/kg;   381 kJ/kg;
             (iii) 2·56 kJ/kg K;   1·97 kJ/kg K;
             (c) 1·475;   (d) −391 kJ;   +229 kJ;   (e) −383 kJ.

14.28 (a) (i) 0·590 kJ/kg K; (ii) 1·472; (iii) 184 kJ/kg; −36 kJ/kg;
(b) (i) 1·472; (ii) 0·860 m³/kg; 570 kJ/kg; 226 kJ/kg;
(iii) 1·84 kJ/kg K; 1·25 kJ/kg K;
(c) 1·475; (d) −391; 2 kJ/kg; (e) −383 kJ.

14.29

	(i)	(ii)
V_2 (m/s)	145	145
$\dot{Q}$ (kW)	2595	7111

14.30 3·60; 3040 kW; 3070 kW.

Chapter 15

15.1 (i) p_{CO} = 13·44 kN/m²; p_{H_2} = 186·56 kN/m²; (ii) 6·43 m³; 3·22 m³/kg;
(iii) V_{CO} = 0·431 m³; V_{H_2} = 5·999 m³;
(iv) n_{CO} = 0·035 7 kmol; n_{H_2} = 0·496 kmol;
(v) 2·210 kJ/kg K; 7·676 kJ/kg K; 5·466 J/kg K; 1·404.

15.2 (i) 309 °C; (ii) 2234 kJ/kg; 1591 kJ/kg; 3·79 kJ/kg K; (iii) 3181 kJ.

15.3 (i) 309 °C; (ii) 2234 kJ/kg; 1591 kJ/kg; 3·79 kJ/kg K; (iii) −3181 kJ.

15.4 (a) 0·015 97 m³; (b) (i) 169·8 kN/m²; (ii) 0·944 kJ/kg K; 0·598 kJ/kg K; 1·579;
(c) 34·3 kN/m²; +2·49 kJ; +0·75 kJ; +0·121 kJ/kg K; (d) 28·0 kN/m²; 2·26 kJ;
+0·004 70 kJ/K; −0·004 70 kJ/kg.

15.5 (i) 15 °C; (ii) p_{air} = 91·72 kN/m²; p_{CH_4} = 8·28 kN/m²; (iii) 0·1216 kJ/kg K.

15.6 (i) 15 °C; (ii) p_{CO} = 64 kN/m²; p_{H_2} = 96 kN/m²; (iii) 1·206 m³/kg;
(iv) m_{CO} = 0·1871 kg; m_{H_2} = 0·0202 kg;
(v) n_{CO} = 0·006 68 kmol; n_{H_2} = 0·010 03 kmol;
(vi) 0·670 kJ/kg K; 2·335 kJ/kg K; 1·665 kJ/kg K; 1·402;
(vii) 0·0935 kJ/K; (b) 15 °C; 80 kN/m²; (c) 15 °C; 80 kN/m².

15.7 93 °C; 2·44 kg.

15.8 (a) (i) p_a = 169·65 kN/m²; p_s = 30·35 kN/m²; (ii) 0·645 m³/kg; (iii) 356 kJ/kg;
(datum states: zero values of internal energy for saturated water at the triple
point and for air at 0·01 °C); (b) as (a).

15.9 (a) m_a = 0·425 kg; m_s = 0·256 kg; (b) 22 °C; 30·2 kN/m².

15.10 (i) 290 kN/m²; (ii) 1·67 kg; (iii) 6 552 kJ; (iv) −18·8 kJ/K.

15.11 1595 kJ/kg; 0·414.

15.12 (i) p_a = 99·70 kN/m²; p_s = 1·63 kN/m²; (ii) 0·849 m³/kg;
(iii) 127·8 kJ/kg; (iv) 41·8 kJ/kg; (v) 0·189 kJ/kg K.

15.13 (i) 0·99; (ii) 0·0101 kg_s/kg_a; (iii) 0·512; (iv) 14·2 °C; (v) 13·6 °C.

15.14 0·512; 0·512; 0·503.

15.15 (a) 18·07 °C; 3·46 × 10⁻⁶ kg; (b) 18·01 °C; 3·34 × 10⁻⁶ kg.

15.16 (a) 0·003 91 lb_s/lb_a; 0·003 91 lb_s/lb_a; (b) 0·779; (c) −10·0 kW.

15.17 (a) 0·0160 lb_s/lb_a; 0·0050 lb_s/lb_a; (b) 1·0;
(c) −22·6 kW; $\dot{m}_{H_2O}$ = 0·303 kg/min.

15.18 (i) 0·0046 kg_{H_2O}/kg_a; (ii) 47 °C; (iii) 47·4 kJ/kg mixture.

15.19 0·6; 1218·2 kJ; 0·188 kg.

15.20 Plot.

15.21 Yes.

15.22 496 m.

Chapter 16

16.1 (a) 3·38 kg;

Products	CO_2	H_2O	SO_4	
m (kg)	3·08	1·26	0·04	
n (kmol)	0·07	0·07	0·000 625	

(b) 14·57 kg;

Products	CO_2	H_2O	SO_4	N_2
m (kg)	3·08	1·26	0·04	11·19
n (kmol)	0·07	0·07	0·000 625	0·4

16.2 (a) 15·13; $n_{CO_2} = 0.0702$; $n_{H_2O} = 0.0789$; $n_{N_2} = 0.415$ kmol;
(b) $V_{fuel} = 0.075\ 7$ cm^3; $V_{air} = 683$ cm^3.

16.3 (i) $n_{H_2O} = 0.080$; $n_{CO_2} = 0.041$; $n_{CO} = 0.029$; $n_{N_2} = 0.366$ kmol;
(ii) CO_2, 9·4%; CO, 6·7%; N_2, 83·9%.

16.4 (i) 13·5; (ii) 24·1%; (iii) 3·37 m^3/s.

16.5 (i) $n_{CO_2} = 0.07$; $n_{SO_2} = 0.000\ 625$; $n_{O_2} = 0.0422$; $n_{N_2} = 0.559$; $n_{H_2O} = 0.0867$
kmol; (ii) 1·561 kg; (iii) 47·94 °C.

16.6 CO_2, 14·4%; O_2, 6·06%; N_2, 79·54%.

16.7 (i) $n_{CO} = 4.36 \times 10^{-5}$; $n_{O_2} = 2.18 \times 10^{-5}$ kmol;
(ii) −282·1 MJ/kmol; (iii) 67·55 kN/m^2; (iv) −283·34 MJ/kmol.

16.8 (a) Plot; (b) +110·034 MJ/kmol CO; (c) 514 °C.

16.9 (i) Sketch; (ii) −44·524 MJ/kg octane;
(iii) −44·360 MJ/kg octane; −44·492 MJ/kg octane;
(iv) −47·743 MJ/kg octane; −46·128 MJ/kg octane.

16.10 1·082 kJ/kg K.

16.11 1 519 °C.

16.12 10·113 MJ/kg CO.

16.13 (a) (i) −1428·791 MJ/kmol C_2H_6;
        (ii) +1428·791 MJ/kmol C_2H_6;
        (iii) 52·022 MJ/kg C_2H_6; 47·626 MJ/kg C_2H_6;
(b) 4525 °C.

16.14 (a) (i) −1428·791 MJ/kmol C_2H_6;
        (ii) +1428·791 MJ/kmol C_2H_6;
        (iii) 52·022 MJ/kg C_2H_6; 47·626 MJ/kg C_2H_6;
(b) 1501 °C.

16.15 (a) 5·837 MJ/m^3; 5·540 MJ/m^3;
(b) 5·770 MJ/m^3 of wet gas; 5·447 MJ/m^3 of wet gas.

16.16 (i) 94·9%; (ii) 96·8%.

16.17 (a) Proof; (b) $\overline{LCV} = 44.80$ MJ/kg; $q_1 = 31.10$ MJ/kg;
$m\bar{c}_p\ (t - t_0) = 7.09$ MJ/kg; $q_u = 1.81$ MJ/kg; (ii) 77·8%.

16.18 (a) 448 kW; (b) 18·9%.

16.19 114·4 MJ/kg H_2.

Index